GLOBAL ISSUES

ENERGY SUPPLY AND RENEWABLE RESOURCES

Regina Anne Kelly

Foreword by Tom Mast

Viva Books

New Delhi | Mumbai | Chennai | Kolkata | Bangalore | Hyderabad | Kochi | Guwahati

To my husband, Brian
for all his hard work in promoting renewable energy.

First Indian Edition 2010

VIVA BOOKS PRIVATE LIMITED

- 4737/23 Ansari Road, Daryaganj, New Delhi 110 002
  E-mail: vivadelhi@vivagroupindia.net, Tel. 42242200
- 76, Service Industries, Shirvane, Sector 1, Nerul, Navi Mumbai 400 706
  E-mail: vivamumbai@vivagroupindia.net, Tel. 27721273, 27721274
- Jamals Fazal Chambers, 26 Greams Road, Chennai 600 006
  E-mail: vivachennai@vivagroupindia.net, Tel. 28290304, 28294241
- B-103, Jindal Towers, 21/1A/3 Darga Road, Kolkata 700 017
  E-mail: vivakolkata@vivagroupindia.net, Tel. 22836381, 22816713
- 7, Sovereign Park Apartments, 56-58, K. R. Road, Basavanagudi, Bangalore 560 004
  E-mail: vivabangalore@vivagroupindia.net, Tel. 26607409, 26607410
- 101-102 Mughal Marc Apartments, 3-4-637 to 641, Narayanguda, Hyderabad 500 029
  E-mail: vivahyderabad@vivagroupindia.net, Tel. 27564481, 27564482
- First Floor, Beevi Towers, SRM Road, Kaloor, Kochi 682 018, Kerala
  E-mail: vivakochi@vivagroupindia.net, Tel: 0484-2403055, 2403056
- 232, GNB Road, Beside UCO Bank, Silpukhuri, Guwahati 781 003
  E-mail: vivaguwahati@vivagroupindia.net, Tel: 0361-2666386

www.vivagroupindia.com

Published by arrangement with

Facts on File, Inc.
132 West 31st Street
New York 10001
USA

ISBN 978-81-309-1418-3

Published by Vinod Vasishtha for Viva Books Private Limited, 4737/23, Ansari Road, Daryaganj, New Delhi-110 002. Printed and bound by Anand Sons, Delhi - 110 092.

CONTENTS

Foreword

One has only to read a newspaper or magazine to realize the importance of energy at this early juncture of the 21st century. It will surely become more critical as world demand for oil continues to increase and its supply peaks and begins to fall.

Industrialized nations have become addicted to an ever-increasing amount of energy, and developing nations are beginning to demand advantages in their business and personal lives that require abundant energy. Competition for scarce resources, primarily oil, is already shaping U.S. foreign policy, military strategy, and international trade balances. Battles over energy sources are not new. U.S. relations with Japan before the bombing of Pearl Harbor deteriorated badly because of Japan's aggression in oil-rich Indonesia. Germany made a huge strategic error in World War II by attacking the Soviet Union to take over its oil fields in Azerbaijan.

Conflicting with our need for energy is our desire to control pollution and global warming. The complex issues surrounding energy are quite confusing to the vast majority of citizens of the world and the United States. Americans very much need to understand energy issues, including crises, shortfalls, affordability, economics, dependence on foreign oil, political instability in oil-producing nations, foreign relations, infrastructure, environmental considerations, and alternative sources. These are exactly the issues explained by this excellent book. Young Americans will be making important national and personal decisions regarding energy during their lives, and they will have to be well informed to make the right decisions. For example, they should form opinions on whether a nation's economic security necessarily depends on oil, what can be done about the environmental impact of energy use, and whether alternatives to oil can be developed quickly enough to have the potential to substitute for dwindling oil reserves.

A careful study of this book will allow the reader to come away better prepared than most Americans to carry on informed conversations on energy topics. The book is also a rich resource, leading anyone interested in quickly learning more about a particular topic to other authoritative sources of information. It is an excellent vehicle for high school students wanting to conduct research. Also, it provides many topics for spirited discussion.

In its introduction, *Global Issues: Energy Supply and Renewable Resources* does a fine job of putting energy into a historical perspective. It tells how humankind used wood for many centuries as the primary fuel or source of energy for heating and cooking. But excessive use of wood for fuel led to deforestation in China and Europe, with the destructive environmental consequence of washing away topsoil. Eventually, a transition to coal for most heating was made, showing that it is possible to change from one source of energy to another with some positive benefits. History also shows how humans harnessed the wind to power ships and grind grain, among other things. Then, in the late 1700s, humans first learned to make an engine that would convert heat energy into work—the steam engine. The Industrial Revolution began, and people harnessed energy for thousands of tasks over the next 200 years. However, technological advancement has not been painless, and there are lessons to be learned from history. Coal pollution became a major problem, so, in the late 1800s, oil began to replace coal as an energy source. Again one source of energy supplanted another, at least to some extent. Use of coal declined because oil was a more convenient and cleaner fuel. Now we are beginning to require other sources of energy to substitute for oil, partly because fossil fuels generate carbon dioxide, which contributes to global warming, but also because we do not have a choice—oil supply will soon be exceeded by world demand.

The introduction also explores renewable and alternative sources of energy, including wind, solar, biomass, and ethanol. Many alternative sources of energy are still in the early stages of development, and the author explains the challenges that need to be overcome before these alternative energy sources become viable, able to reduce our dependency on oil as the primary source of energy in the world. Deciding which alternatives to pursue aggressively—balancing the environmental, efficiency, and scalability aspects—is the challenge humanity is facing today.

The introduction is followed by a very thoughtful chapter on the United States and energy. It talks about U.S. dependence on foreign sources of oil and how it is impossible for the United States to produce enough oil to be self-sufficient. It explains how Americans were affected, including being led into severe recession, by foreign events in the 1970s: U.S. support of Israel leading to the Arab oil embargo and the differences with Iran that are ongoing. This chapter describes how Congress has tried to deal with energy and environmental issues and how, for the most part, U.S. dependence on foreign oil has continued to increase because nothing Congress has done has been very effective. After reading this book you will have some facts on what drilling in the Arctic National Wildlife Refuge would do for U.S. energy needs. You will have some facts to help you decide how close to peaking world oil production might be. You will have a much better understanding of environmental and global warming issues and how they affect energy production and influence the desirability of various sources of energy.

Global Issues: Energy Supply and Renewable Resources also gives the reader an international perspective on energy issues. Sources of energy, such as coal, oil, and natural gas, are commodities that are bought and sold in the global economy. Some nations have more of these resources than others. But balancing the supply and demand of these commodities worldwide is difficult, especially because many of the countries that supply and those that consume energy do not have good relations. The book describes how America uses so much oil that it became an importer in 1970; now it is by far the world's biggest consumer of oil and is dependent on nations such as Venezuela and Saudi Arabia as suppliers.

The book talks in detail about China's very rapid growth as a consumer of energy. China has been very successful in growing its manufacturing industries in recent years. The industrial plants and consumers' automotives and electrical appliances are putting pressure on China's energy supplies and creating pollution. China's determined approaches to solving these problems may well hold some lessons for the United States. Are the United States and China going to butt heads over energy in the years to come? If they do, you can bet that your life will be affected.

Global Issues: Energy Supply and Natural Resources also describes how several other countries have dealt with some of these matters. Germany is a leader in using renewable alternative sources of energy, generating more electrical power from wind than any other country. Iran is an example of an oil-rich country that has mismanaged its resources considerably, subsidizing gasoline consumption and not developing its oil fields wisely. As a result, its economy is weak, and its people are restive. Nigeria also is a major producer of oil and natural gas and should be a rich country, but 70 percent of its people live in poverty. Consequently, sabotage by rebels and vandalism abound, making this major supplier to the United States fairly unstable. This instability makes oil markets nervous and drives up the price of oil.

The world demand for oil will continue to increase as the supply peaks and begins to fall. This trend will lead to an ever-increasing shortage of oil. Over the medium to long term, we can neither drill nor conserve our way out of this dilemma, although doing both will give us more time to develop alternatives to oil. All the alternatives to oil have serious technical and/or social implications that must be addressed before they can replace the huge quantities of oil we use. However, once they become motivated, Americans have historically been very good at solving large and difficult problems. Remember the complete conversion of U.S. industrial might from consumer goods to war matériel in the six months after Pearl Harbor, the rebuilding of Europe after World War II under the Marshall Plan, and the solving of the huge technical problems involved in putting a human on the Moon within 10 years. It is my hope that with their drive, ingenuity, and technology, humans can wean themselves from the addiction to oil.

This book will be a resounding success if it inspires the best and brightest young people to pursue careers in science, engineering, and economics to develop alternatives to the present excessive reliance on oil, coal, and natural gas. It would be delightful if such an effort could have the same sense of purpose, excitement, drive, organization, and success that the National Aeronautics and Space Administration and the entire space program had when they set out to put a human on the Moon.

— Tom Mast
Author of *Over a Barrel: A Simple Guide to the Oil Shortage*

Acknowledgments

I wish to thank Tom Mast for sharing his energy expertise and offering valuable suggestions for making this an accurate and complete manuscript; Claudia Schaab for her editorial insight, professionalism, and guidance; gro! Solar and Sea Bright Solar for their interest and input; my local university libraries for research assistance and a quiet place to work; and especially my husband, Brian, for offering his knowledge of renewable energies, energy policy, and international energy issues.

I also wish to thank Abby and "Miss Wiggins" for showing me what it means never to run out of energy.

PART I

At Issue

1

Introduction

Questions about energy and its sources are major preoccupations in the world of the 21st century. Government leaders, politicians, activists, scientists, economic analysts, researchers, journalists, and citizens—in the United States and beyond—have been asking serious questions about the natural resources the world uses for energy and the way in which the world uses such resources. These questions include the following:

1. How long will the world's supply of its most widely used energy resource, oil, last?
2. What are some ways to prevent an "energy crisis"—that is, an immediate shortfall in energy supply and/or a sudden, sharp increase in the cost of energy resources?
3. How can the affordability of energy resources be assured?
4. Does a nation's energy security necessarily depend on oil as a commodity? And is the stability of its energy system necessarily tied to the availability of foreign energy resources?
5. To what extent have insecurities about energy driven political decisions, controversial government policies, and even wars?
6. How will developing nations' energy needs be met?
7. How extensive is the environmental impact of energy production and use, and what can be done about it? Similarly, to what extent have human beings' energy-related activities contributed to the Earth's changing climate?
8. Can renewable energy sources and oil alternatives be developed quickly and efficiently enough to have the potential for widespread, conventional use? And is nuclear energy really one of these technologies?

These questions can seem overwhelming and even insurmountable, yet they express several basic, underlying energy-related issues that the world confronts today:

- Energy supplies, crises, and shortfalls
- Energy affordability and economics
- Foreign oil dependence, political instability, and foreign relations
- Weak or inadequate energy infrastructure, which causes limitations on energy production
- Energy-related environmental issues and the potential of renewable and alternative sources of energy

Although these basic energy-related issues seem to have a particular urgency today, people have been dealing with similar energy uncertainties for decades—even centuries. That is because energy is one of the most powerful, complex forces behind the development of human civilizations. "Silent and unseen, energy fuels our bodies, our machines, our societies, and our planet. It is the common denominator among widely diverse activities," write the authors of *Biosphere 2000: Protecting Our Global Environment.*[1] Energy is a force upon which human beings throughout the ages have relied to support their progress and survival.

WHAT IS ENERGY?

Energy is the ability to do work, the capacity to cause matter to move or change. Even when you complete the simple act of turning a page in this book, for example, you are using energy. The earliest human beings depended on the natural, biological energy of their own bodies to move from one place to another, to hunt animals or gather vegetation for food, and to build shelter. Energy use in the world today, of course, is quite different. As F. Peter W. Winteringham writes in *Energy Use and the Environment,* "Almost all the activities and products of modern society depend upon the deliberate production, storage, and use of 'artificial' energy: the production, storage, and use of gasoline (petrol) or diesel fuel for transportation; coal and coal-based electricity for heating, lighting, and powered-tool use; etc."[2] Yet dependence on not only these "artificial" forms of energy but all forms of energy has been driven by the same simple principle: As civilizations have developed new ways of generating energy to support their activities, their civilizations have become more advanced. As they have become more advanced, their demand for energy and their need to find additional ways of meeting this demand have escalated. And it is this ever-increasing demand for energy that plays a key role in the energy-related questions and issues of the world today. It has also been a significant force throughout history—a principle at work throughout the development of the world's use of energy, from prehistory to today. As Tom Mast writes in *Over a Barrel: A Simple Guide to the Oil Shortage,* "We have harnessed energy to perform an almost unbelievable list of tasks for us over a very short period in human history."[3] And so many of the world's modern energy-related questions and issues were evident, in varying degrees, in the energy experiences of ages past.

PREHISTORY: EARLY SOURCES OF ENERGY

The human beings who populated the Earth more than 15,000 years ago were food gatherers and hunters. Their energy needs were simple: They needed enough energy to obtain and eat their food, to grow physically, and to reproduce. For these activities they at first harnessed one of the most basic sources of energy, the energy from within their own bodies. All living things (and many nonliving things) have such energy stored within them. It is called *potential,* or stored, energy. When this energy is actually used, it becomes actual, or *kinetic,* energy, which is energy being released to do work. For example, a prehistoric man who used the strength of his body to chase and capture an animal to obtain its meat for food was, in the most basic way, transforming potential energy into actual work.

Although early human energy needs were simple, capturing animals for food was not without its difficulties. Some animals were too dangerous to approach or too quick to catch. So, some of the early human beings—perhaps as they advanced in their awareness of their difficulties and of the possible means for solving them—began to craft simple tools and weapons, which they used to make hunting and capturing animals easier and more effective. Primitive tools and weapons being put to use for energy illustrate a form of energy called *mechanical* energy. Mechanical energy is the energy that an object possesses as a result of its motion or position. Mechanical energy can be either kinetic energy or potential energy. A primitive tool being used to cut through forest brush would be an example of an object that possesses mechanical energy. This energy is *kinetic* because the tool is actually doing *work.* On the other hand, a primitive bow on the verge of releasing an arrow to strike animal prey is an object with mechanical energy that is *potential;* the bow is in the *position* to do work.

Mechanical energy is only one form of energy. Energy also takes the form of light, heat, electrical energy, chemical energy, and nuclear energy. Mechanical energy is a type of energy usually associated with a physical object as opposed to a chemical or biological change or process. Aside from objects, other sources of energy include the Sun, moving water, wind, and substances that can be burned to give off heat.

Turning again to the example of early civilizations, the need to find new sources of energy affected these early peoples rather intensely. To ensure their survival, they needed food. And in order to obtain enough food, and to obtain it in the most efficient way possible, they needed tools and weapons. In order to create their tools and weapons, they needed to use resources from their environment, such as stones or branches—that is, natural resources. So, natural resources were essential to the very continuation of their lives. Natural resources helped to sustain their lives, enabling them to make the tools they needed to obtain food in a more efficient manner. In a rudimentary way, this situation illustrates a problem that lies at the heart of all the energy issues that human beings have ever faced: As each civilization advances, its energy needs multiply, and as these needs multiply, the civilization begins to require more and newer sources of energy to meet these needs.

During this early period, just as now, people also depended on another source of energy: the Sun. During the day sunlight could provide two other valuable forms of energy: heat and light. Heat is also called *thermal* energy, and light is considered a form of *radiant* energy, or energy traveling as electromagnetic waves. The Sun gives off both heat and light. The Sun's energy, known as solar energy, is not limited in supply but is essentially inexhaustible. It is classified as a *renewable* resource, a source of energy that is constantly "renewed," or replaced, by the environment. As long as renewable resources are not used up more quickly than they can be restored, their supply is essentially limitless. Aside from the Sun, wind and water are also renewable resources.

For times when the Sun was not shining, the early human beings discovered the power of fire as a source of heat and light. Perhaps they first found warmth and light through the fires that had occurred naturally—for example, when lightning struck the Earth. Eventually they discovered how to start fires themselves. To do so, however, they needed fuel. Fuel is any material or substance that is burned to provide energy. Wood was the main fuel for early civilizations, although they also used other plant matter, as well as animal waste. Organic matter that is burned, including wood and plant and animal waste, is called biomass. Biomass is also considered a renewable source of energy, since plants can always be regrown and since what living things take in for nourishment they eliminate as waste. (Biomass would actually supply most of the world's energy for thousands of years, until the 1800s, when new energy technologies, such as electricity, along with the discovery of oil, changed the whole energy paradigm.) At this time in history, fire was civilization's greatest energy discovery. It was used for lighting, cooking, heating, and even scaring away wild animals.

ANCIENT CIVILIZATIONS: NEW SOURCES OF ENERGY AND ENVIRONMENTAL CONCERNS

Deforestation in Ancient China

The centuries passed, and the use of fire for energy, along with the widespread use of wood for fuel, began to have a significant consequence. The growing demand for wood as fuel resulted in deforestation, or the cutting down or clearing away of forests. Forests are homes to many species, and they provide a kind of protective blanket over the earth, preventing the soil from eroding and thus from sending excess silt deposits into nearby rivers and streams. Soil erosion and the displacement of forest species from their natural habitat are two of the negative effects of deforestation. Deforestation is known for harming ecosystems, the communities of organisms within a geographical area that depend for their survival and proper functioning on an intricate network of interactions with the environment.

Deforestation is known to have been a problem as early as 770 B.C.E.—when it devastated China. China's early use of wood to fuel fires that generated heat and light led to rapid deforestation, then to a serious wood shortage, and then to a need to turn to charcoal for fuel. The problem of deforestation is one that actually reaches far beyond these early centuries. For example, by the time of the late 1800s, straight through the mid-20th century, deforestation in China was further exacerbated by a burgeoning rice-farming industry that cut into pastures and other tree-lined lands. This activity resulted in soil erosion, a lack of cultivable land, and finally a food shortage and famine. In Europe, too, deforestation became a concern: In the mid-11th to late 13th centuries Europeans cleared millions of hectares of forests and swamps in order to obtain wood for fuel. In 1346 England established a policy to protect forests from further cutting. And in France in the 18th century unprecedented domestic thermal energy consumption and a growing industrial sector dependent on wood for fuel caused serious deforestation. This process then led to escalating wood prices and governmental restrictions on woodcutting. In the 1730s and in 1776 French peasants organized uprisings over these wood price increases and restrictions.

Deforestation remains one of the greatest concerns of environmentalists around the world today. Clearly the world of ages past encountered some of the same energy-related issues that have importance today, such as shortages in energy supply and the environmental impact of energy production and use.

Hydropower in Ancient Egypt and Mesopotamia

Returning to earlier generations, it becomes clear that people's energy needs multiplied further. Animals were another energy resource early civilizations began to tap. At the beginning of the fourth millennium B.C.E., more and more people began to settle in the valleys of the Nile, Tigris, and Euphrates Rivers, within the western and southwestern regions of the Asian continent, in regions such as Egypt and Mesopotamia. The waters of the Nile and other rivers overflowed at certain times of the year and provided a natural irrigation system for the land. Blessed with the natural irrigation produced by these rivers, the Egyptians and Mesopotamians began to grow grains on the land as a source of food. They also began to rely on animal energy to supplement their own human energy. Cattle pulled plows, sheep buried seeds under their hooves, donkeys transported harvested goods, and horses carted carriages.

Just as early tool and weapon production depended on the availability of the natural resources of stones and branches, early irrigated cereal farming activities depended on the availability of other natural resources, such as water for irrigation. The energy that moving water provides is known as *hydropower*. It is so called because water is two atoms of hydrogen plus one atom of oxygen. Like the Sun's, energy from moving water is considered renewable.

With the help of naturally occurring hydropower, then, the Egyptians and Mesopotamians had a flourishing cereal production sector, and more and more people settled around the rivers. Complex governing systems gradually rose up to oversee the land, the harvests, and the peasants and to regulate the distribution and rationing of the grains harvested for cereals. In Egypt the pharaoh was the head of this governing system, while in Mesopotamia temple and palace officials supervised the farmlands and harvests. Regulations required that a portion of the harvests be given to governing officials, who also ordered that retention basins be built along the rivers to accommodate the burgeoning river-valley population. By 2650 B.C.E. the rivers had become a major source of energy to fulfill another need: the transportation and distribution of grains. Thus, the security and well-being of the territories around the rivers began to hinge on the agricultural products they generated. The key to their energy stability was the stability of the political system that controlled them. So, in a sense, these early riverside settlers too met with an energy-related issue that remains a concern for the world today—energy's impact on the economy.

Even during this early age of humankind, dramatic shifts in the use of energy were beginning. In using water for transportation, for example, human beings were for the first time gaining power over a source of energy that did not really need to be supplemented by their own human energy. And even though the efficiency of water as a source of energy for transportation depended on directional currents and winds, this newfound form of energy opened up endless possibilities for transportation through numerous routes along the sea. Further innovations and refinements in shipping vessels enabled people to use the winds and currents in a more efficient way. For example, the addition of sails to shipping vessels allowed them to take advantage of the wind as a source of energy, giving them the ability to steer their vessels in the courses of the wind and travel much more quickly and efficiently. (Their predecessors simply had to float their goods downstream!) In the fifth and sixth centuries B.C.E. Greek shipping vessels in particular had a shipping capacity of up to 150 tons; four centuries later this capacity had increased to as much as 800 tons. By the first century C.E. North Africa and Egypt were shipping vast amounts of wheat along the Tiber River.

Human Energy Crisis in the Roman Empire

The Romans of the third century B.C.E., however, encountered wood shortages and bottlenecks in shipbuilding yards. Human energy from hundreds of thousands of slaves, captured from the many cities Rome had conquered, soon became an important resource in the Roman Empire. Unlike all other energy sources, slaves were capable of doing the work that made the Roman Empire the center of civilization that it was—they built great structures, they labored in mines, they maintained Rome's elaborate water and wastewater systems, and they managed domestic tasks for the elite. However, nearing the time of the Roman Empire's decline in the fifth century C.E., Rome made fewer and fewer conquests and began to experience a shortage in the human energy it once obtained from slaves procured from pillaging other cities.

Rome's human labor shortage exemplifies another issue that has relevance today: energy crises. In this case, the energy crisis was a sudden shortfall of human energy supply that forced the empire to seek out new energy sources. As the authors of *In the Servitude of Power: Energy and Civilization through the Ages* state, "All historical societies have developed an energy thrust-block, a threshold beyond which they could not go."[4] In more modern times, such an energy crisis occurred in the United States in 1973–74 when Arab countries stopped shipping oil to the United States and other Western nations because of their support of Israel during the 1973 Arab-Israeli Yom Kippur War. In 1973, as now, oil was the most widely used energy resource. According to the International Energy Agency (IEA), oil accounted for about 45 percent of primary energy supply in the world in 1973. At this time more than a third of America's oil was imported from the Arab countries of the Middle East. So, when Arab countries stopped shipping, or placed an *embargo* on, oil imports, the United States experienced a severe oil shortage. U.S. oil supply could not catch up to U.S. oil demand, and the price of oil

THE INDUSTRIAL REVOLUTION

The Coal Steam Engine

The Industrial Revolution was a time of major socioeconomic change that began in England in the latter half of the 18th century. One of the driving forces behind it was the invention of a new energy resource: a coal-powered machine devised by the British inventor Thomas Newcomen. This machine had an engine that burned coal in a boiler to create high-pressure steam. The steam drove a shaft that could pump water from coal mines, enabling the British to extract higher-quality coal for heating. With this invention, coal soon surpassed wood as the primary source of thermal energy. But the device's real impact on energy would not be truly felt until a Scottish inventor, James Watt, added improvements to the coal-powered steam engine, beginning in 1765. Watt's steam engine could be used for purposes other than pumping water, and it soon became the energy powerhouse behind the operation of silk mills and other textile factories, first in England in the 18th century and then in the United States in the 19th century. The textile industry, which previously had depended on energy from water mills and windmills, was transformed.

More and more factories sprang up in response to this new development in power-driven machinery. More and more workers settled around the factories that employed them. In this way, urban industrial centers began to develop. Steel mills, steamships, and railways also began to make great use of the coal steam engine. The invention of the coal steam engine thus represents a significant change in energy use, energy needs, and civilization itself, again showing that a civilization's advancement leads to increased energy needs and, in turn, to an increased demand for ways to meet these needs.

The Shift from Renewable to Nonrenewable Sources of Energy

The introduction of the coal steam engine was responsible, in part, for a major shift in the kinds of fuels and resources on which people primarily relied for energy. Whereas mainly renewable biological energy resources, such as water, wind, and biomass, had once dominated energy consumption, nonrenewable energy resources—specifically coal—were increasing in importance.[5]

To understand some of the history behind the rise of nonrenewable resources, it may first help to know a little more about what their components are. Nonrenewable resources are natural resources that are limited in supply and/or cannot be replenished more quickly than they are consumed.

Introduction

Fossil fuels are one type of nonrenewable resource. They are known as fossil fuels because scientists believe that they were formed by the decomposition of plants and animals that died millions of years ago. This decomposing organic material in the Earth's crust resulted in naturally occurring compounds of carbon or hydrocarbon (*hydrocarbon* refers to a combination of the gases hydrogen and carbon). These naturally occurring compounds of carbon or hydrocarbon are *fossil fuels.* They include coal, petroleum, natural gas, oil shale, and tar sands.

OIL

The term *oil* is often used to refer in a general way to the fossil fuel petroleum, or to the other fuels that petroleum yields when it is purified and distilled. Petroleum is an oily, flammable, liquid mixture of hydrocarbons and traces of other substances. It may range in color from almost clear to black. It is also called crude oil. It lies under pressure within porous rock formations in the ground (commonly within specific rock formations known as anticlines). After geologists have identified areas where it is likely to be found, test wells are drilled and oil fields are developed. Oil-bearing rock formations usually lie underneath a layer of natural gas and above a layer of salt water; when these formations are drilled, pressure causes the layers of natural gas and water to expand and push the oil to the surface so that it collects in a well. When insufficient amounts of natural gas remain to allow the oil to be drawn out, drillers add pumps to pull it up. It can take up to a decade before a field of oil is developed and producing significant quantities of oil, but pumps allow the oil to flow quite freely for years or even decades, until the flow decreases and the oil field is depleted.

NATURAL GAS

The natural gas that is found above oil-bearing rock is usually extracted along with oil so that it too can be used as a fuel. Natural gas is a simple molecule made up of four hydrogen atoms positioned around one atom of carbon; it consists mainly of a gas called methane. Most of the time, after natural gas is extracted, it is delivered through pipelines, and its end uses are mainly heating and generating electricity. To transport natural gas overseas, however, it is first chilled so that it liquefies, creating liquefied natural gas, or LNG, a source of energy that can be shipped easily.

COAL

The most polluting of all the fossil fuels—and the one that drove the Industrial Revolution—is coal. Coal is a fossil fuel formed when increased pressure, temperature, and moisture act upon masses of plant matter that have become compacted beneath marshes and lakes. These masses harden into the black or brownish black rock that we call coal. Coal exists as a layer of sedimentary rock in the earth and must be extracted by mining, which traditionally consists of driving shafts into the ground to dig it up. Coal was the first of all fossil fuels to be used extensively. People actually burned coal for thermal energy long before the Industrial Revolution. Early in history (770 to 453 B.C.E.), China, in response to deforestation and a wood shortage, began burning coal. The Chinese used coal in place of wood as a fuel for cooking and domestic heating, as well as for the thermal energy needed to forge metals to make tools. England in the 16th century turned to coal to fulfill the same needs when its population had doubled between 1530 and 1700 and wood became scarce. Many of England's early industries, including glassworks, potteries, and breweries, were powered by coal.

However, it was the coal steam engine of the late 1700s that launched an entirely new and revolutionary consumption of coal. Aside from leading to a proliferation of factories that depended on it for daily production, the coal steam engine was the driving force behind a boom in railway transportation. Railways for steam-powered trains that both were powered by and transported coal sprang up across America and Europe. In 1781 the horse-drawn carriage was the standard means of passenger travel worldwide; within 100 years the coal-powered steam train filled this role. And whereas in 1800 most energy was consumed for domestic purposes, by the end of that century the coal-powered steam engine had created an industrial energy sector whose level of production and manufacturing surpassed anything that had preceded it. Steamships that ran on the coal-powered steam engine also became pervasive; the first steamship companies emerged between 1836 and 1840, and by 1850 they were consuming about 60 tons of coal per day.

Air Pollution

With this rise in coal as a major source of energy there also arose a set of energy-related issues more familiar in nature to those the world experiences today. For example, the environmental impact of energy no longer simply equated with deforestation and soil erosion, but with serious air pollution. This is because coal is a solid, combustible substance consisting primarily of carbon, with sulfur and other impurities mixed in. When it and other fossil fuels are burned for energy, the gas carbon dioxide, or CO_2, is released, along with other gases. Carbon dioxide is a colorless, odorless gas that is also emitted whenever humans and animals breathe, forests are burned, or land is cleared and vegetation left to decompose. The atmosphere also consists of carbon dioxide, plus nitrogen, oxygen, argon, and other gases in trace amounts. And plants absorb carbon dioxide in a process called photosynthesis, which helps to keep them alive. Yet despite all of this, carbon dioxide, of all the gases emitted when fossil fuels are burned, is generally considered to be one of the most harmful to the environment. This is because carbon dioxide collects high up in the atmosphere and reduces the amount of heat from the Sun that the Earth can radiate back into space. Retaining this additional fraction of the Sun's heat can cause the Earth's temperature to rise gradually. So, coal's primarily carbon makeup has given it a reputation as the fossil fuel that is most threatening to the environment. The concentration of carbon dioxide in the Earth's atmosphere has risen sharply since the late 1800s, when the coal-driven Industrial Revolution swept through both Europe and the United States. Energy Information Administration (EIA) estimates say that the concentration of carbon dioxide in the atmosphere today is actually about 30 percent greater than it was in the 1860s, when coal's use for energy was well established and when oil, yet another fossil fuel, was just beginning to be exploited.[6] And, according to the EIA, the world's total carbon dioxide emissions from fossil fuels are projected to reach 40,045 million metric tons by 2025, about a 47 percent increase over 1990 levels.[7] In 2001 the Intergovernmental Panel on Climate Change (IPCC), an organization established by the World Meteorological Organization (WMO) and the United Nations Environment Programme (UNEP), announced that, over the previous 100 years, temperatures had increased between 0.4°C and 0.8°C, with the warmest temperatures all occurring within the preceding 15 years.[8]

During the Industrial Revolution in the 18th century, the increased burning of coal caused air pollution around the industrial centers where workers and factories were clustered. As the British novelist Charles Dickens wrote famously in a description of industrial London in his 1850s novel *Bleak House:* "Fog everywhere. Fog up the river, where it flows among green gaits and meadows; fog down the river, where it rolls defiled among the tiers of shipping and the waterside pollutions of a great (and dirty) city."[9]

Even though its pollution factor is high, however, coal is the cheapest and most plentiful fossil fuel on Earth. During the 1800s, a half-ton of coal could produce as much energy as two tons of wood at half the cost. Even today, when consumption of coal is projected to slow, the burning of coal products is still expected to be great enough to contribute 5,353 million metric tons of carbon dioxide pollution to the Earth's atmosphere through the year 2025, according to the EIA. The EIA also reported that coal accounted for 2,142 million metric tons of carbon dioxide emissions in the United States alone in 2005. Mining for coal, too, has been criticized by environmentalists for its impact on the environment.

Dependency on Foreign Energy Resources

During the Industrial Revolution, England was the most prolific coal producer in the world. It exported one-third of its coal output and kept many parts of the world dependent on it for coal to fuel their own energy activities. For the first time countries had to pay for their energy in a way unknown to them when their energy was derived mostly from renewable sources. Such countries, in their dependency on England for their energy needs, were experiencing one of the energy-related issues that is of great concern to the United States and other countries today: dependency on a foreign country for energy resources. From 1850 to 1914 the presence of coal reserves within a country became a determining factor in whether that country experienced industrial growth. Two countries, however, had major coal reserves and were able to catch up to England's industrial output by 1914. These countries were Germany and the United States. In 1913 Germany produced 195 million tons of coal and exported 45 million tons; in the same year the United States exported 25 million tons of the 500 million tons it produced.[10] Between 1885 and 1950 in the United States, in fact, coal was the most widely used fuel. More countries soon followed England's lead during the Industrial Revolution and boosted both their mining and consumption of coal. As industries, and entire countries, began to depend on coal for their progress and prosperity, and as air pollution from coal increased, more energy-related issues began to rise to the surface, especially energy's impact on the economy and the environment. In the stage of energy history that followed the Industrial Revolution, the environment, in particular, would become a main focus, as well as one of many energy-related concerns that now have heightened significance in the modern world.

THE 19TH CENTURY: EMERGENCE OF CONTEMPORARY ISSUES

Beginning in 1850 the average amount of energy used by each person began to increase steadily, according to the EIA.[11] The growing use of coal was not the only reason for the increase. It also coincided with two important energy developments. The first was an advancement in the technology available for one of the fundamental forms of energy: electrical energy. Electrical energy is energy released from the movement and interaction of negatively charged elemental particles, called electrons, and positively charged elemental particles, called protons. It can occur naturally, such as in lightning or in static electricity, or may be deliberately produced, such as through a generator. The second important development from this period was the discovery of large supplies of a fossil fuel that is now the most widely used energy resource in the world: oil. Developments in electricity and in energy from fossil fuels represent two major trends out of which arose many of the modern-day global issues related to energy supply and natural resources.

Electrical Energy

One of the initial steps in the development of electrical energy was Benjamin Franklin's 18th-century discovery that static electricity and lightning were expressions of one and the same form of energy—electrical energy. This discovery helped to lead to a better understanding of electrical energy. In the 1830s the British scientist Michael Faraday further advanced understanding when he discovered that an electric current could flow in a wire within a magnetic field. His work demonstrated basic principles of electricity production: electromagnetism, induction, generation, and transmission. Technologies that began to make use of electrical energy in the 1830s included the telegraph, a device that electronically transmitted messages over a distance through a wire, and electroplating, the use of an electric current to cause a coating to adhere to a surface.

Other advancements in electrical energy technology quickly followed. In the 1860s James Clerk Maxwell, a Scottish mathematical physicist, demonstrated that electric and magnetic fields travel in waves through space. Maxwell developed a mathematical theory that expressed four basic laws of electricity and magnetism. His equations became the basis for many uses of electrical power in the late 1800s and later for the development of radio and television. The first electric railway was introduced in 1878, and electricity was also found to have applications in the heavy chemicals and metallurgical industries. Furthermore, an electric motor created by the American electrical engineer and inventor Nikola Tesla replaced the coal steam engine in some industries and yielded greater productivity.

But advancements in electrical technology probably had the most revolutionary effect through one particular form of energy: light. This was one form of energy for which demand had greatly increased after the introduction of the coal steam engine. For the many factories and businesses that had sprung up since, the onset of evening represented an obstacle to labor and productivity. In 1878 the American inventor Thomas Edison founded the Edison Electric Light Company. He created the first practical incandescent electric light in 1879. The lightbulb, in which a filament gives off light when heated to incandescence by an electric current, was modeled on an 1878 prototype created by Edison. With this introduction of artificial lighting, one of the remaining limitations on industrial productivity—the onset of evening—diminished. Lightbulbs began to replace the oil lamps used in industrialized nations such as the United States and England. From 1880 to 1914 there was an annual growth of more than 10 percent in electrical consumption in the world's major industrialized countries.

Edison also entered into a partnership with Werner and William Siemens in 1881 to design Europe's first public lighting network in London. In 1882 Edison established the first electricity generating plants in London and New York. They were powered by coal. In 1889 Edison consolidated his various electric businesses into the Edison General Electric Company.

Energy Infrastructure

Another contemporary energy-related issue began to surface at this time: the inadequacy of the energy infrastructure, which caused limitations on energy production. Early electricity generating stations produced electricity in the form of direct current (DC), an electric current that flows in only one direction and cannot easily be transmitted over long distances. Transmission over distances greater than three miles required the use of large, expensive copper cables. Another problem was that early generating stations, along with corporations and factories that were also creating their own electricity, generated power at different voltages and cycles.

When Edison General Electric merged with another U.S. electrical power distribution company in 1892, forming General Electric, diversity among the two companies' existing generation plants, combined with the use of DC, hindered the establishment of effective regional electricity networks. The lack of infrastructure made it impossible to connect the electrical networks of different cities.

However, within a few years, generation plants implemented the use of a rotating coupler, a device that could group their various systems, and alternating current (AC), an electric current that, unlike direct current, flows back and forth. It can also be transmitted over long distances with minimal energy loss. George Westinghouse, a U.S. inventor and manufacturer, first promoted this type of current in the 1890s.

This stage in energy history again illustrates that when societies find new ways to generate energy (in this case, the coal steam engine) they become more advanced and therefore require a greater amount of energy (here, a need for light), and so they seek out additional ways of generating this energy (here, through electric light and electricity in general). Electricity, ironically, also led to an increased use of coal, because coal powered the plants that generated the electricity. (This continues to be coal's primary function in energy production today.)

Energy Monopolies

Outside the United States and England, electrical power took hold in other countries, too. Germany was one such country; it also grappled with the need to establish an infrastructure to unify its electricity networks. In 1926 its Central Electricity Board allocated public funds to unify and centralize the distribution of electricity. Still, right up until the 1940s, Germany and other industrialized countries would continue to struggle with a lack of sufficient regulation and centralization to keep their electrical power networks running smoothly. Compounding this situation was the fact that electric utility companies were able to form monopolies, controlling the related functions of generating electricity, transmitting it over distances, and distributing it to customers. One outcome of electric utility monopolization was that electricity consumers had little or no choice of providers. The monopolizing companies could set prices that were much higher than they might have been had other companies been able to enter the market and initiate competitive pricing. Also, because each monopolizing utility company controlled the generation and distribution of electric energy in a specific service area, interconnection with utility transmission networks in other service areas met with technological hurdles and compatibility issues, affecting energy reliability and efficiency. Additionally, utility monopolization gave electricity providers the power to install facilities they considered necessary without weighing input from consumers, who were all but bound to the electricity provider in their service area.

Consumers encountered such problems with General Electric, which controlled everything from the manufacture of electric motors to the distribution of power. The Commonwealth Edison Company, which was led by Samuel Insull, a former assistant to Thomas Edison, was another monopoly. By 1910 Insull had developed what was the largest electricity distribution network at that time, capable of transmitting electricity over long distances. In his network, electricity was generated as DC at a few central power stations, changed to the more efficient high-voltage AC, and transmitted throughout a city via a network of wires. In substations in different parts of the city, transformers and converters changed the electricity back to low-voltage DC for nearby homes and businesses. Eventually under this system electricity was sent through entire states and then multistate regions. It became the standard way to distribute electricity.

Between 1920 and 1929 electricity output in the United States increased 106 percent; in Europe, 116 percent. By 1939 it increased another 38 percent in the United States.[12] But as U.S. utility companies began to form monopolies, there were calls for reform. The administration of the U.S. president Franklin D. Roosevelt (1932–45) helped to reform the electric sector. His

New Deal program took a direct role in developing U.S. natural resources by establishing the Tennessee Valley Authority in 1933 and the Rural Electrification Administration in 1935. The New Deal administration enacted laws to address problems caused by monopolies and to dissolve public utility holding companies.

The Emerging Oil Industry

The energy-related event that is perhaps most relevant to the global energy-related issues of today is a 19th-century development in which, as Tom Mast writes, "oil shoved aside coal as the world's primary fuel, just as coal had replaced wood."[13] From ancient times human beings had found various uses for petroleum, or oil, which sometimes seeped from the ground. They mainly used it as a fuel for lighting—burning it in torches or distilling it and using it in lamps. Kerosene, a fuel derived from the distillation of petroleum, filled oil lamps as early as the medieval period, and small quantities of oil were produced and sold in the 1800s, mainly for lighting. By 1857 kerosene had almost completely supplanted whale oil, a more expensive and scarce oil that had been used for lamp lighting in earlier generations. Gas, a product derived from the distillation of oil, was also used as a fuel for lighting—mainly to light workshops and factories in early 1800s London. In 1812 the first gas company was founded, and in the 1830s the perfection of the petroleum distillation process and improvements in gas storage units spread the use of gas lighting to other industrial cities.

However, the "oil industry" as we know it today really began in the United States at the end of the 19th century. One can trace its beginnings back to an event that took place in 1859, when the U.S. entrepreneur "Colonel" Edwin Drake struck a large oil deposit 69 feet beneath the earth in Titusville, Pennsylvania. This was the first U.S. oil well, and its discovery soon led to the development of oil fields, places where oil could be extracted and produced for energy. At first horse-drawn carts transported the oil over the rough Pennsylvania terrain. Within six years, however, the first U.S. oil pipeline, a line of pipe with pumps and valves for conveying oil, was constructed to facilitate the oil transportation process. Drake's oil discovery and the subsequent development of oil fields and pipelines are considered founding events of the oil industry—precedents of the energy-intensive, fossil fuel–dependent lifestyle that eventually developed in industrialized nations by the 20th century. Oil is today the most popular fuel in the United States, accounting for 40 percent of U.S. consumption of energy resources in 2005, according to the EIA.[14] The entire world consumed 82,459 million tons of oil in 2005 alone, reports the *BP Statistical Review of World Energy 2006.*[15]

The Invention of the Automobile and the Impact on the Environment

With this intense dependency on fossil fuels has arisen a mounting concern over what the prolific consumption and production of oil have been doing to the environment since the initial 1800s Titusville oil discovery. The burning of fossil fuels releases not only carbon dioxide, but also sulfur dioxide, nitrogen oxides, and ash particles—all pollutants. Some studies by climate scientists show that increases in atmospheric carbon dioxide and these other pollutants have raised the average overall temperature of the Earth by several degrees. Some scientists believe that even a slight change in average global temperature can radically alter weather and climate patterns, as the warmer air melts the ice of the arctic regions, heats up the oceans, and causes more frequent and violent storms. This issue is often referred to as the problem of global warming or global climate change.

The oil discovery happened to coincide with events that greatly raised the value of oil in everyday life. The first of these events was the invention of a new type of engine, the internal combustion engine. This radical new engine was powered by combustion, a rapid chemical process that generates heat. Unlike in the steam engine, where combustion took place inside a separate furnace chamber, combustion in the internal combustion engine took place inside the engine itself. Because of this feature, the internal combustion engine was lighter than the steam engine and therefore more portable and suitable for a greater variety of energy needs. In addition, the internal combustion engine ran best on gasoline, a liquid fuel composed of a mixture of small, light hydrocarbons. Gasoline is one of the products of a process that crude oil undergoes in refineries, where crude oil is heated and broken up into smaller molecules to remove its impurities. Gasoline was a more ideal fuel than coal for powering the new engine, for various reasons. First, it had the desirable quality of being easy to handle, since it was by nature a fluid—not a solid that needed to be broken down, like coal. It was also able to produce enough energy to keep the internal combustion engine running efficiently. And since it could be carried directly within the combustible engine machine itself, it was easily transported, unlike coal. Further, when oil-based fuels such as gasoline were transported in bulk, they did not need to be moved via railways, as coal did; they could be distributed via pipelines. With the newfound use for gasoline, the oil being discovered and extracted in large quantities was now in greater demand. The ease and convenience of oil-based fuels for transportation continue to maintain oil's status as the most consumed energy resource in the world. According to the EIA, oil-based fuels now power about 90 percent of all transportation in the world.[16]

THE EARLY 20TH CENTURY: OIL BECOMES THE PRIMARY SOURCE OF ENERGY

The early 20th century is the period in which contemporary energy issues truly began to emerge; it is the era in which widespread production and consumption of oil really began. In 1901, oil was struck in Spindletop, the first large oil field in Texas, near the town of Beaumont. The Spindletop oil well opened with a "gusher," or a large burst of oil. The oil shot more than 150 feet into the air. After that the field began producing almost 100,000 barrels of oil a day, more than all the other oil fields in the United States combined. The Spindletop gusher set off a veritable race for oil, with oil exploration and drilling taking place all across Texas and other parts of the United States. Since then, according to some estimates, the world's demand for oil has grown by more than 80 million barrels per day, and close to one-half of all the recoverable oil on the planet has been used up.[17] But limits on the supply of oil, which is, after all, a nonrenewable resource, did not—neither in 1859, nor in 1901, nor even within a few decades—pose a pressing concern for most people, who viewed oil as an ideal source of energy. Nor was the environmental impact of burning oil for energy much of a concern yet. In fact, even in the 1860s, long before the Spindletop discovery, there were so many oil producers in the United States that there were not enough businesses to refine the oil or transport it. In addition, because oil was being produced in such large quantities, it cost just 10 cents a barrel in 1862, whereas only three years earlier it had sold for 2 dollars a barrel.[18] In 1863 U.S. petroleum producers even formed their first professional association, the Association of Petroleum Producers. And by 1870 oil had already become the United States' second largest export.[19]

The mass production of automobiles running on the internal combustion engine was another factor (perhaps the greatest) that contributed to the rise of oil. In 1890 the first automobiles utilizing an internal combustion engine were produced in large quantities in the United States, creating a greater demand for the gasoline that was needed to fuel them. In 1899 a machinist and engineer employed by the Edison Company, named Henry Ford, successfully designed his first car and launched the Detroit Automobile Company (which would later become the Ford Motor Company). In 1908 Ford produced his first Model T automobiles. Their popularity resulted in soaring sales of gasoline, and the number of motor vehicles in the United States rose from 8,000 in 1900 to about 1.3 million in 1913. In 1920 the Ford Motor Company began to produce Model Ts by the millions, manufacturing nearly 17 million cars before discontinuing the model in 1928.[20]

In addition, Rudolf Diesel, a German mechanical engineer, developed the first diesel engine in 1895. He originally designed this engine to run on a variety of fuels, including vegetable oil. Aside from being able to run on cheap fuels, his engine had 30 percent efficiency—a 15 percent greater efficiency than the standard combustible engine. After Diesel died in 1913, his engine was modified to run on petroleum—the cheapest fuel available—and oil companies labeled one of the by-products of gasoline distillation *diesel fuel*.

John D. Rockefeller and the Standard Oil Company

As America plunged full-force into oil production and consumption in the late 1800s and early 1900s, driving their gasoline-powered automobiles, the entrepreneur John D. Rockefeller began his domination of the American oil scene. Rockefeller had formed the Standard Oil Company of Ohio, an oil production company, in 1870. Since the production of crude oil was growing faster than the refinery and transport businesses at this time, Rockefeller saw an advantage and began to commandeer the oil refining and distributing operations of weaker or smaller businesses. After seizing control of the transport aspect of the industry—the pipelines—Rockefeller began charging other oil producers, his competitors, high rates for using his pipelines. He applied the same principle to his refining facilities. Soon he was able to eliminate most of his competition and develop a monopoly over U.S. oil production and distribution. In 1900 the Standard Oil Company had control of greater than half of the world's sales of oil, and in 1904 the Standard Oil Company processed more than 84 percent of all U.S. crude oil in its refineries.[21] In 1892, however, the U.S. government dissolved Rockefeller's Standard Oil Trust, and in 1911 a U.S. Supreme Court ruling broke up the holding company to which Standard Oil had transferred its assets. Yet Rockefeller retained his enormous fortune, later making donations to many philanthropic causes.

Government and Industry Collaborate to Develop Global Oil Resources

Oil production and consumption were on the rise in parts of the world other than the United States. In 1890 oil was struck in southeast Asia, and a petroleum company from Britain, British Shell, set up operations there, along with another company called the Royal Dutch Company for the Exploitation of Oil Wells of the Dutch Indies. These two companies would later become part of a cartel of enormous, multinational oil companies that reigned over the oil industry. By the early 20th century the Middle East also became the locus of various competing oil companies, as well as governmental interests, with Germans, Russians, Britons, and Americans all seeking control of the Persian oil fields. It was Britain that seemed to gain the strongest foothold in oil production, taking advantage of its landholding rights in the Ottoman Empire to surpass any other countries that produced oil in that region.

In the early 20th century politics, business, and oil began to overlap: The U.S. federal government and the Standard Oil Company began to collaborate on a program for obtaining oil in the Middle East. Such cooperative efforts between governments and oil companies multiplied, setting the stage for an interdependence among governments, the economy, and the energy industry that still shapes world politics today.

The Struggle for Dominance in the Oil Market

Some contemporary energy issues began to emerge as countries became entangled in wars and conflicts. As early as World War I, there were troubling shortages in energy supply. By that time society's energy needs were entirely centered around oil, and one of the aftermaths of World War I was that access to some oil pipelines and distribution systems were destroyed, causing a blockage in the transport of oil. This resulted in a glut of oil in oil-producing regions, but a shortage of oil in those countries where oil was needed most, where oil was consumed most rapidly. At the same time increasing oil discoveries in countries outside the United States gradually began to create a U.S. dependency on foreign sources of oil. In 1925 the potential for oil production reached beyond the Middle East into Latin America, as oil wells were discovered in Venezuela and Mexico. Oil production in Venezuela started in 1921; by the outbreak of World War II in 1939 production in that region reached almost 30 million tons of oil per year. U.S. companies began to step up their investment in Venezuela's oil, setting the stage for an eventual U.S. dependency on oil imports from Venezuela.

Also at this time, large petroleum corporations formed and began to dominate the market. They competed with one another for lower prices and thus a greater share of the market. Yet even when one of these companies needed to increase prices, it was still able to maintain its profitability. Britain, for example, dominated the Middle Eastern oil scene in the early part of the 20th century and was the main supplier of oil to the European market. While Britain sought out cheaper supplies of oil, it kept its oil prices elevated, yet never lost its dominant position. In 1928 an agreement among the major oil corporations, known as the Achnacarry Accords, set production limits and standards for pricing and marketing among the major global oil corporations, keeping profits high.

Meanwhile, major corporations from Western European countries and North America reigned over the production of oil outside their own borders, causing other countries to become dependent on them and on their control of the existing fossil fuel energy infrastructure. Smaller, less industrialized countries' energy sectors relied heavily on oil and coal from other parts of the world.

While the abundance of energy propelled industrialized areas, such as North America and Western Europe, into greater economic and industrial development, developing countries grappled with scarcity and energy shortages because the more highly developed countries now owned all the energy surpluses. Thus, foreign oil dependence became an issue even as the international oil industry was just taking shape, and developing countries had the additional problem of lacking a developed energy infrastructure to support their energy needs.

THE SECOND HALF OF THE 20TH CENTURY

The Seven Sisters

The scramble to fill the increasing demand for oil let to further exploration and production and to further consolidation in the industry. "[Oil] output was regularly expanded—from nine to ten per cent yearly—and exploration uncovered sufficient new reserves to supply mass consumption and the conversion of a large part of the economy to oil."[22] By the end of World War II there were seven large, multinational oil companies, based in the United States and Britain, that held a near monopoly over deliveries of oil in the industrialized world. These seven large companies formed a cartel that is often called the Seven Sisters. The Seven Sisters included the British companies Royal Dutch Shell and British Petroleum (BP) and the American companies Exxon (former Standard Oil of New Jersey), Gulf, Texaco (the latter two created after the Splindletop discovery), Mobil (former Standard Oil of New York), and Standard Oil of California (later Chevron).[23] They controlled 92 percent of reserves and 88 percent of production until the 1950s.[24] These companies, mostly privately owned, became partial owners or controllers of the assets of oil-producing countries. They obtained contracts from producer countries that enabled them to control the exploration and production of oil resources in these countries, while paying royalties to the countries on any profits they made from the successful exploration and production of the countries' oil reserves. The companies essentially controlled the world's oil trade, even the oil refining and transportation aspects. Through the 1950s the Seven Sisters dominated the world oil market, with three-fourths of the world's refinery capacity and distribution, nearly one-third of the oil tanker fleets, and a good share of maritime traffic, and they were able to keep prices high for all their oil-related activities without much competition.

OPEC and the Oil Crisis of 1973–1974

Industry analysts sometimes cite the dominance of these oil companies in the middle of the 20th century as a contributor to many of the energy-related issues that resonate in the world today, such as the oil crisis in America in

1973–1974. The roots of this oil crisis are often traced back to 1959 and 1960, when new oil companies began to challenge the dominance of the Seven Sisters cartel, bringing about pricing reductions. This development destabilized the relationships among the seven large companies and between the seven companies and the countries in which they produced oil. This shift in balance of power from the oil-producing companies to the oil-producing countries occurs naturally over time: Initially the host government needs the oil industry to develop the resources that it cannot develop on its own; the host government has little to offer except future profits. Once the industry is in place, however, the oil company cannot simply pick up and go; it needs to remain where the natural resource is, and thus the host government gains a bargaining chip. When one of the Seven Sisters, Standard Oil of California, decided to defer an oil price cut by reducing the royalties it paid to oil-producing countries for developing oil reserves in these countries, the oil-producing countries formed the Organization of Petroleum Exporting Countries (OPEC), a multinational organization aimed at coordinating oil production policies and pricing among oil-producing nations. Functioning in a similar way to the Achnacarry Accords between the major oil companies at the time, OPEC standardizes the amount of oil exported by its members, thus regulating the price of oil. When this organization was formed, its highest priority was to establish arrangements between major oil companies and oil-producing countries that were more favorable to the latter. When it was founded in 1960, its original members were Saudi Arabia, Iran, Iraq, Kuwait, and Venezuela. Today member nations, which together produce 40 percent of the world's oil, are Algeria, Indonesia, Iran, Iraq, Kuwait, Libya, Nigeria, Qatar, Saudi Arabia, the United Arab Emirates, and Venezuela. It is the founding of OPEC, say analysts, that connects the Seven Sisters to the oil crisis of the early 1970s, for it did not take long before OPEC began to exert significant influence over the oil market.[25] In 1970 the Middle East owned most of the world's oil reserves, with the Seven Sisters controlling 70 percent of the Western world's oil production. By 1973 oil had commanded nearly half of the world's total primary energy supply, double its share in 1949.[26] And in 1970 the Seven Sisters decided to reduce their investments in the Middle East to keep prices down and deflect competition. This policy resulted in sudden oil shortages in the United States in the winter of 1970–1971. But the real crisis began when a coalition of Arab nations in the Middle East became entrenched in a political conflict in 1973, engaging in a brief war with Israel known as the Yom Kippur War. The war started in October of that year. Because the United States and other Western nations had shown support for Israel during this war, the Middle Eastern oil-exporting countries that were part of OPEC decided to stop exporting oil to the United States and the other Western nations to exert political pressure.

This gave OPEC the opportunity to carry out a coup on oil prices, raising them from $3.0111 a barrel in October 1973, at the start of the embargo, or stoppage, of oil exports, to $11.651 in January 1974.

The oil embargo ended in March 1974, but it had great implications for the U.S. economy and American sentiments about energy. American consumers waited on long lines for gasoline that cost four times the amount it cost just a few months earlier. The situation led to greater conservation in oil-consumer countries such as the United States; in fact, between 1979 and 1982 overall energy consumption in the United States decreased 10 percent, and consumption in the industrial sector dropped by 20 percent.[27] The shock to the American energy system also resulted in the establishment of government policies to prevent the United States from being caught short on oil should another oil emergency occur. The Energy Conservation and Policy Act of 1975, for example, established the Strategic Petroleum Reserve (SPR), an emergency stockpile of oil, which was recently ordered filled to capacity. Also, the U.S. president who took office in 1977, Jimmy Carter, approved additional funding for the development of alternative energy resources to offset the country's dependency on oil.

Shortage of Oil and Other Nonrenewable Energy Sources

Thus, during this period, energy-related issues such as energy crises and shortfalls, energy affordability and economics, and foreign oil dependence, political instability, and foreign relations all converged upon the United States and other oil-consuming countries. In 1970 U.S. production of crude oil reached its highest level, at 9.4 million barrels per day (11.7 million including both petroleum and natural gas), as reported by the EIA.[28] Since then production in the lower 48 states has been generally declining. Thus, the peak in U.S. oil production occurred in 1970, and there are ample data to suggest that world oil production will peak in the early part of the 21st century.[29] Yet the world's population is now four times higher than it was when that first oil field was struck in Titusville, and the demand for oil around the world shows no signs of slowing. The largest oil-consuming countries, China and the United States, appear to be demanding more oil than ever. China is the world's second largest consumer of oil, behind the United States, with a total demand of 7.4 million barrels per day projected in 2006 by the EIA.[30] China's oil demand is projected by EIA to reach 14.2 million barrels a day by 2025. Meanwhile, average U.S. petroleum consumption stood at about 20.6 million barrels a day during 2006, which was less than the 2005 average of 20.8 million.[31]

There are still many who contend that the future oil supply situation is not as dire as some predict. The U.S. Geological Survey and the EIA both report that oil and fossil fuel production will peak much later, about 50 to 100 years into the future, and that they will remain dominant energy sources for at least another 25 years.[32] Other commentators point to improvements in oil extraction technology and the discovery of new supplies of oil, including vast undersea pools of oil in the Gulf of Mexico and off the coasts of West Africa and Brazil.

Whatever the case, it is true that oil is the dominant source of energy for the entire world. It replaced coal as the world's primary energy source in 1946; the world now relies on oil-based fuels for about 90 percent of its energy for transportation, using more than 26 billion barrels of oil each year. Furthermore, the manufacture of 95 percent of all goods and food products worldwide involves the use of petroleum, which is the basis for fibers, polymers, packing materials, plastics, and more.[33] In addition, the world's consumption of other hydrocarbon fuels aside from oil has proceeded at a rapid pace. Natural gas, for example, is the resource of choice for both heat and electricity in the modern world.

Natural gas usage is of particular concern, because its consumption is especially on the rise; the number of central electrical power plants that use natural gas has been growing each year. Some experts project that the production peak for natural gas is only a decade or two later than that of oil. According to BP, the world's resources-to-production (R/P) ratio for natural gas declined to 65.1 years in 2005 (which, however, is still well above oil's R/P ratio of 40.6 years). Coal is another resource whose supplies are waning, though coal is recognized to be more available than other hydrocarbon fuels, with estimates of remaining reserves ranging from 200 to 300 years.[34]

NUCLEAR ENERGY

The 20th century also presented issues related to the development of another form of energy—nuclear energy. In 1896 Antoine Henri Becquerel, a French professor and scientist, made a discovery that was foundational to the development of nuclear energy. He discovered the radioactivity of uranium. Radioactivity is a property possessed by some elements, by which energy can be spontaneously emitted during disintegration of these elements' atomic nuclei. Becquerel shared the 1903 Nobel Prize in physics with his colleagues Marie and Pierre Curie, who had conducted further research into the phenomenon. In 1939 scientists discovered that it was possible to maintain a fission reaction capable of releasing enormous amounts of energy. Fission is the splitting apart of the atomic nucleus of a radioactive element, such as uranium, by the impact of a subatomic particle, or neutron, which results in the release of large amounts of energy.

Introduction

In 1942 U.S. scientists produced nuclear energy in a sustained nuclear reaction, and in the 1940s nuclear power was used in the creation of atomic bombs. It was an atomic bomb that destroyed the cities of Hiroshima and Nagasaki in 1945. Electricity was generated for the first time by a nuclear reactor in 1951—in the United States at an experimental station in Idaho. In 1957 the first commercial nuclear power plant began operating in England. In nuclear power plants, electrical energy is produced by controlled fission reactions that result in the production of steam that turns turbines. But for several reasons nuclear power plants have never reached the energy-producing output that they were once expected to achieve. The uranium and plutonium that are used in fission reactions are nonrenewable resources, and the cost of setting up a nuclear power plant is very high. But even beyond this, the production of nuclear energy carries many risks. The greatest is the possibility of the leakage of the radioactivity that is produced when a nuclear reaction takes place. In 1979 at a nuclear power plant in Three Mile Island, Pennsylvania, a small amount of radioactive material leaked out into the air and water as a result of equipment failure and human error. Far worse than this accident was one that occurred in 1986 in Chernobyl, in the Ukraine, which was then part of the Soviet Union. A reactor in a nuclear power plant overheated, and the reactor's lid was blown off, spewing a cloud of radioactive waste into the atmosphere that affected all of Europe. Approximately 135,000 people were evacuated from surrounding areas immediately after the accident, and a total of 30 fatalities resulted directly from the accident. The high levels of radiation covering the 20-mile radius surrounding the accident site have had lasting and devastating effects on the lives of the people in the region in countless other ways. For example, an enormous increase in rates of thyroid cancer—from four to six cases per million Ukranian young children in the years preceding the accident to 45 cases per million from 1986 to 1997—has surfaced since the accident.[35] In addition, in the years since the accident first occurred, more than 300,000 people have had to be resettled from areas contaminated by radiation.

Aside from operational safety issues, there are other problems associated with nuclear power that have incited many emotionally charged arguments about its usefulness as a form of energy. For example, there is still great uncertainty about the safest way to handle, store, and dispose of nuclear waste, which has the potential to expose people and the environment to harmful radiation. Also, there is a concern that nuclear power plants may be prime targets for terrorist attacks, since an attack on a nuclear power plant can result in a massive, uncontrolled radioactive explosion. Because of all the risks involved in nuclear power production, stringent government intervention and regulation of nuclear power plants are necessary.

ALTERNATIVE SOURCES OF ENERGY

Besides nuclear energy, there are many other alternatives to fossil fuels that are being explored in the world today, but doubts about whether these alternatives can ever achieve the ease and convenience of fossil fuels have hindered their widespread implementation. Descriptions of some of these alternative energy sources follow.

Biomass

Biomass is both a renewable and an alternative source of energy. It is primarily burned for heat and electricity. Wood chips, sawdust, black liquor, and by-products of pulp and paper are forms of biomass that may be burned in large power plants instead of coal. Biomass energy can be far less polluting than coal energy, but some level of greenhouse gas emissions is still involved in its use.

Biomass may also be used as an alternative car fuel. It can be combined with gasoline to form a substance known as gasohol, or it may be converted into the alternative auto fuels ethanol and methanol. Ethanol is typically produced from corn in the United States. Both of these fuels are better for the environment than gasoline, but automobile engines need to be specially designed to accommodate their use.

Another alternative biomass fuel is biodiesel, which can be used in any standard diesel engine. It is derived from vegetable oil and other plant fats, as well as animal fat. Biodiesel is reportedly highly energy efficient, yielding more than three times the energy needed to produce it. There are more than 500 biodiesel filling stations in the United States, and the National Aeronautics and Space Administration (NASA) and the U.S. military also use biodiesel in selected vehicles and equipment.[36]

Animal waste has also been used as a biomass fuel, but it is associated with a high pollutant effect. Also, traditional fuels need to be used in the growing, processing, conversion, and transportation of many biomass sources of energy, minimizing their value as fossil fuel alternatives.

Solar Power

Solar power is another alternative and renewable source of energy. Solar energy can be harnessed passively or actively. An example of a passive use of solar energy would be designing a home with most of its windows facing south so as to receive the rays of the daylight Sun and with thick walls and floors that could absorb the heat and release it at night when the Earth cools down. An example of an active use of solar energy would be to install solar panels on a home for heat or electricity. The solar panels that are used for electricity are flat panels made up of photovoltaic cells, which are tiny, crystalline semiconductor structures made from silicon, with boron and phosphorus added. The boron and phosphorus create a positively charged region and a negatively charged region within the cell that produce a current when stimulated by the Sun's rays. Solar panels that are used for heating do not have cells but are specially designed to absorb heat from the Sun. The energy generated is then used for hot-water heating.

When used to generate electricity, a solar system needs to be connected to the electrical power grid and thus will automatically shut down when power to the grid fails, unless a battery backup system is used to store a portion of the solar-generated electricity. Batteries can be expensive, and solar panels are also costly to install, but many states offer rebate programs. On the other hand, solar power plants to generate electricity for large areas, as opposed to individual homes and businesses, are expensive to build and require a great deal of space. Nonetheless, there are many advantages to having solar-generated electricity or hot water heat in a home or business, including reduced (or absent, if the solar system is large enough) utility bills. At the same time, however, solar energy systems are not for everyone, as they require a certain amount of space to allow enough panels to be installed to make the system worthwhile, should be exposed to little or no shading from trees or surrounding buildings, and generally need to be installed in an east-, west-, or south-facing area. Additionally, cloud cover, the winter season, and northern latitudes limit the energy that solar systems can produce.

Geothermal Energy

Geothermal energy is another type of alternative energy source. It harnesses the temperature of the Earth's core as a source of energy. Geothermal systems involve the drilling of wells into underground areas that contain water that is heated by the Earth's core. The hot water or steam that travels up the wells can be used to generate electricity or to heat or cool a structure. When geothermal energy is used for electricity, the steam or hot water is harnessed as a force to turn turbines that generate electricity. When geothermal energy is used for heating and cooling, the pressure of the steam or hot air is harnessed to distribute hot or cold air within a structure. Geothermal energy has little environmental impact, but geothermal power plants can be unwieldy, geothermal energy has limited efficiency, and geothermal systems are also highly location dependent; reserves of geothermal energy are not available everywhere.

Hydropower

Water power, or hydropower, is another form of renewable energy. It has been used too long to be considered alternative; today it is the most widely used renewable resource in the world and accounts for almost half of all renewable energy used in the United States. However, while it is renewable and relatively benign, it may also be associated with substantial environmental risks. When hydropower derives from falling water that runs through a dam and turns turbines, thereby generating electricity, it is called hydroelectric power. Dams and reservoirs of large hydroelectric power systems have been known to cause significant flood damage in portions of the United States and elsewhere in the world, wiping out local ecosystems.

Hydrogen

Another alternative source of energy is hydrogen, which is extracted from a compound such as water, natural gas or other hydrocarbon fuels, or ethanol. It can be harnessed for power in the form of a fuel cell, a device similar to a battery that contains a mixture of hydrogen and oxygen that engage in a continuously flowing chemical reaction that produces water and energy. Fuel cells are stacked together to produce energy. Limitations of hydrogen are that the process used to extract it is very energy intensive and there are difficulties involved in storing it. It must be stored in liquid form below −423°F, a temperature that demands extreme insulation (and caution). It may also be stored as a compressed gas at a pressure of a few thousand pounds per square inch. Another downside is that automobiles and other equipment must be altered to allow the use of energy from hydrogen via fuel cells. Nonetheless, hydrogen is nonpolluting and versatile, and it is being researched for applications in electric power plants and to replace gasoline in motor vehicles.

Wind Power

Wind power is used to generate electricity by turning turbines. It is the fastest-growing renewable energy source in the United States. As water power has, wind power has been used far too long to be truly considered alternative. Complaints associated with wind power include that wind is irregular and unpredictable and that wind turbines can be loud, large, unsightly, and hazardous to birds. Wind turbines are also limited in efficiency because of the sporadic nature of wind. However, wind is an inexhaustible resource, and wind farms, sites consisting of multiple wind turbines that generate electricity, are capable of producing enough electricity to rival that of conventional, coal-fired power plants. The United States has some of the largest wind farms in the world. The wind power market is also highly developed in Western Europe, particularly in Denmark, Germany, and Spain.

The history of energy production and consumption has placed at the forefront many issues and challenges that are still being played out in the world today.

But it is mainly in the past 30 years that the world has confronted its most looming energy-related challenges: disruptions in oil supply, price fluctuations and vulnerable economies, environmental dilemmas resulting from polluting fossil fuel emissions, the ominous possibility of running short of oil and other sources of energy at a time when the world is most dependent on them, and shakeups in relationships among countries that depend for their economic and political stability on an ever-flowing supply of oil and natural gas. These problems have had implications both in the United States and throughout the countries of Asia, the Middle East, Europe, Africa, and Latin America. The next two chapters will explore how the United States, as well as China, Germany, Iran, Saudi Arabia, Nigeria, and Venezuela, have confronted these issues.

---

1 Donald G. Kaufman and Cecilia M. Franz. *Biosphere 2000 . . . Protecting Our Global Environment,* 3d ed. Dubuque, Iowa: Kendall/Hunt, 2000, p. 194.

2 F. Peter W. Winteringham. *Energy Use and the Environment.* Chelsea, Mich.: Lewis, 1992, p. 2.

3 Tom Mast. *Over a Barrel: A Simple Guide to the Oil Shortage.* Austin, Tex.: Hayden, 2005, p. 18.

4 Jean-Claude Debeir et al. *In the Servitude of Power: Energy and Civilization through the Ages.* London: Zed Books, 1990, p. 13.

5 According to the Energy Information Administration (EIA) the popularity of nonrenewable resources persists in the world today, as only about 13 percent of the world's primary energy supply was from renewable resources, including hydropower, in 2004, and the rest from nonrenewable energy sources. In the United States in particular, renewable energy accounted for just 6 percent of energy consumption in 2005. The remaining percentage was mainly from fossil fuels—that is, from coal, oil, and natural gas. The relevant statistics can be found in the chapter of the EIA's *International Energy Annual* titled "World Energy Overview." Available online. URL: http://www.eia.doe.gov/iea/overview.html and in *Annual Energy Review 2005,* p. 280. Available online. URL: http://www.eia.doe.gov/oiaf/aeo/index.html. Both accessed August 2, 2006.

6 See the EIA report *Emissions of Greenhouse Gases in the United States 2005,* p. 5. Available online. URL: http://www.eia.doe.gov/oiaf/1605/ggrpt/index.html. Accessed March 15, 2007.

7 See the EIA report *International Energy Outlook 2006,* p. 73. Available online. URL: http://www.eia.doe.gov/oiaf/ieo/index.html. Accessed August 2, 2006.

8 Presentation of Robert T. Watson, chair, Intergovernmental Panel on Climate Change (IPCC), at the Sixth Conference of Parties to the United Nations Framework Convention on Climate Change, November 13, 2000. Available at the IPCC Web site. URL: http://www.ipcc.ch/press/sp-cop6.htm. Accessed August 3, 2006.

9 Charles Dickens. *Bleak House.* New York: W. W. Norton, 1977, p. 5.

10 Jean-Claude Debeir et al. *In the Servitude of Power,* p. 111.

11 See entry for 1850–1980 in the Energy Information Administration's Energy Kid's Page, section "History: Timelines." Available online. URL: http://www.eia.doe.gov/kids/history/timelines/index.html. Accessed August 3, 2006.

[12] Jean-Claude Debeir et al. *In the Servitude of Power*, p. 118.

[13] Tom Mast. *Over a Barrel*, p. 15.

[14] See the EIA's *Annual Energy Review 2005*, p. 8.

[15] *BP Statistical Review of World Energy 2006*, p. 11. Available online. URL: www.bp.com/statisticalreview. Accessed March 15, 2007.

[16] The relevant statistics can be found in the EIA's *International Energy Annual:* "World Energy Overview." Available online. URL: http://www.eia.doe.gov/iea/overview.html. Accessed August 2, 2006.

[17] Paul Roberts. *The End of Oil: On the Edge of a Perilous New World.* Boston: Houghton Mifflin, 2004, p. 15.

[18] Jean-Claude Debeir et al. *In the Servitude of Power*, p. 113.

[19] Jean-Claude Debeir et al. *In the Servitude of Power*, p. 114.

[20] Douglas G. Brinkley. *Wheels for the World: Henry Ford, His Company, and a Century of Progress.* New York: Penguin Books, 2004, pp. 333–351.

[21] Jean-Claude Debeir et al. *In the Servitude of Power*, p. 114.

[22] Jean-Claude Debeir et al. *In the Servitude of Power*, p. 130.

[23] Today the Seven Sisters have become four: ExxonMobil, Chevron-Texaco, BP, and Royal Dutch Shell.

[24] Jean-Claude Debeir et al. *In the Servitude of Power*, p. 136.

[25] See M. A. Adelman. *The Economics of Petroleum Supply: Papers by M. A. Adelman, 1963–1993.* Cambridge, Mass.: The MIT Press, 1993, pp. 329–357.

[26] Richard Kerr. "OPEC's Second Coming." *Science*, August 21, 1998, p. 1,129.

[27] See entry for 1979–1982 in the Energy Information Administration's Energy Kid's Page, section "History: Timelines." Accessed August 3, 2006. See also EIA data for 1979 and 1982 in Table 1.3, Energy Consumption by Source, and Table 1.5, Energy Consumption, Expenditures, and Emissions Indicators, 1949–2005. Available online. URL: http://www.eia.doe.gov/emeu/aer/overview.html.

[28] See entry for 1970 in the Energy Information Administration's Energy Kid's Page, sections "History: Timelines." See also EIA's report *Annual Energy Review 2005*, p. xxiii.

[29] See the EIA's online resource "Energy Plug: Long-Term World Oil Supply: A Resource Base/Production Path Analysis." Available online. URL: http://www.eia.doe.gov/emeu/plugs/plworld.html. Accessed August 3, 2006.

[30] The relevant statistics can be found in the EIA's *Country Analysis Briefs: China* (August 2006), p. 2. Available online. URL: http://www.eia.doe.gov/emeu/cabs/contents.html.

[31] See the EIA's *Monthly Energy Overview* (February 2007), p. 43. Available online. URL: http://www.eia.doe.gov/emeu/mer/overview.html.

[32] See "Energy Plug: Long-Term World Oil Supply: A Resource Base/Production Path Analysis." Available online. URL: http://www.eia.doe.gov/emeu/plugs/plworld.html. Accessed Augu: : 3, 2006.

33 Chris Skrebowski. "Joining the Dots." Paper presented at 2004 Energy Institute Conference: Oil Depletion: No Problem, Concern or Crisis? London, November 10, 2004.

34 David Goodstein. *Out of Gas.* New York: W. W. Norton, 2004, p. 21.

35 Mykola D. Tronko et al. "Thyroid Carcinoma in Children and Adolescents in Ukraine after the Chernobyl Nuclear Accident: Statistical Data and Clinicomorphologic Characteristics." *Cancer,* 86, no. 1 (July 1, 1999), pp. 149–156.

36 The Tribune Co. "Foreign Fuel Subsidy Burns Taxpayers." *Tampa Tribune,* December 26, 2005, p. 14.

2

Focus on the United States

THE DEMAND FOR OIL

On July 28, 2005, members of the U.S. House of Representatives were debating a new U.S. energy policy. This policy, the Energy Policy Act of 2005, would be passed by the U.S. Congress the next day and signed into law 10 days later. The debate on July 28, however, raised some interesting, if not alarming, questions about the future of U.S. energy. The chairman of the House Committee on Energy and Commerce, Congressman Joe Barton, remarked, "Here is the fundamental problem we face on our mobile energy sources. We consume 21 million barrels of oil every day in this country, and we only produce eight. You subtract the eight out of 21 and you get 13. So, we are importing about 13 million barrels of oil a day. On our best day, the United States of America did not produce more than 10 million barrels of oil a day—on our best day. There is nothing we can do that is going to generate another 13 million barrels of oil produced in the confines of the United States of America. It cannot be done."[1]

In his concern that the United States does not have the resources to keep up with its energy demands, Congressman Barton was not alone; nor was he unsupported by national energy data. According to EIA statistics, each person in the United States used an average of 337 million Btu of energy in 2005. (A Btu is a British thermal unit, a measurement of energy that is gauged on the amount of heat required to raise the temperature of one pound of water by one degree Fahrenheit.) This level of energy usage equates to about two times the energy per capita of people living in Japan and Europe, and 10 times that of the average world citizen.[2] In the meantime, growth in U.S. energy consumption is projected only to increase, outpacing U.S. energy production over the next 25 years.

Yet the overwhelming concern, as Congressman Barton observed, is not that the United States will not be able to meet all of its energy demands but that it will not be able to keep up with its demand for *oil,* its primary source of energy for more than a half-century. Energy is consumed in four broad categories: residential, commercial, industrial, and transportation. According to the EIA, the transportation sector accounted for 28 percent of energy usage in the United States in 2005. In addition, almost 100 percent of all transportation—whether by land, sea, or air—was fueled by oil. Of all the ways of getting from one place to another (electric railways, tramways, diesel railways, buses, trucks, and individual cars), people in the United States overwhelmingly favor the least energy-efficient form of transportation: individual cars, powered by gasoline. During 2005, motor gasoline accounted for 44 percent of all the oil consumed each day in the United States. According to David Garman, U.S. under secretary for energy, science, and environment, "The most urgent need is to address our transportation sector, which consumes two-thirds of all U.S. oil and is still growing. Petroleum imports already supply more than 57 percent of U.S. domestic needs, and those imports are projected to increase to more than 68 percent by 2025 under a business-as-usual scenario."[3]

Oil Imports

Oil's status as the primary U.S. energy source has been firmly established since at least the mid-1900s, when decades of oil discoveries and a booming transportation sector greatly increased the nation's reliance on gasoline. Today the United States consumes more oil than any other region in the world. EIA data for early 2007 reported a U.S. demand of 22 million barrels of petroleum per day in February, and U.S. crude oil production is now at a 50-year low.[4]

In terms of oil production, the United States' "best day," as Barton referred to it, came and went in 1970. That year U.S. production of crude oil reached 9.4 million barrels a day—the highest level of oil production ever achieved in the nation. In 1970 the United States produced about 3.1 billions of barrels of oil annually. Since then oil production in the lower 48 U.S. states has been declining, according to the EIA. Quantities of oil produced in the United States in late 2006 were about 50 percent below 1970 quantities.[5] In 1956 a renowned U.S. geophysicist named M. King Hubbert calculated that U.S. oil production would reach its peak in the 1970s and decline thereafter, basing this projection on his conclusion that oil production rises and falls according to a bell curve.[6] Needless to say, his projection regarding U.S. peak oil production turned out to be accurate.

The United States produced sufficient quantities of oil to meet its oil demands until about 50 years ago. During World War II there were even adequate supplies of U.S. oil for the country's operations both home and abroad. Around this time, however, American habits, especially in relation to transportation, began to require large amounts of oil, and consumption soon began to outpace production. To meet increased oil demand, the nation started importing more energy in the 1950s and has relied steadily on imports since then. In the early 1990s the United States for the first time imported more oil and refined oil products than it produced, and greater quantities of imports have been added each year as U.S. demand for petroleum grows and domestic production declines. In 2005, according to EIA data, 30 percent of all energy consumed in the United States was imported from other countries, and the EIA projected that the United States would import about 17.7 million barrels of oil each day in 2030—about 61 percent of total U.S. oil demand.[7]

Oil Production in the United States

Part of the predicament, as many see it, is that there do not appear to be any major U.S. oil discoveries forthcoming; nor do there appear to be untapped fields capable of producing enough oil to make a substantial impact on U.S. oil demand. Even the Arctic National Wildlife Refuge (ANWR), a 19-million-acre tract of federally protected land near an oil-rich site in Alaska and the center of a controversial oil-drilling proposal from the administration of the U.S. president George W. Bush, is estimated to be capable of meeting U.S. petroleum demands for only six months to, at the most, several years, according to the U.S. Geological Survey (USGS).[8]

Despite improvements in the technologies used to find oil, oil discoveries have steadily declined over the past 50 years. U.S. "oil rushes" seem to be phenomena of the past, confined to the first half of the 1900s. These flurries of U.S. oil exploration and production followed discoveries such as the 1901 Spindletop gusher. One of the last major examples of an incident like this occurred in the 1930s, when a significant source of oil was discovered in East Texas. Columbus Marion ("Dad") Joiner, convinced that land near Tyler, Texas, held oil, leased the land so that he could drill it for oil. On October 5, 1930, beneath 140,000 acres of land, Joiner struck possibly the largest pool of oil ever found in the United States. An oil entrepreneur named H. L. Hunt bought Joiner's leases and sold them to oil companies, and the land eventually yielded 4 billion barrels of oil. Even with such a massive discovery of oil domestically, the United States in the 1930s was already branching out into other countries to obtain the greatly coveted energy resource. In the 1930s Chevron, Texaco, Exxon, and Mobil bought the rights to extensive oil fields in Saudi Arabia—today one of the top three suppliers of crude oil to the United States.

Since the 1930s the United States has seen only a modest overall increase in the amount of its proven oil reserves, the quantities of oil that can actually be recovered under existing economic conditions. There were 18.5 billion barrels of proven oil reserves in 1939 and 21.7 billion barrels in 2005, a difference of only 3.2 billion barrels in 65 years. Some research indicates that U.S. oil discoveries actually peaked in the 1930s, and many researchers point to U.S. proven oil reserves today as a direct measure of U.S. ability to meet future oil demands. Historically estimates of what is left in oil fields—that is, the reserves—become more precise as the end of oil fields' production draws near. Most of the United States' 500,000 or so producing oil wells are generally producing only a few barrels of oil each day. Therefore, researchers have concluded that estimates of proven U.S. oil reserves can be considered quite accurate.

There were 21.7 billion barrels of total proven oil reserves in the United States at the end of 2005, according to the EIA.[9] This represented only about 3 percent of the total reserves in the world. More than 80 percent of these U.S. oil reserves were located in four states where oil production has been declining in recent years, including Texas, where 22 percent of total U.S. oil reserves are concentrated. In fact, U.S. oil reserves figures have, as have oil production figures, declined overall since 1970, when U.S. proven oil reserves peaked at 39 billion barrels, bolstered by a reserve of oil tapped in the North Slope region of Alaska. The 21.7 billion barrels of proven oil reserves in the United States in 2005 represent a 44 percent decline in reserves since 1970. Since 1990 U.S. proven oil reserves have already diminished more than 17 percent, with the largest single-year decline, 1.6 billion barrels, in 1991.

Global Oil Reserves

Even so, the likelihood of an oil shortage would not be such a cause for concern for the United States were there an infinity of oil supplies to import from other countries. Yet research has generated global oil reserves figures that are also disconcerting to many people. On a worldwide scale, about 5 billion barrels of new oil reserves were discovered in 2005, while 30 billion barrels of oil were consumed worldwide, according to the International Energy Agency (IEA). In June 2006 the *BP Statistical Review of World Energy* reported a worldwide oil reserves-to-production (R/P) ratio of 40.6 for 2005.[10] R/P ratios, calculated by dividing the proven oil reserves in the ground at the end of the current year by total oil production in that same year, provide a measure of the number of years that oil reserves would last if production were to continue at the level of the current year. At 2005 production rates, worldwide oil supplies would run out in about 40 years, by BP estimates.

M. King Hubbert used his bell curve model to calculate not only when U.S. oil production would peak but also when worldwide oil production would peak. He concluded that the world's oil production would reach its peak around the year 2000.[11] A former colleague of Hubbert, the petroleum geologist and Princeton University professor emeritus Kenneth S. Deffeyes, has stated that, on the basis of Hubbert's calculations, the world's oil production either is peaking now or has already peaked.[12] Colin J. Campbell, a British geologist, independent oil-industry consultant, and member of the American Association of Petroleum Geologists (AAPG) since 1959, joins Hubbert and Deffeyes in the opinion that world oil production will reach its peak in the first decade of the 21st century. Campbell has suggested that grave economic and political consequences will result as nations around the world experience energy scarcity and sustained increases in the price of oil.[13]

The Debate over Oil Demand and Supply

However, oil reserves as a measure of the ability to meet current oil demands have frequently been the subject of debate. The EIA, for example, projects that world oil production will peak as late as 2037, assuming an average growth in demand of 2 percent per year based on previous years' data.[14] Some researchers say that the industry experts who cite ever-improving recovery technology for finding oil as a promise of greater reserves should feel validated by a 2000 USGS report stating that the world has 20 percent more oil waiting to be discovered than originally estimated in 1994.[15] And then there are some commentators who dismiss "peak oil" theories altogether. Mark Jaccard, a British author and sustainable energy expert, was a recognized supporter of peak oil calculations for 20 years. In 2005, however, Jaccard surprised the energy industry by concluding in his book *Sustainable Fossil Fuels: The Unusual Suspect in the Quest for Clean and Enduring Energy* that there is enough oil, gas, and coal to sustain the world at least for another 500 years. He acknowledged that oil and gas are in limited supply but pointed to energy efficiency technologies that may possibly extend fossil fuels' availability indefinitely. He argued that modifying the *way* we use fossil fuels is an essential part of any plan to achieve energy sustainability. He especially highlighted coal's potential for generating hydrogen, a clean fuel, as well as carbon dioxide.[16]

M. A. Adelman was another detractor from peak oil theories. A professor of economics emeritus at the Massachusetts Institute of Technology (MIT) and an oil economics authority, Adelman contended that the world is not running out of oil. He argued that calculations of peak oil production based on the world's proven oil reserves are flawed because oil reserves rise and fall in a continuous cycle of replenishment and depletion.[17]

Other experts point to improvements in oil exploration, extraction, and production techniques that over the years have led to more extensive recovery of oil in the United States and elsewhere. They argue that there is no reason to think that further improvements and, subsequently, greater reserves of oil are not in the United States' future. For example, when the U.S. oil industry was in its infancy, around the year 1900, oil companies routinely relied on surface geology, or a basic evaluation of land formations, to find likely spots for oil. Visible features of the Earth's surface, such as seeping oil or natural gas, craters caused by escaping gas, or the inverted cup–shaped hills that are characteristic of anticlines, provided straightforward evidence of oil reservoirs. By 1925, however, geologists were routinely using seismic exploration, sending high-intensity sound waves into earth or water (at first, mainly through the use of explosives) and then interpreting the resulting echoes for evidence of the presence of oil reservoirs. Today, however, most oil exploration relies on highly sophisticated technology: hydraulic surveying, three-dimensional digital imaging, undersea seismic exploration using compressed air, and more. Similarly, drilling techniques have undergone continuous improvements since U.S. oil production began: from rotary drilling to offshore drilling barges, from deepwater drill ships to semisubmersible oil-drilling rigs, and, more recently, to horizontal drilling. Many commentators suggest that there are virtually no areas where oil exploration cannot be successfully carried out, presuming geological studies have indicated a good possibility of finding oil.

In the 1970s Peter Odell, an energy consultant and professor of international energy studies at Erasmus University, Rotterdam, the Netherlands, challenged the view that worldwide oil supplies were waning, citing unexpected increases in proven oil reserves during prior years, new technologies for recovering oil, and oil price increases that encouraged oil companies to seek out more oil than they did when oil was less expensive. He argued that the oil industry would continue to expand into the mid-21st century.[18] As has Odell, Michael C. Lynch, a prominent oil economist at the Massachusetts Institute of Technology, has countered the claims that the world has reached or is reaching peak oil production. He has asserted that these claims underestimate the oil-producing potential of nations outside OPEC as well as the technological changes that can result in better oil recovery.[19]

Nevertheless, the notion that demand for oil will continue to rise over the next 20 years is widely accepted, and it is a known fact that oil and other hydrocarbon fuels are nonrenewable resources, by nature limited in supply. Many researchers agree that the world has used such resources too aggressively. Some experts claim that the world has burned approximately one-half of all the recoverable oil that was created over a period of 100 million years.[20]

U.S. Energy Crises

How has the United States addressed this issue of dwindling supplies of its primary energy resources? The U.S. strategy for responding to the problem has had two dimensions: (1) finding short-term solutions to immediate energy crises and (2) making long-term provisions for the projected decline in primary energy resource supplies.

Some commentators point out that in the United States it is the short-term solutions that have received the most emphasis, perhaps too much emphasis. This emphasis is due, in part, to energy crises that triggered knee-jerk reactions in the United States. Since the 1970s the United States has met with a number of immediate shortfalls in energy supply and/or sudden, sharp increases in the cost of energy resources—in other words, energy crises.

THE OPEC OIL EMBARGO IN 1973

The first of these crises resulted from the October 1973 Arab oil embargo, when Arab member countries of OPEC cut off oil imports to the United States and other Western nations because of their support for Israel during the Arab-Israeli conflict known as the Yom Kippur War. A simultaneous quadrupling of oil prices—the outcome of a concomitant OPEC decision to use its oil price-setting leverage to its advantage—compounded the crisis. It immediately became clear that overdependence on oil from Middle Eastern countries put the United States in a vulnerable position. The embargo lasted until March 1974 and left Americans idling for hours in long lines at gas stations to buy gas that was four times as expensive as it had been just six months earlier.

IRANIAN REVOLUTION IN 1979

In 1979 another oil crisis occurred, again in the wake of political instability in the Arab world, this time because of a change in Iranian leadership. The troubles were rooted in a 1953 change in leadership: Mohammad Reza Pahlavi had assumed power as the shah of Iran in 1953 after the overthrow of the Iranian prime minister Mohammad Mossadegh with the help of the U.S. and British governments. Whereas Mossadegh's oil policies had been antagonistic to the United States and Britain (he had nationalized the oil industry), Pahlavi was an American-friendly leader who opened Iranian oil reserves to Western companies. However, in 1979, Islamic fundamentalists who resented his pro-Western policies drove him out during the Iranian Revolution. A new Islamic republic emerged, based on fundamentalist Islamic principles and led by the national spiritual leader, Ayatollah Khomeini. This new regime reduced oil exports and raised the price of oil. Saudi Arabia and other OPEC nations increased their oil production in response, but Americans, for whom the 1973–74 oil crisis was still a fresh and bitter memory, panicked. The United

States experienced a widespread panic-driven oil demand, which drove the price of oil even higher and caused more long lines at the gas pumps.

THE IRAN-IRAQ WAR AND THE PERSIAN GULF WAR

Yet another crisis occurred in 1980 after Iraq, under the leadership of Saddam Hussein, its president from 1979 to 2003, invaded Iran. The conflict nearly drove Iranian oil production to a halt and greatly slashed oil production in Iraq. The situation impacted oil imports to the United States but did not affect the country as severely as the previous two crises had. In 1990, however, during the U.S. presidency of George H. W. Bush, Iraq invaded neighboring Kuwait under the leadership of Hussein, possibly to commandeer Kuwaiti oil operations. In response a coalition of U.S. and international military forces launched Operation Desert Storm, or the Persian Gulf War, to force Iraq out of Kuwait. Not only did the conflict threaten oil imports to the United States, but during the war Iraqi soldiers set oil fields ablaze in a gesture of defiance of Western countries.

OTHER EVENTS IMPACTING OIL PRICES

Another event that threatened the stability of U.S. oil prices and supplies was the September 11, 2001, terrorist attack on the United States. After the attack the United States experienced record-high oil prices and a general insecurity over oil imported from the Middle Eastern region. Other challenges that have affected U.S. energy supply in immediate and substantial ways are hurricanes and other unpredictable natural events. In August 2005 Hurricane Katrina struck the Gulf of Mexico, damaging oil production facilities, terminals, pipelines, and refineries off the coasts of Texas and Louisiana, which are responsible for a quarter of all domestic oil production. Oil production throughout the region declined significantly, causing a severe drop in U.S. oil output. The disruptions in the supply of gasoline and other refined oil products that followed the hurricane caused oil prices to spike nationwide. The price of oil reached a record high of $70.82 per barrel on August 30, 2005.

SHORT-TERM SOLUTIONS

Strategic Petroleum Reserve

The United States has used a number of different measures to address these kinds of sudden oil shortfalls and price spikes. One is the Strategic Petroleum Reserve (SPR), the world's largest emergency stockpile of oil, which now contains about 700 million barrels. Held in underground caverns along the coast in the Gulf of Mexico, the SPR has a 727-million-barrel capacity. The SPR rose out of the Energy Policy and Conservation Act of 1975, a U.S. energy law whose purpose was to reduce U.S. dependency on high-priced oil imported from politically unstable countries, to contend with energy shortage conditions, and to improve energy efficiency and conservation. This energy law authorized the stockpiling of up to 1 billion barrels of petroleum to tap into during energy emergencies. The impetus for establishing such an emergency oil stockpile was the 1973–74 oil crisis; the U.S. government saw the SPR as a way to sidestep similar future crises. The Energy Policy and Conservation Act gave the president of the United States the authority to withdraw crude oil from the SPR in response to an energy emergency and distribute it to oil companies by competitive sale.

The SPR has been drawn down twice. The first emergency drawdown of the SPR was authorized in January 1991, during the Persian Gulf War. In the middle of the effort to counter Iraq's invasion of Kuwait, the U.S. Department of Energy (DOE) implemented a plan to draw down and sell 33.75 million barrels of crude oil from the SPR because oil supplies were threatened. The drawdown proceeded on schedule, but world oil supplies and prices stabilized before the final sale, so the United States reduced the drawdown to 17.3 million barrels, selling them to 13 companies.

The second emergency drawdown was a response to the oil production disruption caused by Hurricane Katrina. President George W. Bush, acting in conjunction with the IEA, authorized a drawdown of crude oil from the SPR in September 2005. Secretary of Energy Samuel W. Bodman immediately authorized the sale of 30 million barrels of crude oil to U.S. markets. The DOE accepted the bids of five companies for a total sale of 11 million barrels.

Drawing from the SPR is one way the United States has attempted to minimize its vulnerability to oil shortages. However, because filling the SPR reduces the amount of oil available for current consumption, the SPR is controversial. For example, the Bush administration ordered the SPR filled to capacity in November 2001 in response to the threat to oil security posed by the September 11 terrorist attacks that year. When gas prices soared, 50 members of the U.S. House of Representatives requested that the president stop the filling of the SPR so that the amount of oil available for current use would increase and gas prices would stabilize. The president denied their request.

More recently, the secretary of energy was authorized to fill the reserve to its previously prescribed 1-billion-barrel capacity. This authority was granted by the Energy Policy Act of 2005, a statute more than 1,700 pages long that was passed by the U.S. Congress on July 29, 2005, and signed into law on August 8, 2005. Aside from authorizing the filling of the SPR, the law has been touted by its supporters as a comprehensive energy policy that tackles shortages in energy resources through such measures as developing alternative and renewable energy technologies to supplement the use of fossil fuels for energy. In January 2007, during the State of the Union address, President Bush asked Congress to expand the SPR even further than the Energy Policy Act of

2005 prescribed—to double its current capacity. In February 2007 the DOE announced expansion plans for two existing SPR storage locations and named a site in Mississippi where a new SPR storage location would be developed.

International Energy Agency

Another outcome of the 1973–74 oil crisis was that the United States joined other Western nations, including the United Kingdom, Germany, Canada, France, Spain, and Australia, in forming the IEA. The initial goal of the IEA was to coordinate the efforts of member countries during an energy crisis to maintain energy supply. Today the IEA's stated goal is to ensure the reliability, affordability, and environmental soundness of the energy supply in its member countries. The IEA uses energy intensity as an index to measure energy efficiency; an *intensity* is the energy used per unit of activity—for example, the gasoline used per mile driven by a car. The IEA has reported that as a result of declining energy intensities, significant energy savings began around the world in 1973, when the first oil "price shock" hit, and lasted through the mid- to late 1980s. The IEA concluded that the changes caused by the 1970s oil crises, including the revisions in energy policy that resulted from the crises, had a substantially greater impact on reducing energy consumption than energy conservation and efficiency policies implemented in the 1990s. Declining energy intensities, according to the IEA, led to significantly reduced energy costs since 1973, but the reductions in energy intensities have been much more modest since the late 1980s, so the rate of energy savings has slowed since then.[21] According to the IEA, energy use for cars is much greater in countries with low fuel prices and there is a correlation between higher fuel prices and lower vehicle fuel intensity and travel per capita.

Price Regulation

Another measure that the U.S. government has taken to deal with energy emergencies is to regulate the price of energy resources. In response to the 1979 oil crisis, for example, the Carter administration instituted price controls. However, the chief outcome of this decision was snaking lines at gas stations around the country. It has been proposed that Americans wasted up to 150,000 barrels of oil each day while their engines idled at gas stations.[22] The Carter administration also attempted to encourage citizens to conserve energy at home. In a symbolic effort to encourage conservation, President Carter had solar power panels and a woodburning stove installed at the White House. His administration also proposed removing price controls imposed during the administration of the U.S. president Richard Nixon in response to the 1973 oil crisis.

LONG-TERM STRATEGIES

Other U.S. policies attempt to address the issue of limited energy supply on a longer-term basis. Such policies have involved increasing energy supply, reducing energy demand, or attempting to do both. U.S. strategies to increase supply have involved attempts to tap into further sources of fuel domestically or to stimulate existing domestic production by offering tax incentives or other advantages. Another strategy has been to increase the amount of energy derived from renewable sources. This latter strategy has involved government programs to subsidize renewable energy production or to offer tax incentives to those who generate energy from renewable sources. U.S. measures to reduce demand have included efforts to encourage energy conservation and efficiency by regulating the levels of energy produced and consumed—for example, by requiring that vehicles meet a certain level of fuel efficiency. No matter what approach the United States has taken to increase energy supply or to reduce energy demand, however, nearly every strategy has sparked some type of controversy.

Increasing Energy Supply

ARCTIC NATIONAL WILDLIFE REFUGE

For example, the massive energy bill introduced by President George W. Bush in 2001 and later enacted into law as the Energy Policy Act of 2005 originally included a controversial plan to drill for oil on 2,000 federally protected acres within ANWR. As passed in August 2005, however, the energy policy did not include the ANWR plan, mainly because of lack of support. The proposal to drill in ANWR instead became a separate bill in the Senate and part of the president's proposed 2007 budget. But the Senate rejected the ANWR drilling plan in early 2006. By mid-2006, however, as gasoline prices climbed back up to post–Hurricane Katrina levels, members of Congress were expected to resurrect the debate over drilling in the ANWR site. The ANWR drilling plan has been controversial because the level of environmental damage that might result from drilling in the refuge and the quantity of oil that such drilling would actually produce are matters of great dispute. The USGS estimated that between 5.7 and 16 billion barrels of oil could be recovered from the site, while the Alaska Coalition, which opposes drilling in ANWR, has claimed that 3.2 billion barrels is a more realistic estimate. Even the USGS conceded that the most technically realistic estimate for the proposed drilling on the site was about 7.7 billion barrels.[23] Opponents of the ANWR plan add that the environmental risks of drilling would probably outweigh any benefits to domestic oil supply anyway. On the other hand, supporters of the ANWR plan say that the environmental impact of drilling on the site would be minimal because new technology makes possible the drilling of multiple wells with a single rig. They also point out that ANWR oil production could reduce fuel prices significantly because it would increase the amount of crude oil available to satisfy current consumption rates. The DOE, meanwhile, has indicated that drilling in ANWR would cause only a 4 percent reduction in the quantity of imports the United States is projected to take in between the present time and 2025.[24] The EIA has also suggested that the impact of ANWR oil on gasoline prices would be minimal, since ANWR oil would represent just a fraction of worldwide crude oil supply, which averaged 84 million barrels per day in 2006.[25]

INCREASING IMPORTS

Another U.S. strategy for increasing energy supply has been to step up imports of energy sources from other countries. According to the *BP Statistical Review of World Energy 2006*, more than 60 percent of the world's remaining oil reserves are located in the Middle East, a region whose relationship with the United States and other Western nations is rife with adverse ideological and political differences. Increasing energy imports from such regions, of course, is highly controversial. Many policymakers, researchers, and commentators have implicated U.S. dependency on oil imports in entangling the United States in too many political conflicts and economic instability.

Imports of oil, in general, are a hot-button issue. In January 2007 U.S. imports of crude oil and petroleum products were at higher levels than in any previous January. The EIA reported total imports of 14.3 million barrels per day for the week ending January 12, an increase of 4 million barrels per day above levels from five years earlier. In January 2007 imports supplied more than 65 percent of U.S. daily oil demand, 10 percent more than five years earlier.[26] Canada, already a top supplier of oil to the United States, has become the focus of U.S. efforts to meet its growing domestic demand for natural gas. Overall the United States depends on natural gas for about 22 percent of its total primary energy requirements (oil accounts for around 41 percent and coal for 23 percent). More than 80 percent of U.S. natural gas imports are already from Canada, mainly from the western provinces of Alberta, British Columbia, and Saskatchewan. The United States has connected Canadian natural gas supplies to U.S. consumers through the construction of a pipeline from Canada to Chicago through the upper Midwest. This pipeline was put onstream in late 1999 and was further extended into Indiana in 2001. Two additional pipelines, running from Canada's Sable Island to New England, went onstream in January 2000, with further extensions into the Boston area completed in 2003. Another proposed pipeline would have run 3,500 miles from the North Slope into Alberta and on to markets in the U.S. Midwest.

Construction of this pipeline was estimated at $30 billion, so major energy companies in early 2007 were considering a scaled-down version of the project that would involve terminating the pipeline in Alberta. Overall, however, the United States in recent times has introduced more policies aimed at *decreasing* dependency on energy imports rather than facilitating it.

Diversifying the Sources of Energy

Another way of dealing with the energy shortage, one that has gained support among environmentalists and activists in the United States, is that of seeking energy from alternative and renewable sources. Initially the administration of President George W. Bush and supporters of his energy policies argued that the United States must increase its domestic supply of oil, natural gas, and coal and that incentives for fossil fuel exploration and the removal of environmental protection regulations are necessary to help accomplish this goal. In addition to promoting drilling in ANWR the administration's proposals have included instituting greater incentives for corporations to undertake fossil fuel exploration and repealing environmental protection rulings that prevent oil production in certain areas. President Bush's administration also phased out earlier statutes or laws that had helped to create more diversity in U.S. energy supply, such as the Public Utility Holding Company Act (PUHCA) of 1935, which was repealed with the enactment of the Energy Policy Act of 2005. The Energy Policy Act of 2005 also contained provisions to expand the use of nuclear energy for electricity and committed $2 billion over 10 years to support research into environmentally "cleaner" " ways of using coal to generate electricity.

Early 2006, however, marked what seemed to be a shift in the Bush administration's approach to the U.S. energy supply issue—at least in what was *said* about the issue. In his State of the Union address on January 31, 2006, President Bush, a former oil executive, stated, "America is addicted to oil, which is often imported from unstable parts of the world. . . . I announce the Advanced Energy Initiative—a 22 percent increase in clean energy research at the Department of Energy."[27] Among the plans he introduced was a proposal for a greater U.S. investment in solar and wind energy technologies. At the same time, however, President Bush's Advanced Energy Initiative also called for greater investment in both coal-fired power plants (with reduced polluting capacity) and advanced nuclear energy designs focused on "clean and safe" technologies. During the president's address he also announced plans to increase research and development funding for the manufacture of batteries for hybrid and electric cars, pollution-free hydrogen-fueled cars, and ethanol fuels derived from corn, wood chips, stalks, or switch grass.[28] Ethanol is a fuel usually made from biomass; it is an alcohol that can be used in internal combustion engines that have been modified to run on it. The president's January 23, 2007, State of the Union address reiterated many of the same themes; he called for a mandatory fuels standard requiring 35 billion gallons of renewable and alternative fuels by the year 2017.

Many of the president's proposals, including those outlined in the Advanced Energy Initiative and the policies in the Energy Policy Act of 2005, are criticized for including measures that act as incentives for the fossil fuels industry or that lack real appropriation of funds or concrete plans. For example, the Energy Policy Act includes forms of liquefied natural gas and derivatives of petroleum in its definition of "alternative fuels." Nonetheless, there was an increased emphasis on renewable energy technology in some of the U.S. energy initiatives introduced in 2005–07. The Energy Policy Act of 2005, for example, included a measure aimed at tripling, by 2012, the current required amount of biomass-derived fuel (or biofuel), such as ethanol, that must be mixed with gasoline sold in the United States. It also provided for a tax credit of up to $3,400 for owners of hybrid vehicles, cars that are powered by both an electric motor and a gasoline engine. Yet this tax credit is controversial because it is based on a formula that allows higher tax incentives for hybrid versions of models that are by nature fuel inefficient, such as sport utility vehicles (SUVs), in comparison with compact hybrids with the greatest fuel efficiency. In addition, some industry analysts have observed that tax incentives for hybrids have been shifting the focus on fuel conservation away from its most important task: continuing to demand and achieve greater standards in fuel efficiency among all vehicle models, hybrid or not. And the way to do this, they point out, is to continue to gradually increase the government's fuel economy standards for vehicles.

In 2005 the U.S. government also introduced a federal tax credit for solar and other energy-saving systems installed and operating on a home or a commercial building between January 1, 2006, and December 31, 2007. The tax credit (up to $2,000 per system) was made available to residential users of solar electricity, wind power, and solar water heating (excluding pool heating). A tax credit of up to 30 percent plus depreciation was made available to businesses for solar or wind power systems. In addition, some U.S. initiatives to encourage reliance on renewable energy sources have been instituted at the state level. These have included the following:

- the implementation of "green pricing," a program in which state utilities give consumers the option of paying more to have their electricity generated by renewable resources
- net metering, in which consumers and businesses can produce their own energy from renewable sources, such as solar, and then sell any excess power they produce back to their utility company

- state rebates given to consumers and business for installing solar electric systems or for purchasing gas-electric hybrid or ethanol-using cars
- state renewable energy credits (also known as SRECs, green tags, or green credits), financial credits granted by public utility commissions, which require that a certain percentage of energy be from renewable sources (a concept known as a "renewable portfolio standard"); the credits are granted to those who produce energy from renewable sources, and consumers and others who produce renewable energy can sell their renewable energy credits to entities who use more than their share of energy and do not currently produce enough renewable energy to meet the state's renewable portfolio standard

However, many commentators have noted that even with such programs, U.S. oil consumption continues to grow and does not appear to be slowing. Many say that it is obvious that not enough is being done to diversify the sources from which Americans get their energy. They also point out that the current presidential administration needs to improve continually upon the energy policies implemented by previous administrations—not repeal them altogether, as has been done, or allow them to become completely outdated.

Energy Conservation and Efficiency

One of the major problems that the United States faces in relation to energy is an inability to balance energy demand with energy supply. The nation will not get very far with its efforts to conserve fuel and improve energy efficiency as long as demand for energy continues to increase and even surpass energy supply. Energy demand is soaring, as personal computers, mobile phones, and various other electronic devices take on greater importance in daily American life and as the transportation sector continues to grow. The real problem with energy thus reaches beyond oil reserves, gasoline prices, fuel efficiency, and similar issues; it is a problem based in a way of life, one in which convenience, speed, and technology have driven American attitudes about energy and lifestyle. As many commentators note, short-term solutions instituted by the federal government, such as rebates or tax incentives, have had little effect on the growing problems with U.S. energy supply. Rather, they argue, more consumer education in the value of conservative energy use may help to influence American citizens to begin living in a less energy-intensive way.

CORPORATE AVERAGE FUEL ECONOMY

An example of such a policy is the Corporate Average Fuel Economy (CAFE) program, which was established by the Energy Policy and Conservation Act of 1975. CAFE was part of this law's strategy for improving energy efficiency and conservation in the wake of the Arab oil embargo. CAFE sets minimum requirements for the number of miles per gallon (mpg) of gas that cars and light trucks should be able to achieve. Under the CAFE program automobile manufacturers are required to average a minimum specified mpg for each category of cars they make: The initial minimum for passenger cars was 18.0 mpg in the 1978 model year; this was later changed to a minimum of 27.5 mpg in 1985. Also in 1985, the minimum for light trucks was set at 19.5 mpg (a vast improvement over the 10.5 mpg that most light trucks averaged prior to CAFE's implementation). In 1996 the light trucks average was increased to 20.7 mpg. However, around the same time, notoriously gas-inefficient SUVs began to soar in popularity as passenger vehicles, yet they were classified by CAFE as light trucks and thus did not need to meet the mpg standard for passenger vehicles, which was their primary role. The largest SUVs classified as heavy trucks rather than as passenger vehicles, so they too were subject to a less stringent mpg requirement. So, as many commentators have pointed out, there are some defects in the CAFE system that need to be corrected in order to achieve true fuel efficiency across the board. In April 2003 the National Highway Traffic Safety Administration agreed to revise the CAFE standard upward by 1.56 mpg by 2007. As of this writing, a draft bill known as the Vehicle and Fuel Choices for American Security Act was under review in Congress; it proposed that loan guarantees and other incentives be given to U.S. automobile manufacturers for retooling their assembly lines to increase sharply the production of cars that run on an alcohol-gasoline combination, hybrid vehicles, and plug-in hybrid vehicles. The bill's sponsors claim that this strategy will help to reduce oil consumption by 2.5 million barrels a day by 2015 and by 7 million barrels a day by 2025. Yet there has not been much support for this bill from the Bush administration.

In May 2006 the House of Representatives Energy and Commerce Committee gave the Bush administration the authority to set per-gallon mileage targets for passenger cars in a new system that is based on vehicles' size and weight. Some lawmakers called for this authorization to include a stipulation that higher CAFE standards be established by 2015 and a fleetwide average of 33 mpg for light trucks and passenger cars be set by 2015, but the Energy and Commerce Committee rejected that proposal. This is just one example of a proposal that many say the United States could have taken to address the energy supply issue better—but chose to reject, continuing to follow a path of oil depletion.

Many of the possible strategies for addressing energy crises and shortfalls, if implemented, could help the United States to cope with other energy-related issues, too. For example, measures to promote energy efficiency, energy conservation, and renewable energies may also help to address energy-related environmental issues. And measures designed to decrease reliance on nondomestically produced energy may also help to address concerns about energy affordability and economics. Energy affordability and economics, in fact, are two issues that historically have been directly linked to many of the U.S. energy crises and shortfalls just reviewed.

U.S. ENERGY AFFORDABILITY AND ECONOMICS

Energy has a direct impact on the U.S. economy. Energy shortages, in particular, can have significant economic implications. With an energy crisis, there generally is a period of decreased spending and reduced confidence in the economy. For example, during a period of high oil prices or oil shortfalls, gasoline prices increase, and higher gasoline prices translate into higher transportation costs. Higher transportation costs affect not only the day-to-day living expenses of most citizens but the daily operational expenses of businesses. This effect leads to an overall increase in the price of goods and services and to decreased consumer spending, which then, in many cases, can lead to a recession, a period of slowed economic activity.

Energy-Related Recessions

There have been several examples of energy-related recessions in the United States. The 1973–74 oil crisis, for example, led to a severe recession in the United States and most of the Western world, specifically in the countries that were dependent on Arab oil. The situation was especially bad after the implementation of a U.S. policy requiring people who drove vehicles with odd-numbered license plates to purchase gasoline on only odd-numbered days of the month and those who drove vehicles with even-numbered plates to purchase gasoline on only even-numbered days. The second oil crisis, in 1979, caused decreased spending and heightened economic insecurity in the United States. During this second crisis many people believed that oil companies artificially created the oil shortage to drive up prices.

Yet another example of a shortage-related economic crisis occurred in 2003, when a massive electrical blackout in North America in August resulted in $6 billion in financial losses. Other factors can also bring on increased energy costs and lead to economic problems. In 2003, for example, crude oil prices in the United States rose to a 29-month high as a result of a cold winter and a strike by Venezuelan oil workers at the end of the previous year. In February 2003 the price of oil was $49 a barrel, almost double the previous February's price. This situation was compounded by a pending war with Iraq; in March 2003, after Iraq's president, Saddam Hussein, failed to cooperate with weapons inspections supervised by the United Nations

(UN), the United States led an invasion of Iraq, contending that Iraq held weapons of mass destruction. The U.S. invasion of Iraq put Iraqi oil production on hold. The economic impact of this stoppage could have impacted the U.S. economy more severely, but Saudi Arabia's minister of petroleum and mineral resources, Ali bin Ibrahim al-Naimi, asserted Saudi Arabia's commitment to stepping up its own oil production to fill the void in Middle Eastern oil exports. Saudi Arabia and other oil-producing countries then increased production to circumvent the shortfall. The remainder of 2003, however, was marked by reduced economic activity in the United States, with a stagnant stock market, especially during the early part of the year.

Another example of the interplay between energy supply and the economy followed Hurricane Katrina's landfall in late August 2005. Reacting to the hurricane-induced damage to the oil rigs in the Gulf of Mexico, many gas stations increased prices by a considerable amount; in most regions of the United States gasoline cost more than three dollars per gallon. By September 1, 2005, many gas stations were actually running out of fuel because consumers, panicked over possible mass U.S. oil shortages, began stocking up on gasoline. As demand increased, so did price; Atlanta, Georgia, reported gasoline sales of nearly six dollars per gallon. However, prices eased as the true extent of the damage to Gulf oil facilities became known. Saudi Arabia, too, stepped up oil production to fill the gap in U.S. supply.

Impact of Supply and Demand on Price

Many of these economic troubles are related to the fact that the United States has a free-market economy. In such an economy, the basic economic principle of supply and demand for the most part governs the price of energy supplies, including oil, gas, and electricity. This principle theorizes that as demand for a product or service increases, supply decreases, thus triggering an increase in price, which keeps demand in check so that supply can keep up with it. Similarly, the theory goes, if the supply is greater than the demand, lower prices are maintained, helping to prevent demand from dropping down to the point where the good or service is not purchased at all. Because this principle generally drives the market for energy, changes in either supply or demand can trigger sudden fluctuations in price.

Yet a free market is an abstraction that is implemented to different degrees in real markets. Types of factors that limit the freedom of markets include regulations (such as trade agreements or minimum wages), government interventions, and lack of economic competition. For example, even in countries that have a free-market economy, the government sometimes intervenes to prevent the economy from spiraling into a recession or to lessen the hardship that price increases may cause for citizens. The even- and odd-numbered license plate restrictions placed on gasoline buying during the 1979 oil crisis provide one example; in this case, however, the measure backfired, exacerbating the energy crisis and its negative effects on the economy.

In the case of oil, the freedom of the market is also limited by the fact that several nations who control significant reserves of oil formed a cartel in 1960. Founded to gain leverage over the major oil-producing companies at the time, OPEC coordinates oil production policies among its member nations, standardizing the amount of oil its members export. Since OPEC produces 40 percent of the world's oil, it has the potential to exert a good deal of influence over the price of oil worldwide by regulating the supply side of the market. Some critics say that OPEC artificially inflates its proven oil reserves figures to generate an R/P quota that trends downward, suggesting scarcity, thus justifying higher prices. OPEC has also been accused of producing oil from older oil fields that are nearing depletion, enabling the organization to report lower production rates, which then can also allow for higher prices. Yet there are limits to OPEC's ability to regulate prices. The nations of OPEC are highly dependent on oil exports for their economic stability, so driving prices too high by restricting oil supply too much could generate a decrease in demand that would adversely affect their economies.

In the United States government subsidies are another limitation to the free determination of oil prices. Subsidies are government funds allocated to lower the price of a particular good or service, usually with public interests in mind. Many people criticize U.S. oil subsidies for their potential to mask the true cost of oil and artificially keep it just low enough so that demand for oil will continue at the present level or even increase. Oil subsidies, along with the hidden "environmental costs" of oil (such as air pollution caused by motor vehicle usage), give oil an artificially low price. Some commentators say that the main purpose of oil subsidies is to maintain the financial health of large, influential oil companies. They cite the "artificial underpricing" of fossil fuels as the primary reason for U.S. fossil fuel overconsumption.[29] Many commentators have suggested that instead of an artificial lowering of the price of oil, the price should be artificially increased by increasing gasoline taxes so as to stave off further upswings in U.S. gasoline consumption and spur greater efforts toward the development of fuel alternatives. In 2004 the General Accounting Office (GAO), a nonpartisan, independent accounting office of the U.S. Congress, estimated that a new gasoline tax of 46 cents per gallon could help reduce oil consumption by 10 percent over the next 14 years.[30] No such legislation yet exists, however. And one criticism of the Bush administration's Energy Policy Act of 2005 is that it encourages additional subsidies for the oil industry, such as reduced corporate taxes, reduced average gasoline sales taxes, and federal grants for programs that are beneficial to oil companies.

On the other hand, in the earliest days of the U.S. oil industry, the tendency of oil prices to drop *too* low presented more of a problem to the U.S. economy. For example, around the time of the Great Depression, the huge leap in U.S. oil reserves that resulted from the Joiner discovery in 1930 caused a glut of supply that then drove the price of oil down to 10 cents a barrel in 1931. This wreaked havoc on the economic stability of the oil industry and had ripples in the economy as a whole. The New Deal administration, however, helped to restore some measure of stability, and World War II stimulated the oil business enormously.

Nonetheless, as oil companies' profits today continue to soar along with the price of oil and gas, oil companies have become the subject of intense scrutiny in the United States. The world's five largest oil companies, including Exxon Mobil and BP Plc, earned record profits in the first quarter of 2006—about $29 billion, or $4.46 for every person on Earth.[31] At the same time, by mid-2006 gasoline prices reached about three dollars per gallon and oil was nearly $65 per barrel.

Oil price hikes usually trigger outrage among U.S. consumers and politicians, especially when oil companies continue to report record profitability. In the first half of 2006, some members of the U.S. Congress searched for legislative avenues through which to transfer some of oil companies' economic gains back to average U.S. citizens in order to make up for the high price of oil. For example, in May 2006 the U.S. House of Representatives approved a Democratic plan to renegotiate contracts granted to oil companies in 1998 and 1999 for leasing land in the Gulf of Mexico for oil production. The original leases excused several oil companies, such as Exxon Mobil Corp. and ConocoPhillips, from any additional royalty fees they would have had to pay to account for increases in oil prices. (The agency responsible for the contracts claimed that the price thresholds for royalty relief were omitted by mistake.) The Democratic plan also proposed to take back $10 billion in tax incentives granted to oil companies. The approval of this plan marked the first time that either house of Congress passed a measure designed to address tax incentives and subsidies granted to the oil industry. However, another proposal from Congress, aimed at taxing oil companies for their record profits, met with opposition from the president.

In addition to members of Congress, political activists have pressured the U.S. government to do something about rising energy costs—and something more than the provisions laid out in the Energy Policy Act of 2005. For example, a leader of the Alliance to Save Energy, a nonprofit coalition of business, government, environmental, and consumer groups that advocate energy efficiency programs and reduced energy use, addressed the U.S. House Committee on Energy and Commerce, Subcommittee on Energy and Air Quality, in November 2005. Brian Castelli, the Alliance's chief operating officer and executive vice president, criticized what he viewed as loopholes in the Energy Policy Act of 2005 that would only result in higher oil prices. Castelli's testimony highlights certain energy–price-related concerns that other Americans have also expressed, such as a need to revise CAFE standards according to realistic emissions testing and to classify SUVs and minivans as passenger vehicles instead of light trucks. Castelli also lamented the "startling and immediate effects of Hurricane Katrina and Rita" on energy prices, which, he says, show that we need to balance energy demand and supplies.[32] Other activists have looked for evidence that U.S. oil companies are intentionally reducing domestic oil-refining capacity to drive up their profits. The Foundation for Taxpayer and Consumer Rights (FTCR), for example, a U.S. consumer watchdog group, has claimed that internal business memos of Mobil, Chevron, and Texaco show that different tactics have been used to drive independent refiners out of business and create increased-demand conditions that justify higher prices for oil. The FTCR's president claimed that large oil companies artificially shorted the gasoline market for a decade to drive up prices.[33] Accusations of oil company price gouging are nothing new; they were suspected during the 1970s oil crisis. They have also been refuted by the many oil economists who observe that high oil prices often occur as the natural outcomes of uncontrollable events such as Hurricane Katrina.

And then there are others who flatly predict that the combination of rising oil prices and a mounting shortage of oil is precipitating a major economic crisis in the United States from which it will never recover. As Jeremy Leggett, an expert in renewable energy and chief executive of a large independent solar electric company in the United Kingdom, writes, "When it becomes clear that there is no escape from ever-shrinking supplies of increasingly expensive oil, there will be a paroxysm of panic. Human society will face an energy crisis of unprecedented proportions, and that, plus the panic, will spark an economic collapse of unparalleled awfulness."[34] Whether such doomsaying can be believed, one thing is clear: Energy, particularly oil, has substantial and lasting effects on the U.S. economy that only appear to be increasing in intensity. And many Americans would agree that problems with the price of oil and oil-related economic issues are very much tied to the fact that so much of U.S. oil is imported (more than 65 percent as of January 2007).[35] Thus, overdependence on foreign oil is another energy issue that the United States has had to address, and with a particular urgency, in recent years.

U.S. FOREIGN OIL DEPENDENCE, POLITICAL INSTABILITY, AND FOREIGN RELATIONS

Foreign Oil Dependence

U.S. consumption of imported oil has increased dramatically since 1970, particularly from the mid-1980s onward. Through the first 11 months of 2006 the top suppliers of crude oil to the United States were Canada (1.8 million barrels per day), Mexico (1.6 million barrels per day), Saudi Arabia (1.4 million barrels per day), Venezuela (1.1 million barrels per day), and Nigeria (1.0 million barrels per day). By 2050, only 30 percent of the 28.3 million barrels of oil per day that the United States is projected to consume are likely to be from domestic sources, according to the EIA.

That obtaining most of one's oil from foreign sources can inflict severe economic damage became painfully clear to the United States during the 1973–74 oil crisis. The oil price spikes instituted by the countries of OPEC wreaked havoc on the U.S. economy. Yet the dependency on oil makes the U.S. economy vulnerable no matter the source, say many commentators. According to the DOE, for every million barrels of oil taken out of production each day, regardless of the region where they originate, world oil prices will increase by three dollars to five dollars per barrel. An increase of $10 per barrel has the potential to cut U.S. economic growth by 0.2 percent and increase consumer prices by 0.4 percent. Considering that oil is a worldwide commodity sold at a world market price, obtaining it from foreign sources merely multiplies the economic vulnerability of the United States, according to many who have studied the issue.

U.S. reliance on foreign oil has been associated with other costs, too. For example, the trade deficit tends to increase when oil prices and imports are high. Both the quantity of oil imported and the price of oil per barrel increase the portion of the U.S. trade deficit that is attributed to the value of oil. According to the U.S. Census Bureau, the United States had a trade deficit of $449 billion in 2000, and 20 percent of this deficit, reported the EIA, was the value of imported oil.[36] As greater quantities of oil are imported into the United States at the world price, the U.S. trade deficit inches up further. Census Bureau data also indicate that petroleum products accounted for one third of the overall trade deficit increase in 2000, a year when the price of oil reached its highest level above any prior year since 1990. In the same year petroleum products also contributed to a worsening trade balance between the United States and Canada, Latin America, the Middle East, and the European Union (EU), the union of 25 European states that includes Germany, the United Kingdom, and France.

Oil Dependence and Foreign Policy

Historically U.S. foreign oil dependence has been associated with other costs, as well. U.S. military spending generally increases as U.S. forces are called on to respond to political instability in the oil-producing nations on which the United States relies for energy supplies. Critics of U.S. military spending and foreign oil dependence have suggested that the 1990s U.S. military intervention in Iraq's invasion of Kuwait was driven by a desire to protect oil production in Kuwait and to prevent Iraq from further threatening U.S. oil imports by invading Saudi Arabia, a major supplier of U.S. oil.

Others have alleged that the United States–led invasion of Iraq in March 2003 had less to do with suspected Iraqi holdings of weapons of mass destruction than with an apparent Iraqi plan to begin accepting euro dollars (the currency of the EU) for its oil exports. There were also reports that Iraq intended to open an international oil exchange market for trading oil in euros—a move that might have been devastating to the value of the U.S. dollar.[37]

Whether these claims are true, there are undeniable connections among U.S. foreign relations, oil issues, and the stability of the U.S. economy. For example, political instability in one of the United States' top suppliers of oil, Venezuela, has had noteworthy effects on the U.S. economy. Venezuela sends close to 70 percent of all its oil exports to the United States, according to the EIA. In 2003 U.S. oil prices soared after Venezuelan oil workers went on strike at the end of 2002 as part of a national revolt against the nation's leader, Hugo Chávez. Some politicians have cautioned that U.S. dependence on oil from the Middle East and politically unstable countries such as Venezuela may give political leaders in these regions undue leverage over U.S. foreign and defense policy.

Strategies to Lessen Foreign Oil Dependence

What are some of the U.S. strategies for coping with the real and hidden costs of its reliance on foreign oil? One possible solution that has been proposed by U.S. legislators from time to time is to establish a tax on energy imports as a way to stimulate domestic production and conservation. However, such a tax has never been implemented. Ronald Reagan, U.S. president from 1981 to 1989, refused to implement taxes on energy imports even when pressured to do so. The Reagan administration also sharply cut funding for federal energy programs, particularly alternative energy research and development (another proposed solution to the problem).

In 1991 the U.S. president George H. W. Bush unveiled an energy policy that promised to scale down foreign oil dependence by increasing domestic oil production, producing oil from environmentally sensitive areas in the

United States, and encouraging domestic pipeline construction. Of course, these proposals met with opposition from environmentalists and renewable energy supporters.

In more recent times, the U.S. president George W. Bush has promoted the diversification of oil import sources and included provisions to facilitate this in the Energy Policy Act of 2005. But critics have said that diversification of oil imports does nothing to minimize the severe trade deficit caused by foreign oil imports.

The Energy Policy Act of 2005 also contained another measure designed to address the issue of U.S. foreign oil dependence: the Set America Free Act of 2005, which calls for the creation of a "United States Commission on North American Energy Freedom" to study the particulars of U.S. foreign oil dependence and to "make recommendations for a coordinated, comprehensive, and long-range national policy to achieve North American energy freedom by 2025."[38]

Elements of the president's Advanced Energy Initiative, announced in his January 31, 2005, State of the Union address, were also intended to manage the U.S. foreign oil dependence issue. In the January 31 speech, Bush promised alternative energy breakthroughs and investments in new technologies that would help the nation "replace more than 75 percent of our oil imports from the Middle East by 2025."[39] A centerpiece of the president's new strategy was ethanol. In 2006 ethanol that was derived from corn, which currently accounts for only a tiny fraction of fuel supply in the United States, required U.S. subsidies for its production. The Advanced Energy Initiative contained plans to subsidize ethanol production even further. This was also a matter of contention in the United States, as critics said that the amount of land and water and level of energy required to produce ethanol make it useful as only an add-on fuel to complement primary, nonrenewable oil. Others suggest that the significant amount of traditional fuels expended in the growing, processing, conversion, and transportation of biomass fuels, including ethanol, greatly reduces their value as alternatives to oil.

Aside from promising to boost government spending on the production and development of ethanol fuels, the Advanced Energy Initiative stressed the development of new or renewable sources of energy as a way to scale back consumption of fossil fuels, especially in its section "Changing the Way We Fuel Our Vehicles."[40] Commentators have said that the Advanced Energy Initiative's emphasis on renewable energies betrays a misconception that renewable sources of energy are more consistent and efficient than they really are. These critics highlight the fact that many renewable energy sources, such as wind or solar sources, provide energy only intermittently, are unequally disbursed in supply across the country, require enormous space for production, and still require the support of traditional energy sources in their manufacture and transport. In addition, some industry analysts have suggested that the United States needs to expand its pursuit of "alternative energies" beyond renewable energies such as wind, solar, and biomass. They point to new possibilities for carbon sequestration, a method of capturing the carbon content of fuels like coal and enabling it to be taken up and stored by soil, oceans, or forests. Energy experts also raise the possibility of using more nuclear fission or hydrogen energy technologies.[41]

Other industry analysts also lament that many of the measures put forth as "U.S. energy policy" usually amount to nothing more than proposals that never translate into real action. The truth of the matter is that oil consumption and oil imports continue to increase in the United States, despite developments in U.S. energy policy. Many U.S. citizens, politicians, researchers, and economists are very concerned that the United States will become even more "addicted to oil!," and even more dependent on getting it from unstable parts of the world. Some even contend that until the United States builds an energy infrastructure that allows it to become entirely self-sufficient, dependency on foreign oil will always be a threat to national security, economic stability, and peaceful foreign relations.

U.S. INFRASTRUCTURE-RELATED LIMITATIONS ON ENERGY PRODUCTION

Another energy issue of significance in the United States is the need for a sufficiently developed infrastructure to support escalating energy demand and to deliver energy to the regions most in need of it. The existing U.S. energy infrastructure consists of a complex web of oil and natural gas pipelines; oil refineries and other facilities that convert natural resources into energy products; gasoline stations and other services for selling energy to the end consumer; electricity marine transmission lines; power plants; rail lines, truck lines, and transportation networks for delivering energy products; and waste disposal systems for removing the by-products of energy production. Often flaws in this unwieldy infrastructure become apparent when portions of its networks fail to deliver energy in a reliable, consistent, and safe manner.

The 2003 Blackout

One example of such an infrastructure failure was the massive North American electrical blackout on August 14, 2003. Affecting an estimated 50 million people over an area of 9,300 square miles across Canada and eight U.S. states, the blackout was attributed to the failure of an electrical utility supplier, FirstEnergy Corporation, to trim trees in part of its Ohio service area. According to reports from the task force that investigated the causes of the blackout, a generating plant in a suburb of Cleveland, Ohio, went offline in the midst of high electrical demand, putting a strain on high-voltage power lines. These lines came into contact with overgrown trees, short-circuited, and automatically went out of service. Because of a glitch in intersystem communications, other control centers were not warned about the event, and a cascading series of shutdowns were triggered until ultimately more than 100 power plants were offline. In the days after the blackout the nationwide electric power transmission grid was scrutinized. The event highlighted the need for a more modernized electrical energy infrastructure.

The U.S. electricity industry was once vertically integrated, meaning that a single corporation controlled all or nearly all aspects of production, including generating plants, long-distance transmission lines, and electricity distribution centers. While such an arrangement can streamline the coordination of generation, transmission, and distribution activities, it can also result in monopolies. U.S. deregulation laws and policies to discourage monopolies have led to more defined divisions among generation, transmission, and distribution activities. Many people blamed deregulation for the North American blackout, asserting that it had created an electricity market with very little accountability for outdated transmission lines and malfunctioning security systems.

Another factor cited for the inadequacies in the U.S. electrical power infrastructure is the very way electricity is generated. Electricity cannot be stored with much stability for extended periods and is quickly consumed after being produced. So an electrical power grid must be able to meet the demand placed on it with both adequate supply and great efficiency. Also, the dynamics of power grids themselves can cause infrastructure problems. For example, power grids automatically disconnect when sudden changes in demand on power lines and generators are detected, because such imbalances can cause costly damage, and power lines are automatically disconnected when a short circuit is detected. Fluctuations in power caused when a power line goes out of service can cause failure in other parts of the system when the changes in power are detected. Normally, during a built-in delay period before further shutdowns kick in, the cause of the power failure is communicated between systems and electrical power is redirected through appropriate alternate transmission lines. The blackout of 2003 was, in part, blamed on a defect in such a communication process.

Flaws in the U.S. Oil Infrastructure

Infrastructure flaws have also befallen the U.S. oil industry. The primary transportation system for oil and petroleum products in the United States is a network of 2 million miles of pipelines. After oil is pumped from the ground, it travels through gathering lines to these pipelines. The pipelines deliver the oil to refineries, where the oil is transformed into petroleum products such as gasoline. One average pipeline can transmit the equivalent of 750 tanker truck loads of oil per day.

Many of the existing pipelines in the United States are fairly old and require regular safety and environmental checks. Insufficient pipeline capacity has been known to disturb the flow of oil and petroleum products from one region of the country to another, and energy supply shortages can create operational difficulties in the pipelines themselves. By the nature of the cargo they carry, pipelines can be dangerous. In August 2000 a corroding natural gas pipeline near the Pecos River in southeastern New Mexico ruptured, causing an explosion so large that 11 nearby campers were killed by the blast and a fireball could be seen as far as 20 miles to the north in Carlsbad, New Mexico. Failures such as these are significant threats to the safety of individuals and communities. According to the U.S. Department of Transportation, there were 429 hazardous pipeline incidents in 2004 and 396 in 2005.

Dangerously outdated infrastructure has also plagued other aspects of the U.S. oil industry. Oil refineries have suffered infrastructure problems as a result of ongoing industry consolidation ever since the oil shocks of the 1970s. Environmentally destructive oil spills from oil tankers that were not double-hulled for safety were also a major problem, until the Oil Pollution Act of 1990 mandated double-hulling.

Government Response

In the United States there have been several attempts to stimulate improvements in America's energy infrastructure. One was the Pipeline Safety Improvement Act of 2002, which outlined specific pipeline system checks to ensure the integrity and security of existing pipelines. It also set a goal of reducing the number of hazardous pipeline incidents by 5 percent each year, with a target number of 295 for 2005. (Unfortunately, the United States in 2005 was more than 100 incidents above this goal.) On the electrical energy front, the Energy Policy Act of 2005 also established reliability standards for electricity utilities to follow and initiatives for modernizing the electrical grid. The act also extended daylight savings time in the United States by four weeks on the presumption that Americans will use less electricity if daylight is available for a longer period.

In addition, in 2001 the U.S. vice president, Dick Cheney, a former chief executive officer of the energy contractor Halliburton Company, led a task force called the National Energy Policy Development (NEPD) Group, which made recommendations to the president for a new national energy policy. Among its recommendations were suggestions for updating America's energy infrastructure, including the following:

- That the secretary of energy work with the Federal Energy Regulatory Commission to improve the reliability of the interstate electricity transmission system and develop laws to enforce compliance
- That the secretary of energy examine the benefits of a national electrical grid and identify transmission bottlenecks and ways to remove them
- That the renewal of the lease of the Trans-Alaskan Pipeline System (which is responsible for a fifth of all oil transported throughout the United States) be expedited to ensure an uninterrupted flow of Alaskan oil into the U.S. West Coast
- That the federal government, the state of Alaska, and Canada work together to expedite the construction of a natural gas pipeline from Alaska and Canada to the lower 48 states, repealing any portions of the Alaska Natural Gas Transportation Act of 1976 that discourage investment in new infrastructure.[42]

The report offers background and recommendations on U.S. energy infrastructure issues. However, the group's overall findings were controversial because environmentalists and watchdog agencies, pointing to the oil industry background of Vice President Cheney, suspected that the interests of the fossil fuel industry had too much influence on its recommendations. The NEPD Group was dismantled after making its recommendations.

U.S. ENERGY-RELATED ENVIRONMENTAL ISSUES

Many in the United States who criticized the NEPD Group's report, and many others who disapprove of energy policies that show any favorability to the fossil fuels industry, base their objections on one major factor: the environment. Fossil fuels have been linked to various types of environmental degradation, such as air and water pollution and destruction of land and natural habitats. Perhaps most troubling to those concerned about the environmental impact of fossil fuels is the fact that scientific evidence has linked the use of fossil fuels for energy to a change in the climate of the entire Earth. This so-called global warming effect of fossil fuels has been said to disrupt storm systems, ocean currents, and natural balances and life cycles on Earth.

Global Warming

The apparent cause of global warming is something called the greenhouse effect, a process in which gases in the Earth's atmosphere, especially water vapor and carbon dioxide, absorb energy from the Sun, resulting in increased heat in the Earth's atmosphere, oceans, and land. Some of this warming is actually natural, the outcome of longwave, or infrared, radiation from the Sun being reemitted from the Earth's surface back into the cooler atmosphere above by way of trace gases. Scientific research has shown that the temperature of the Earth has been increasing by small increments for at least 200 years. However, the burning of fossil fuels for energy emits carbon dioxide and other so-called greenhouse gases (methane, water vapor, ozone, nitrous oxide) into the Earth's atmosphere. And research has shown that as the concentration of these gases in the Earth's atmosphere has increased through human energy consumption, the balance of outgoing, infrared radiation and incoming radiation has been disrupted, trapping more radiation in the atmosphere. This, say many scientists, has caused the Earth to become warmer than it might have been without the use of fossil fuels for energy. Just how much warming is the direct result of human activities or how much will occur in the future, however, is disputed by the scientific community. Most scientists, however, do agree that human energy consumption has had an impact on the overall increase in the Earth's temperature, and many (but not all) scientists agree that most of the warming over at least the past 50 years is probably due to an increased concentration of greenhouse gases.

Recently Robert Correll, a senior fellow of the American Meteorological Society and a U.S. authority on global climate change, led the International Arctic Science Committee, a scientific team commissioned by the United States and seven other countries, in a study of the impact of climate change on Greenland, above the Arctic Circle. This is a region where glacial ice has been melting at a rate that some scientists find alarming. Correll estimated that about 100-million-plus acres of the region's ice has melted over 15 years. Reportedly, 98 percent of the world's mountain glaciers are also melting, so quickly that, according to Corell, in 100 years the runoff water will push sea levels around the world three feet higher.[43]

In 2004 Correll's team produced a report, the Arctic Climate Impact Assessment. The report found that harmful effects on the food chain, ocean currents, storm systems, and indigenous life have resulted from the rapidly melting glacial ice. Correll also reported that the polar bear population in Greenland has dropped by more than 200 bears since the mid-1990s as a result of the loss of habitable ice. The melting ice is attributed to an increase in global temperature, which, in turn, has been traced back to greenhouse gas emissions from fossil fuels. In February 2007 a report by the United Nations' Intergovernmental Panel on Global Climate Change (IPCC), authored by 600 international scientists, unequivocally concluded that the Earth's temperatures were increasing. It also stated that the rise in global temperatures was almost certainly caused by increased carbon dioxide and other greenhouse gases in the atmosphere and that the increase in these gases was more than 90 percent likely the result of human activities. The United States endorsed the IPCC report.[44] Additionally, the U.S. Supreme Court ruled in April 2007 that greenhouse gases are polluting and ordered federal environmental officials to consider limiting emissions of these gases from cars and trucks.

According to EIA statistics the United States emits more greenhouse gases than any other nation. Of all the greenhouse gases, carbon dioxide is considered to have the greatest potential for causing pollution and global warming, and fossil fuels, by far, contribute the greatest share of carbon dioxide emissions to the Earth. The United States is the largest contributor of carbon dioxide emissions per capita among countries in the Organisation for Economic Cooperation and Development (OECD), an international organization of countries with market economies and pluralist democracies. The high index of U.S. carbon dioxide emissions when compared with that of other OECD countries is attributed to the United States' strong demand for energy, particularly energy from a carbon-intensive mix of sources. According to EIA data, the United States consumed a total of 91.09 quadrillion Btu of energy during the first 11 months of 2006, and 85 percent of this total, or 77.85 quadrillion Btu, was from fossil fuels. In the United States, of all the energy derived from fossil fuels, nearly half, or 36.87 quadrillion Btu, was from petroleum alone during these first 11 months, and EIA projections forecast that petroleum consumption will increase almost 75 percent by the year 2030.[45]

Between 1990 and 2005 emissions from greenhouse gases increased by 17 percent in the United States, according to *Emissions of Greenhouse Gases in the United States 2005*, a report released by the EIA. Petroleum accounted for 2,614 million metric tons of carbon dioxide emissions in 2005 in the United States, while natural gas and coal accounted for 1,178 and 2,142 million metric tons, respectively. Total carbon dioxide emissions reached just beyond 6 billion metric tons in 2005.[46]

Carbon dioxide is the primary product of burning oil and coal for fuel. Natural gas burns much more cleanly but still produces carbon dioxide emissions. Because it is mostly methane gas, natural gas can also leak traces of methane into the environment. Of the carbon dioxide emissions attributable to petroleum, motor gasoline accounted for the greatest share, 1,170 million metric tons. Carbon dioxide emissions from fossil fuel use in the United

States are also projected to be 54 percent higher in 2020 than in 1990, despite a reduction in the share of coal in total power generation.

Environmental Impact of Fossil Fuels

COAL

Coal contributed 35 percent of U.S. carbon dioxide emissions in 2005, according to the EIA. Aside from releasing harmful carbon dioxide emissions into the atmosphere, the use of coal as a fuel has been known to damage the environment in the United States in other ways. The methods used to extract coal from the ground, for example, have been criticized for their environmental hazards. When seams, or layers, of coal are near the surface, coal is extracted by strip, or surface, mining, by which an open pit or strip is used to expose and extract the coal. Strip mining generally devalues the land through which it passes. When coal seams are deeply underground, supports are usually used to hold up the roof of the coal mine, and a deep cavity is dug, into which the coal can collapse. Historically deep underground mining has involved dangerous mine roof collapses, explosions, and exposure to lung-damaging gases. In addition, all forms of coal mining have been criticized for their tendency to leave behind heaps of coal, sometimes having a high sulfuric content. Nearby bodies of water can become contaminated with acidic, toxic metal–rich runoff from these coal heaps. In 2006 coal accounted for 23 percent of total energy consumption in the United States, and it was mostly used in power plants to generate electricity.

OIL

Oil, which, of course, is the primary fossil fuel used in the United States (and around the world), contributed 43 percent of U.S. carbon dioxide emissions in 2005, according to the EIA. Burning oil releases fewer carbon dioxide emissions than burning coal but more than natural gas. Aside from its notorious carbon dioxide–emitting effects, oil has been associated with many other environmental hazards in the United States. These have included pipeline explosions, oil spills from tanker ships, and destruction of marine habitats caused by offshore drilling and dredging (a type of digging to extract oil). The "gushers" of oil that were characteristic of many early oil discoveries in the United States were also environmentally damaging in that they spewed oil into the areas surrounding the oil fields. Oil spilled from tanker ships is an especially thorny environmental issue because the oil can remain on the shoreline or water surface for years after the spill, continuously releasing its toxic hydrocarbon components into the marine environment and poisoning marine creatures. Oil, as previously mentioned, is primarily a fuel for transportation.

NATURAL GAS

Natural gas, a less polluting fuel than oil or coal, contributed 20 percent of U.S. carbon dioxide emissions in 2005, according to the EIA. Environmental hazards related to natural gas in the United States have included LNG spills and pipeline explosions as well as explosions caused by leaks of natural gas used for heating and cooking. Two environmental risks that have been associated with the recovery of natural gas include subsidence, a settling of the ground that may occur when the extraction of natural gas from an oil field leads to decreased pressure in the oil reservoir, and the emission of a toxic, acid gas containing hydrogen sulfide. In the past, before natural gas was viewed as a viable energy source, it was flared, or lit on fire, when it was extracted along with petroleum, generating undesirable carbon dioxide emissions. Natural gas is mainly used for heating, but greater quantities of it are also being used to generate electricity in the United States, mainly because the environmental pressures against coal are so great.

Strategies to Reduce the Impact of Fossil Fuel Energy on the Environment

The use of environmentally damaging fossil fuels as a primary energy source has proved to be a difficult issue for the United States to address. As previously discussed, oil-based fuels have some built-in convenience, density, and portability features that make them particularly easy to transport and especially suitable as a form of energy to power transportation activities. Nonetheless, the United States has taken some important measures to minimize the impact of fossil fuel energy on the environment.

GOVERNMENT REGULATIONS

For example, the U.S. Federal Surface Mining Control and Reclamation Act of 1977 was a law enacted to restore topsoil and vegetation to areas damaged by coal surface mining. The Clean Air Act is a U.S. law that addresses another environmental issue related to fossil fuel energy: air pollution. The law was instituted to improve air quality in the United States and to minimize the impact of harmful pollutants, including those emitted by electric power plants and automobiles. First passed as the Air Pollution Control Act in 1955, the law was reinstituted in 1970 as the Clean Air Act with the essential features that it has today. It was revised in 1977, 1990, and 1997. Under the act the U.S. Environmental Protection Agency (EPA) sets limits on the amount of emissions allowable anywhere in the United States from pollutants considered to be primary causes of ozone depletion and harm to the environment. These pollutants include sulfur dioxide, ground-level ozone, nitrogen oxides, carbon monoxide, particulates, and volatile organic compounds. The law gives each state the authority to develop its own guidelines regarding how that state will clean up polluted areas and improve air quality, basing such regulations on the needs of that particular state's industries, geography, population patterns, and so on.

The administration of President George W. Bush, however, has been criticized for waiving one of the requirements of the Clean Air Act of 1970. Known as the new source review, it mandated inspections of new or remodeled coal-burning power plants to ensure that they were utilizing the cleanest-burning technologies possible to limit emission of sulfur dioxide. This requirement did not apply to older, unremodeled plants, which were already a subject of criticism by many who believed that this measure only encouraged the power industry to keep old, heavily polluting plants in operation. When President George W. Bush waived the new source review requirement for remodeled plants altogether in 2003, environmentalists viewed this decision as one that would further increase pollution from coal.

THE KYOTO PROTOCOL

The United States has also been criticized for its refusal to ratify the Kyoto Protocol, an international treaty aimed at reducing emissions from fossil fuels for the purpose of reducing global climate change. The Kyoto Protocol is an agreement that calls for the world's 38 most industrialized countries to reduce fossil fuel emissions to an average of 5 percent below 1990 levels by 2012. The United States committed to the Framework Convention on Climate Change in 1992, which called for stabilizing greenhouse gas concentrations in the atmosphere, helped to draft the Kyoto document in 1997, and signed it in 1998, agreeing, at least initially, to a 7 percent cut in emissions. However, President George W. Bush refused to ratify the protocol; ratification would make it mandatory for the United States to abide by the agreement. Bush has objected to the protocol's exclusion of developing countries from the reduction, which was based on the fact that industrialization was considered vital to their economies. The president has said that the reduction in emissions would seriously harm the U.S. economy and the protocol is too lenient on developing countries whose consumption of these fuels is on the rise. Remaining steadfast in the decision not to ratify the Kyoto Protocol, the United States has entered into other global agreements that attempt to address the global climate change issue. For example, the United States joined with Australia, India, Japan, China, and South Korea to form the Asia-Pacific Partnership on Clean Development and Climate, or AP6. Member countries, which are responsible for about 50 percent of global greenhouse gas emissions, entered into a non-treaty agreement to cooperate on developing and implementing technology to reduce greenhouse gas emissions. The agreement, first announced July 28,

2005, at an Association of South East Asian Nations (ASEAN) Regional Forum meeting, allows countries to set their own goals for reducing emissions, and there is no mechanism for mandating compliance. On January 12, 2006, the member countries agreed to a formal charter, communiqué, and work plan for addressing climate change, energy security, and air pollution. Because AP6, unlike the Kyoto Protocol, imposes no mandatory limits on greenhouse gas emissions, environmentalists and nations that have ratified the Kyoto Protocol have called AP6 meaningless. There has also been a grassroots movement among mayors in the United States to fill the gap left by the U.S. refusal to sign the Kyoto Protocol. Mayors from more than 220 U.S. cities, representing more than 40 million Americans, have signed the U.S. Mayors Climate Protection Agreement, an agreement to meet or exceed the U.S. target for reducing greenhouse emissions as specified in the Kyoto Protocol (a reduction of 7 percent below 1990 levels by 2012), despite the federal government's refusal to ratify the Kyoto Protocol.

RENEWABLE AND ALTERNATIVE SOURCES OF ENERGY

Another approach that the United States has taken to reduce the environmental impact of fossil fuel use is directing government efforts and funds toward renewable and alternative energy development. There are many alternatives to fossil fuels that involve using resources from the environment that are renewable. In recent years many of these alternatives have shown great promise in terms of their efficiency and their beneficial impact on the environment.

However, it is also clear that many of these alternatives to fossil fuels, even when they are renewable and minimally damaging to the environment, have their share of limitations. Such limitations lead some commentators to regard as improbable the possibility that alternative energy sources such as solar, wind, or geothermal could ever truly compete with fossil fuels—or at least conclude that the current energy policies of the United States do not provide enough initiatives to allow them to compete. The Energy Policy Act of 2005, nonetheless, contains some provisions to promote research into and development of these alternative energy sources. These provisions include the following:

- A tax credit for owners of hybrid vehicles
- Authorization of a loan guarantee for energy technologies that avoid greenhouse gases (including "clean coal" and nuclear energy technologies)
- The tripling of the required amount of biofuel that must be mixed with gasoline sold in the United States

- Authorization of $2 billion over 10 years toward the development of clean coal technologies, as well as repeal of a 160-acre cap on coal leases and reassessment of available coal resources on federal lands
- Subsidies for producers of wind energy and other alternative energies
- Authorization of $50 million annual biomass grant
- Tax breaks for energy conservation improvements made to homes
- A requirement that a federal fleet capable of operating on alternative fuels be operated exclusively on these fuels

Such Energy Policy Act of 2005 measures, however, have yet to meet with widespread approval, for several reasons. One is that the term *authorization* simply indicates that funds have been earmarked, not *appropriated;* in other words, the monies have yet to be translated into actual spending. So, when it comes to the authorization of funds for renewable energy technologies, many commentators point out that the measures to promote renewable resources do not mean anything until monies are actually appropriated. Additionally, many U.S. energy industry analysts see the limitations of the novel energy sources emphasized in the Energy Policy Act of 2005, such as ethanol and fuel cells, and complain that the act does not go far enough to make these energy sources viable enough to compete with fossil fuels. They say that rebates and subsidies passed along to the end user or producer do not amount to much when one considers that what is *really* needed is the massive funding of research into ways to overcome the limitations inherent in these sources of energy.

Finally, many object to the fact that the Energy Policy Act's emphasis on alternative fuels is heavily geared toward the development of environmentally "cleaner" ways to use coal, which is still, after all, a fossil fuel, and nuclear energy, a type of energy that many regard as unsafe and impractical.

Over the years, and especially since the Three Mile Island incident in 1979, the United States has attempted to address some of the concerns associated with nuclear energy by making sweeping changes in emergency response planning, reactor operator training, engineering, radiation protection, and other aspects of nuclear power plant operations. The U.S. Nuclear Regulatory Commission has also been directed to tighten and increase its regulatory authority. Nonetheless, it has been about 30 years since any utility has ordered the building of a new nuclear plant, and new nuclear plants ordered after 1973 were never built, according to the EIA. In addition, nuclear plants in the 1970s, on average, produced less than 60 percent of the power they were capable of generating. Nuclear production has never, as it was hoped to, caught up to fossil fuel production to make it a viable alternative.

The Energy Policy Act of 2005, however, contains many measures to reignite the use of nuclear power and stimulate improvements in efficiency and safety. The following is a summary of some of the act's nuclear-specific provisions:

- The extension of a law of the U.S. Nuclear Regulatory Commission known as the Price-Anderson Act, which was originally signed in 1957 to provide for the payment of public liability claims in the event of an accident at a nuclear power site; the extension limits liability in the event of an accident.
- Authorization of up to $2 billion for up to six new nuclear power plants.
- Authorization of a nuclear production tax credit of up to $125 million annually.
- Authorization of $1.25 billion for the building of a nuclear reactor that can generate both electricity and hydrogen.
- Permission for nuclear plant employees and certain contractors to carry firearms.
- Prohibition of the sale, export, or transfer of nuclear materials or confidential nuclear technologies to a region that sponsors terrorist activities.
- A requirement that the Department of Energy analyze and report on how to dispose of high-level nuclear waste.

Since the passage of the Energy Policy Act of 2005 nine companies have proposed plans to build 19 new reactors. The EIA is projecting that nuclear energy consumption will grow 65 percent over 2004 consumption levels by 2035. Still, there are many questions about the viability of nuclear energy and what it will take (financially and technologically) to make it a primary energy source. The same questions surround the development of other fossil fuel alternatives. The quantity of nuclear energy consumed during the first 11 months of 2006 did not even equal a tenth of that consumed in the form of fossil fuels, while hydropower and other renewable sources only accounted for only about an eighth of fossil fuels' share, according to EIA statistics.

To many Americans it seems that the most effective approach to limiting the impact of fossil fuel energy on the Earth has yet to be undertaken by the U.S. government. Perhaps the IEA summed up the issue best: "In some areas, the U.S. policy debate is too narrowly based on current economic benefits and costs. Insufficient weight is given to external environmental

costs. Adjusting energy prices to reflect environmental costs is a key means of achieving cost-effective changes in energy end-use and of encouraging the development of new and cleaner energy sources, including renewables. A transition to sustainable development will be made more difficult if environmental costs are not valued by the market."[47]

---

[1] Congressman Joe Barton. Congressional testimony, July 28, 2005.

[2] Paul Roberts. *The End of Oil: On the Edge of a Perilous New World.* Boston: Houghton Mifflin, 2004, p. 15.

[3] David Garman. "Testimony before the Joint Economic Committee, U.S. Congress, July 28, 2005." Available online. U.S. Department of Energy. URL: http://www1.eere.energy.gov/office_eere/congressional_test_072805_congress.html. Accessed August 2, 2006.

[4] For data on the 22-million-barrels-per-day high of 2007, see entry for February 16, 2007, in the EIA's "U.S. Weekly Petroleum Products Supplied" tables for 2007 at http://tonto.eia.doe.gov/dnav/pet/hist/wrpupus2w.htm. Accessed March 15, 2007. For data on the 50-year low, see p. xxiii of the EIA's *Annual Energy Review 2005.* Available online. URL: http://www.eia.doe.gov/oiaf/aeo/index.html.

[5] Percentages based on EIA data ("U.S. Crude Oil Production") for October 1970 and December 2006. Available online. URL: http://tonto.eia.doe.gov/dnav/pet/hist/mcrfpus1m.htm. Accessed March 15, 2007.

[6] Hubbert announced these findings in *Nuclear Energy and the Fossil Fuels,* Shell Development Company publication number 95, Houston, Texas, June 1956, presented before the Spring Meeting of the Southern District, American Petroleum Institute, Plaza Hotel, San Antonio, Texas, March 7–9, 1956.

[7] The relevant data can be found in the EIA's *Annual Energy Outlook 2007,* pp. 10, 11, and 78. Available online. URL: http://www.eia.doe.gov/oiaf/aeo/pdf/0383(2007). Accessed March 15, 2007.

[8] See the USGS report on ANWR oil reserves, which have been estimated to represent, at the low end, a six-month supply of oil, in *Arctic National Wildlife Refuge, 1002 Area, Petroleum Assessment, 1998, Including Economic Analysis.* Available online. URL: http://pubs.usgs.gov/fs/fs-0028-01/fs-0028-01.htm. Accessed August 2, 2006.

[9] See the entry for 2005 in the EIA's "Crude Oil Reserves, Reserves Changes, and Production" table at http://tonto.eia.doe.gov/dnav/pdf/pet_crd_pres_dcu_NUS_a_htm.

[10] *BP Statistical Review of World Energy 2006,* p. 6. Available online. URL: www.bp.com/statisticalreview. Accessed March 15, 2007.

[11] See *World Petroleum Availability 1980–2000,* a report Hubbert coauthored with Richard Nehring in 1980. Available online. URL: http://www.wws.princeton.edu/cgi-bin/byteserv.prl/~ota/disk3/1980/8023/8023.PDF. Accessed August 2, 2006.

[12] See Kenneth S. Deffeyes. *Hubbert's Peak: The Impending World Oil Shortage.* Princeton, N.J.: Princeton University Press, 2001.

[13] See C. J. Campbell. *The Coming Oil Crisis.* Essex, England: Multi-Science, 2004.

[14] See the EIA's online resource *Energy Plug: Long-Term World Oil Supply: A Resource Base/Production Path Analysis.* Available online. URL: http://www.eia.doe.gov/emeu/plugs/plworld.html. Accessed August 2, 2006.

[15] Richard Kerr. "USGS Optimistic on World Oil Prospects." *Science,* July 14, 2000, p. 237.

[16] These arguments are summarized in Marc Jaccard's book *Sustainable Fossil Fuels: The Unusual Suspect in the Quest for Clean and Enduring Energy.* Cambridge: Cambridge University Press, 2006.

[17] These arguments are summarized in *The Economics of Petroleum Supply: Papers by M. A. Adelman, 1963–1993.* Cambridge, Mass.: MIT Press, 1993.

[18] This position is a major theme of Odell's essays from 1961 to 2000 as reprinted in *Oil and Gas: Crises and Controversies, 1961–2000.* Vol. 1, *Global Issues.* Brentwood, England: Multi-Science, 2001.

[19] This position is another main theme of Odell's essays from 1961 to 2000 as reprinted in *Oil and Gas.*

[20] Kenneth S. Deffeyes. *Hubbert's Peak,* p. 173.

[21] See International Energy Agency (IEA) Report: *Energy Prices and Taxes, 1st Quarter 2004:* "Thirty Years of Energy Prices and Savings," pp. 1–2. Available online. URL: http://www.iea.org/Textbase/Papers/2005/cost.pdf.

[22] John C. Mowen. "A Perfect Storm: Gasoline Prices and Consumer Behavior." September 7, 2005. Spears News Archive, William S. Spears School of Business, Oklahoma State University. Available online. URL: http://spears.okstate.edu/secondary/creator.php?c=./info_pages/news/pr/fall2005/mowen-gas.php. Accessed February 17, 2006.

[23] For the USGS estimates, see the USGS report on ANWR oil reserves, which have been estimated to represent, at the low end, a six-month supply of oil, in *Arctic National Wildlife Refuge, 1002 Area, Petroleum Assessment, 1998, Including Economic Analysis.* For the Alaska Coalition figures, see p. 4 of the report *Broken Promises: The Reality of Big Oil in America's Arctic,* released by the Alaska Coalition. Available online. URL: http://www.alaskacoalition.org/Arctic_Oil.htm. Accessed August 2, 2006.

[24] See p. vii of the document *Impacts of Modeled Provisions of H.R. 6 EH: The Energy Policy Act of 2005,* released in July 2005 by the EIA Office of Integrated Analysis and Forecasting, U.S. Department of Energy. Available online. URL: http://www.eia.doe.gov/oiaf/servicerpt/hr/pdf/sroiaf%282005%2904.pdf.

[25] See the online EIA report *Annual Energy Outlook 2007.* Available online. URL: http://www.eia.doe.gov/oiaf/aeo/pdf/0383(2007).pdf. Accessed August 2, 2006.

[26] See relevant entries for January 2007 and January 2002 in the EIA's "U.S. Weekly Total Crude Oil and Petroleum Products Imports" tables. Available online. URL: http://tonto.eia.doe.gov/dnav/pet/hist/wttimus2w.htm. Accessed March 15, 2007.

[27] U.S. President George W. Bush, State of the Union address, January 31, 2006.

[28] Ibid.

[29] For an example of an activist organization's position on this issue, see Global Energy Network Institute's "Global Issues—Policy Options: Subsidies." Available online. URL: http://www.geni.org/globalenergy/policy/renewableenergy/subsidies/index.shtml.

[30] As cited by Elisabeth Bumiller in "Bush's Goals on Energy Quickly Find Obstacles." *The New York Times,* February 2, 2006, Energy. Available online. URL: http://www.nytimes.com/2006/02/02/politics/02energy.html?ex=1154836800&en=d0b 283fb3af4da6c&ei=5070#.

[31] As reported by the Bloomberg News Service at http://www.bloomberg.com/apps/news?pid=10000103&refer=news_index&si d=aZ.dPxUR6nbw.

[32] "Capitol Hill Hearing Testimony: Statement of Brian Castelli, Executive Vice President and COO, Alliance to Save Energy, Committee on House Energy and Subcommittee on Energy and Air Quality, November 2, 2005." In *Congressional Quarterly*, Federal Document Clearing House Congressional Testimony. November 2, 2005. Excerpts from the document are reprinted in this book starting on page 159.

[33] See "Internal Memos Show Oil Companies Intentionally Limited Refining Capacity to Drive Up Gasoline Prices." Press release of the Foundation for Taxpayer and Consumer Rights (FTCR) released September 7, 2005. Available online. URL: http://www.consumerwatchdog.org/energy/pr/?postId=5110. Accessed August 2, 2006.

[34] Jeremy Leggett. *The Empty Tank: Oil, Gas, Hot Air, and the Coming Global Financial Catastrophe*. New York: Random House, 2005, p. xiv.

[35] Based on data for January 2007 from the EIA's "U.S. Weekly Total Crude Oil and Petroleum Products Imports." Available online. URL: http://tonto.eia.doe.gov/dnav/pet/hist/wttimus2w.htm.

[36] As cited by the National Environmental Trust (NET) on p. 33 of *America, Oil, and National Security: What Government and Industry Data Really Show*. Available online. URL: www.net.org/security/america_oil.pdf.

[37] Reuters. "U.N. to let Iraq sell oil for euros, not dollars," CNN.com. October 30, 2000. Available online. URL: http://archives.cnn.com/2000/WORLD/meast/10/30/iraq.un.euro.reut. Accessed August 2, 2006.

[38] *Energy Policy Act of 2005* (42 U.S.C. 15801 and sections thereafter; P.L. 109–48), signed into law on August 8, 2005. (See page 105.) Also accessible online. URL: http://frwebgate.access.gpo.gov/cgi-bin/getdoc.cgi?dbname=109_cong_public_law&do cid=f:publ058.109.

[39] U.S. President George W. Bush, State of the Union address, January 31, 2006.

[40] The Advanced Energy Initiative was released by the National Economic Council in February 2006 and announced by U.S. President George W. Bush during his January 31, 2006, State of the Union address. It is accessible online on the Web site of the White House. URL: http://www.whitehouse.gov/stateoftheunion/2006/energy/energy_booklet_pdf. Accessed August 2, 2006.

[41] For a thorough review of both the merits and limitations of alternative energy sources such as hydrogen in their ability to limit foreign oil dependence, see "Testimony of David Garman, Under Secretary for Energy, Science, and Environment, before the Joint Economic Committee, United States Congress, July 28, 2005." Available online. U.S. Department of Energy. URL: http://www1.eere.energy.gov/office_eere/congressional_test_072805_congress.html.

[42] The NEPD Group's report on the condition of the U.S. energy infrastructure and recommendations for improving it can be found in *Reliable, Affordable, and Environmentally Sound Energy for America's Future: Report of the National Energy Policy Development Group*, May 2001, chapter 7, America's Energy Infrastructure. Available online. URL: http://www.whitehouse.gov/energy.

[43] Scott Pelley. "Global Warning!" *60 Minutes*. Produced and directed by Bill Owens. CBS Television, February 19, 2006. Television news report.

[44] For more information on the IPCC's assessment of global climate change, see this organization's Web site at http://www.ipcc.ch/.

[45] See pp. 3 and 7 of the EIA's *Monthly Energy Review* (February 2007) and p. 8 of the EIA's *Annual Energy Outlook 2007.*

[46] See pp. ix and xv of the EIA's *Emissions of Greenhouse Gases in the United States 2005.* Available online. URL: www.eia.doe.gov/oiaf/1605/ggrpt/index.html. Accessed March 15, 2007.

[47] International Energy Agency (IEA). *Energy Policies in IEA Countries: 2004 Review.* Paris: OECD, 2005, pp. 163–164.

3

Global Perspectives

INTERNATIONAL ENERGY ISSUES

Nations throughout the world are, like the United States, largely dependent on oil for energy, and at times their economies, governments, and people experience difficulties and crises related to this dependence.

Environmental Issues

Global warming, or global climate change, is one example. As its very name indicates, of course, it is of particular concern not only to the United States but also to the entire global community. As of this writing, more than 165 countries have ratified the Kyoto Protocol, agreeing to achieve, by 2012, at least a 5 percent reduction in greenhouse gas emissions below 1990 levels. The Kyoto Protocol went into effect on February 16, 2005.

Energy Security

Most countries are also greatly affected by issues related to oil supply. Energy security has become a major issue, as countries with enough energy and countries without begin to compete for a piece of what is left of world oil and natural gas reserves, to develop technologies to help their nations keep pace with a skyrocketing demand for energy, to lock in agreements with other nations to address global energy-related environmental issues, and to arrange mutually beneficial trade pacts that exploit existing energy sources or developing technologies. In the United States *energy security* is now also a term that suggests the need to protect oil refineries, pipelines, and nuclear power facilities, yet for the United States and the entire international community, the term now increasingly suggests the need to maintain a level of fuel and electricity production and affordability that will secure a country's economic stability and political power for generations.

The potential of renewable energies when society is already dominated by fossil fuel energies is another hot-button issue for the world, just as it is for the United States. Renewable energy's share of total world energy supply (among countries of the IEA) grew by an annual rate of only 2.2 percent from 1973 to 2004, according to the IEA.[1] Wind and solar were the fastest growing renewable energies among IEA countries but still only accounted for less than 1 percent of the 13.1 percent of energy that was derived from renewable sources. Nuclear electricity production among IEA countries, on the other hand, contributed 6.5 percent of total energy supply in 2004.[2] Yet the demand for fossil fuels is still alive and well. According to the IEA, oil's contribution to total primary energy supply among IEA countries was 34.3 percent in 2004.[3] Total demand for oil among IEA countries fell by 20 percent between 1979 and 1983, but since then it has been increasing continuously, returning to 1973 demand levels in 2000.[4] After 1994 natural gas surpassed coal as a primary contributor to energy supply among IEA countries. In 2004 natural gas accounted for 20.9 percent of total primary energy supply, while coal's share was 25.1 percent.[5]

Foreign Dependence

Also of major concern to the United States and other countries alike are issues related to the need to import oil from other countries, especially from the politically volatile Middle East. Global oil demand just keeps increasing, but proven oil reserves in oil-producing countries outside the Middle East, such as Nigeria, Venezuela, and Mexico, are 70 percent lower than the reserves available in the Middle East. Many commentators have pointed out that even OPEC countries are appearing to exhaust their oil supplies, and oil consumption is outpacing new oil discoveries. In fact, by 2025 all the regions with major oil reserves are projected to possess levels of reserves that are lower than current levels. According to the EIA, there will be a 50 percent increase in global oil demand by 2025, a growth in demand that is double the rate of oil production since 1980.

Six countries, in particular, have grappled with the specific energy issues that have beset the United States, and their experiences with these issues provide interesting case studies, illustrating, in some cases, alternative approaches to the most pressing energy challenges. Also, through their interactions with the United States on specific energy issues, these countries have impacted U.S. economic and political security in substantial ways. These countries are China, Germany, Iran, Saudi Arabia, Nigeria, and Venezuela.

CHINA

Energy Demand

As the United States has, the People's Republic of China has seen its demand for energy skyrocket. But unlike in the United States, which is a fully industrialized country, in China, a developing country, energy demand increased over an extremely short period. Coal is China's dominant fuel; its coal resources are rich, and coal consumption is deeply embedded in its history. China is both the largest consumer and the largest producer of coal in the world. However, China's rapid economic growth in the past decade has resulted in an increasing demand for oil. In 2006, according to the EIA, China's oil demand was expected to soar above 2005 levels by a considerable half million barrels per day. In 2006 alone China generated more than half of a worldwide increase in oil demand, according to the *BP Statistical Review of World Energy 2006.*[6] China is also the second largest consumer of petroleum products in the world and the third largest net importer of oil after the United States and Japan, reported the EIA.[7] In 2006 China's total demand for oil was estimated to reach 7.4 million barrels per day, a demand that the EIA projected would increase significantly by 2030.[8]

It was in 2003 that China, the world's most densely populated country, surpassed Japan as the world's second-largest consumer of petroleum products. And it was only in 2000 that the surge in China's energy consumption began. The roots for this rapid increase in oil consumption can be traced back to the beginning of the millennium, when an upswing in economic investments in China spurred massive economic growth. For the first time many of China's citizens began to experience a level of prosperity that allowed them to purchase cars, home appliances, and other energy-intensive goods. China's mostly human energy–dependent way of life was transformed into an increasingly fuel-dependent lifestyle. The IEA reported that in 2004 for every 1 percent increase in China's gross domestic product (GDP), its energy demand grew by more than 1.5 percent.

Reliance on Imports

Just as in the United States, one outcome of elevated energy demand in China has been an increased reliance on energy imports. China reported a 37 percent leap in oil imports in 2004.[9] Before 1993 China imported no oil at all; now its oil imports are at a level second only to that of the United States. As the United States does, China funnels most of its total domestic oil production, which consists of about 3.8 million barrels per day, into its own country, rather than exporting it, according to the EIA.[10] The demand for oil is particularly strong in China's financial sector, which is already well established. The surge in the number of workers employed by that growing sector has led to an increase in personal car ownership and a growth in transportation overall, as well as a need for oil-powered backup generators to meet this sector's increased demand for electricity. China's budding oil import program and heavy oil demand make it a key player in world energy markets.

One of the major issues that China now faces is the need to protect its economic progress, which has resulted in a better quality of life for its people, while avoiding the consumer-driven, energy-intensive, nonsustainable, and environmentally taxing habits that often accompany industrialization. Some commentators have noted that China's lightning-fast economic boom and corresponding rise in energy demand are not uncharacteristic of underdeveloped countries in recent times. These countries, which contain about 80 percent of the world's population, have been adopting energy-intensive industrialization at a rapid pace. This, in fact, is one of the reasons that was cited by President George W. Bush for the U.S. refusal to sign the Kyoto Protocol. The Kyoto Protocol excludes developing countries, including China, from greenhouse gas emission reduction requirements, allowing them to continue to develop their economies, unchecked by consideration for the global community, whereas other more industrialized countries might lose economic ground while trying to satisfy global environmental requirements. However, supporters of the Kyoto Protocol counter that the United States, while accounting for less than 5 percent of the world's population,[11] is responsible for as much as 25 percent of the world's greenhouse gas emissions,[12] so its emissions-reduction requirement under the Kyoto plan is appropriate to the level of polluting emissions that it outputs. Kyoto Protocol supporters also argue that the United States, with its fully industrialized economy, is better able to invest in the expensive technology that can help to improve fuel efficiency and reduce emissions than developing countries, which might undermine their efforts to compete in the global economy if they attempted to do so.

Counterstrategies

ELIMINATING SUBSIDIES FOR DOMESTIC PETROLEUM

China's leaders, nonetheless, have shown concern about the country's dependence on oil, at least on oil that is imported, and China has implemented strategies to address its mounting demand for oil and inability to meet this demand fully with its own oil production. China has kept domestic petroleum affordable for its citizens by subsidizing it. As a result, the demand for cheap domestic fuels just keeps increasing. A series of state-mandated price increases was instituted in an effort to raise the price of China's oil to a level more closely reflective of the world oil price and shake off some of the domestic demand. However, this effort proved insufficient to keep China at pace with the world oil market; in fact, demand for Chinese exports of certain petroleum products increased in the first half of 2005 as the gap between domestic prices and world prices widened.[13] However, the Chinese government (unlike the U.S. government) has lately expressed intentions to eliminate subsidized prices altogether.

RESTRUCTURING DOMESTIC OIL AND GAS INDUSTRIES

Another approach that China has taken to diversifying its energy sector is embarking on a major restructuring of its state-owned oil and gas entities, beginning in 1998. The purpose of the restructuring effort was to simplify the energy system in China, making the state-owned firms more like the vertically integrated oil corporations in other countries. These state-owned oil companies, known for their massive overstaffing, are the China National Petroleum Corporation (CNPC), the China Petrochemical Corporation (Sinopec), and the China National Offshore Oil Corporation (CNOOC). Under China's oil-company restructuring program, CNPC, Sinopec, and CNOOC began shedding unprofitable ancillary businesses and instituting massive layoffs. CNPC, which had engaged mainly in oil and gas exploration and production, and Sinopec, which had been engaged in refining and distribution, were reorganized so that CNPC focused on oil activities in the north and west of China, and Sinopec in the south. In 2003 China also created a regulatory agency to oversee its energy industry, the State Energy Administration (SEA).

INCREASING FOREIGN OIL SUPPLY

China's leadership has also taken other steps to address China's oil demand. Hu Jintao, president of China since March 2003, led the nation into a deeper dependency on oil from the Middle East. Under Hu's leadership, China has taken measures to increase the amount of the oil it imports from Iran. In October 2004, China's large oil company, Sinopec Group, signed a $100 billion, 25-year oil and gas trade agreement with Iran—China's largest energy deal with the country that is the number-two oil producer in OPEC. Iran's former minister of petroleum, Bijan Namdar Zanganeh, said that Iran is China's major oil supplier and that Iran wanted to become China's long-term business partner.

A few years prior, in 1998, China's former president, Jiang Zemin, made the first visit by a Chinese head of state to Saudi Arabia. During this visit China and Saudi Arabia formally approved certain oil cooperation agreements and discussed plans for using Saudi oil in a large Chinese petrochemical complex. Earlier, in 1995, China agreed to import 3.5 million tons of crude oil from Saudi Arabia annually.

Commentators have pointed to these and other arrangements with Middle Eastern oil-producing countries as a sign of China's limited ability to meet its oil needs. The increased reliance on oil from the Middle East demonstrates a major difference between the United States and China in their approach to domestic oil shortfalls, as the strategy of the United States has involved at least an attempt to cut back on Middle Eastern oil imports.

China has also made massive investments in infrastructure projects in Africa in exchange for securing contracts in Africa's energy sector. This Chinese strategy for ensuring continued oil supply has incited the disapproval of the United States and other Western nations because some of China's dealings are with African states that are notorious for their human rights abuses.

DIVERSIFYING THE SOURCES OF ENERGY

China's nuclear power program has gained momentum in recent years, as China has been investing in the construction of several nuclear power plants. Another Chinese approach to handling its accelerating oil demand has been an effort to convert the country to alternative sources of energy. The reasons for making such a switch include not only the concern that China will soon be unable to meet its immense energy demand but also the serious problems with pollution and environmental degradation caused by China's swift industrialization and extensive reliance on oil and coal. China is the second-largest emitter of greenhouse gases after the United States. And China's level of energy efficiency is one-quarter that of industrialized countries like the United States, according to the United Nations Development Programme (UNDP). Furthermore, rapid economic development in China has led to not only harmful fuel emissions but power shortages. Such energy-related issues are compounded by the fact that China has limited available natural resources; agricultural activities have stripped China's land of its protective plant canopy and exhausted the land's fertility. Moreover, China's booming population, coupled with agricultural encroachment upon scarce arable land (particularly through irrigated rice farming), has generated soil erosion, deforestation, grassland destruction, and soil and water pollution. China reportedly lost 20 million hectares of cultivable land between 1957 and 1977. The UNDP has also reported that only 75 percent of China's citizens are assured access to safe drinking water.

According to the EIA, China's renewable energy consumption in 2003 accounted for only 3 percent of total energy consumption. In 2005, China passed a significant renewable energy law, known as the Renewable Energy Law, which has won praise from some U.S. environmentalists, who say it is a good model for the United States to follow in formulating future U.S. energy policies. This law is expected to boost China's capacity to use renewable energy to 10 percent by the year 2020. Through passing this law, the country aimed to institute measures that would protect the environment, prevent energy shortages, and reduce dependence on energy imports. Effective in 2006, the law stipulates the following:

1. That electricity power grid operators purchase resources from approved renewable energy producers
2. That national financial incentives be available to foster state and local development of renewable energy resources, including solar electricity, solar water heating, and renewable energy fuels
3. That loan and tax discounts be given for renewable energy projects, such as the construction of commercial renewable energy facilities
4. That specific penalties be imposed for noncompliance with the law

The law includes in its definition of renewable energy hydroelectricity, wind power, solar energy, geothermal energy, and marine energy. In addition, China's National Development and Reform Commission (NDRC) will establish specific renewable energy targets for China as the framework for the implementation of this law. In relation to the law, China has also unveiled plans to increase environmental spending from 0.7 percent of GDP in 1996 to 1.7 percent in 2010 and 10 percent in 2020.

There are a number of other environmental laws and regulations that China has instituted in order to improve its environmental situation and address its energy issues. Xie Zhenhua, head of China's State Environmental Protection Administration (SEPA), pledged in 2005 that China would maintain its commitment to environmental protection and efficient, sustainable use of natural resources. The UN secretary-general, Kofi Annan, awarded Xie the UN's Sasakawa Environment Prize in 2003.

Some significant environmental protection programs that China launched in the past have included a 1970s biodigester waste recycling program and a reforestation policy, but ensuring compliance with these programs by China's agricultural producers proved difficult. However, SEPA has also established a number of successful measures to improve lake, river, and coastal water quality; urban and agricultural environments; and forest and grassland coverage.

In addition, Li Peng, China's former prime minister and later chair of China's National People's Congress (NPC), proposed a plan in 1994, while still prime minister, to create an alternative-fuel "people's car," the production and assembly of which China has since arranged through Volkswagen and other

Western auto companies. Other environmental projects Li promoted include China's Three Gorges Dam, the world's largest hydroelectric plant, designed to generate 18.2 gigawatts of power. In March 2002 the China Yangtze Three Gorges Electric Power Corporation was established, and in June 2003 the reservoir created by the dam began to fill, with its initial turbines operating a month later. As many dams have, the Three Gorges Dam has raised concerns over its potential to cause huge floods that could wipe out local ecosystems and result in human fatalities. The Yangtze project is also controversial because of the million-plus people who are expected to be forced from their homes as the reservoir fills to capacity—a capacity that is said to span the area of 13 cities, 140 towns, 1,352 villages, and 650 factories.[14] The cost of the project is also a matter of contention, with an anticipated total bill of $29 billion.[15] The water level, too, which is supposed to reach 577 feet, will submerge local archaeological sites and unique natural landscapes that have been a draw for tourists.

While China is still considered to be a major contributor of fossil fuel emissions and a country where rapid economic growth has the potential to cause real national energy crises, efforts are under way to reduce the impact of an ever-dwindling, always-in-demand supply of oil.

GERMANY

Energy Demand

As is the United States, Germany is a fully industrialized country with a large appetite for oil. Germany has a highly developed market economy—the largest economy in Europe, and the fifth largest in the world. As a result Germany has considerable purchasing power in the world market. With its advanced economy Germany is also a voracious consumer of energy, the world's fifth largest consumer for several consecutive years. Germany consumed 14.7 quadrillion Btu of energy in 2004, according to the EIA.[16]

Many of the EIA data for Germany's level of energy consumption are comparable to those for the United States, especially in terms of fossil fuel consumption. In 2004 oil met 37 percent of Germany's energy demand, while coal and natural gas each provided 24 percent of energy. Germany is the fifth-largest consumer of oil in the world and Europe's largest consumer of electricity; in 2004 electricity consumption totaled 524.6 billion kilowatt hours (a kilowatt hour, often abbreviated as *kWh,* is a measure of a unit of electricity expended per hour). This represented a significant portion of the 566.9 billion kWh of electricity Germany produced that year. Most of Germany's 2004 electricity production, 61 percent, was powered by conventional sources of energy—mainly coal. The IEA has reported that brown coal accounted for the largest share of conventional thermal electricity generation, 42 percent; hard coal and natural gas accounted for 37 percent and 16 percent, respectively.[17] Germany is actually the world's seventh-largest coal producer and fourth-largest coal consumer, and it has the largest quantity of coal reserves of all EU nations. Brown coal is Germany's most important domestic energy source, representing about 40 percent of its total domestic energy production and meeting about 24 percent of its total energy needs in 2004.

Germany also consumes a sizable quantity of natural gas, consuming the second-highest levels of natural gas among EU nations. At the same time, Germany is the EU's third-largest producer of natural gas, having 9.1 trillion cubic feet of proven natural gas reserves, according to *Oil and Gas Journal* in 2006.[18] The EIA reported that in 2004 Germany produced 730 billion cubic feet of natural gas but also suffered from a lack of new natural gas discoveries in recent years, which could impede future growth in natural gas production.

Reliance on Imports

Yet Germany's energy habits are most similar to those of the United States in terms of importing energy. Although the country is able to produce significant amounts of coal and natural gas, Germany, as is the United States, is a huge importer of oil because it does not possess significant oil reserves and has a meager domestic oil production sector in comparison with that of the rest of the world. With an insufficient level of domestic oil production to satisfy the needs of its large, developed economy, Germany's energy needs rely heavily on oil imports. In fact, after the United States and Japan, Germany is the world's third-largest importer of oil. Most of Germany's energy is imported, according to the EIA. Oil imports supply more than 90 percent of the 2.7 million barrels of oil that Germany consumes each day.[19] The German economics statistics agency reported that Russia was the largest supplier of crude oil imports to Germany in 2004, followed by Norway and the United Kingdom. Additionally, a lack of new coal discoveries in Germany has recently been reflected in a downturn in Germany's domestic coal production. To meet domestic coal demand, then, Germany has begun to depend, in part, on coal imports, mainly from South Africa.

Environmental Problems

Germany's heavy reliance on coal for electricity generation, combined with its advanced industrialization, has led to air pollution, forest damage caused by acid rain (rain contaminated by emissions of sulfur dioxide), and environmental degradation. According to the EIA Germany is the sixth-largest emitter of carbon dioxide in the world and the third largest among OECD countries. It emitted 862.2 million metric tons of carbon dioxide in 2004.[20]

Counterstrategies

RENEWABLE ENERGY

However, Germany has made a bold effort to respond to insufficient domestic energy supply, foreign oil dependence, and energy-related environmental damage by ratcheting up its development of renewable energy technologies. Germany has become a world leader in renewable energy. It generates the largest amount of electricity from wind power in the world, and in 2004, according to the IEA, Germany possessed 390 megawatts of installed solar power capacity and 14,600 megawatts of installed wind capacity. This represents a respective 43 and 40 percent of total installed renewable energy capacity among all OECD countries. Although renewable energy still has a modest share of overall energy consumption in Germany—hydropower and other renewables accounted for only about 6 percent of Germany's total energy consumption in 2004—Germany's capacity to generate energy from renewable sources, including solar, wind, biomass, hydropower, and geothermal energy, in order to meet its energy needs has increased annually. Germany has become the world's largest supplier of biodiesel.

Germany's energy policies continue to emphasize conservation and further development of renewable energy sources. For example, on July 21, 2004, Germany passed the Renewable Energy Sources Act (Act Revising the Legislation on Renewable Energy Sources in the Electricity Sector). One of the aims of this law, which went into effect in August 2004, is to increase the percentage of Germany's electricity derived from renewables—from 6.7 percent in 2000 to 12.5 percent by 2010, 20 percent by 2020, and 50 percent by 2050. These goals do not seem unrealistic for Germany, which, as of 2005, had already begun generating 10 percent of its electricity from renewable sources. The Renewable Energy Sources Act calls for improved financial incentives for the continuing development of not only wind power for electricity, but also hydropower, now the second-largest renewable source of electricity in Germany after wind. It also calls for more electricity derived from solar, biomass, and geothermal power. The law seemed to spur almost immediate action, as Shell's five-megawatt solar electricity plant in the city of Leipzig, Germany, went online in September 2004 and another corporation was expanding a 4-megawatt solar plant in Gottelborn, Germany, to 8 megawatts in 2006. Furthermore, in late 2003 Germany finished building its first geothermal plant, in Neustadt-Glewe. Germany's energy industry as a whole has agreed to invest 30 billion euros in constructing new power plants and other forms of energy infrastructure by 2012, and 40 billion euros in the expansion of renewable energy use.

NUCLEAR ENERGY

However, one form of renewable energy has been just as controversial for Germany as it has been for the United States: nuclear energy. Yet Germany's reaction to the controversy is distinctly different. While Germany was the fourth-largest generator of nuclear power in the world in 2003, possessing more than 15 operating nuclear power plants, nuclear energy's unpopularity among environmentalists in Germany, led by members of the Green Party, has resulted in a policy of turning away from this form of energy. The Green Party was a member of the governing coalition from 1998 to 2005 and unlike the U.S. government, whose Energy Policy Act of 2005 contained plans for reviving nuclear energy production, the German government instituted in 2001 a plan to phase out all nuclear energy production by 2022. More receptive to the concerns of industry groups, Germany's chancellor since November 2005, Angela D. Merkel, has cautioned that the nuclear energy phaseout will cause a gap in energy production that creates an urgent need for the development of other energy technologies. Merkel has promised that Germany's commitment to renewable energy and environmental protection will continue. Merkel, who supported the U.S. invasion of Iraq, has been viewed as sympathetic to U.S. energy interests. She is also Germany's former minister for the environment and reactor safety and German's first female chancellor.

THE KYOTO PROTOCOL

Another way in which German energy policy differs from U.S. energy policy is that Germany has fully supported the Kyoto Protocol, which it ratified on May 31, 2002. According to an EU arrangement, Germany and other EU countries share the burden of meeting the protocol's carbon dioxide emission reduction requirements. Because Germany's level of carbon dioxide is significantly higher than that of some other EU countries, Germany is required by the EU to cut its carbon dioxide emissions by 21 percent relative to the 1990 baseline. Germany demonstrated compliance with this requirement, cutting a significant amount of its carbon dioxide emissions by 2006, mainly through an increase in the usage of renewable energy. In addition, Germany has recently instituted environmental regulations that have curtailed exploration for natural gas.

ADDITIONAL STRATEGIES

Under the leadership of Chancellor Merkel, the German government has also announced plans to formulate an energy strategy for further reducing dependency on energy imports, preventing energy price increases, and tackling the country's energy-related environmental challenges through the year 2020. Germany's minister of education and research, Annette Schavan, announced that by 2009 a 2-billion–euro funding increase would be implemented to support research and development in the energy sector. Horst Koehler, president of Germany since July 2004, promoted Germany's Renewable Energy Sources Act. It was Horst's administration that set the goal of having 12.5 percent of Germany's electricity generated by renewable energy sources by 2010 and 20 percent by 2020.

In addition, the former chancellor of Germany, Gerhard Schroeder, who held office from October 1998 to November 2005, spearheaded a partnership with the United States to cooperate on issues related to energy supply, energy efficiency, and renewable energies. In August 2005 Schroeder and the U.S. president, George W. Bush, met in Berlin for the Working Group on Energy, Development, and Climate Change, a U.S.–German effort that has continued under the leadership of Chancellor Angela Merkel. Germany and the United States also announced a partnership to capture wasted methane from oil and natural gas systems, coal mines, and landfills and use it as an energy source.

Aside from partnerships between the United States and Germany, Germany's energy policy, with its emphasis on renewable energy, has the potential to influence U.S. energy policy in significant ways. Many in the United States hope that Germany's progress with using renewable energies to combat foreign energy dependence, environmental degradation, and shortages in domestic energy supply will become an example to the United States of the real viability of renewable energies as a long-term solution to fossil fuels.

THE MIDDLE EAST

Iran

OIL RESERVES

Iran is the second-largest OPEC oil producer. Oil production is extremely important to Iran, accounting for almost all of Iran's energy production. Iran's economy relies heavily on oil export revenues. The EIA has reported that those revenues account for about 80 to 90 percent of Iran's export earnings and about 40 to 50 percent of Iran's budget.[21] Iran's oil export revenues were further boosted by increases in crude oil prices in the first few years of the new millennium.

Iran holds about 10 percent of the world's oil reserves, according to *Oil and Gas Journal* in 2006. In October 1999 Iran announced what it deemed to be its biggest oil discovery in 30 years. And in July 2004 Iran's oil minister, Bijan Namdar Zanganeh, also reported that new oil discoveries in the region of one of Iran's existing oil fields had boosted Iran's proven oil reserves to about 132 billion barrels. Iran also has the world's second-largest reserves of natural gas, according to the EIA.[22]

ECONOMIC PROBLEMS

Yet despite high revenues from energy exports (or perhaps because of them), Iran is beset by a number of economic problems, many of them energy related. According to the EIA, Iran's oil production and exports are both declining, with some of the downturn attributed to guerrilla attacks on oil installations and some to a shortage of available gas to inject into existing oil wells to draw up more oil as depletion of the wells draws near. At the same time Iran's rate of domestic oil consumption is increasing rapidly, about 7 percent per year.[23] A plunge in oil prices during 1998–99 was also devastating to Iran's economy, although the effects were not long lasting, as oil prices soon rebounded. However, Iran has a large budget deficit, mainly because of government subsidies of domestic gasoline, other fuels, and food products. Gasoline in Iran cost less than 40 cents per gallon before April 2003 (in the United States, the average price per gallon of gasoline was $1.56 in 2003). In 2003 Iran increased gasoline prices by 30 to 35 percent and announced that it might need to ration gasoline to control gasoline subsidy expenditures. Iran's government spent approximately $4.7 billion on domestic fuels in 2004, according to the EIA,[24] and in January 2005 Iran's parliament refused to implement an additional increase of the price of domestic gasoline and other fuels, which might have helped Iran's economy. Because gasoline prices remain low, gasoline consumption in Iran keeps rising, driving subsidy expenditures up even further. To pay for the increase in subsidy expenditures, in November 2005 the parliament approved a measure to spend an additional $3 billion on subsidies during 2005–06, withdrawing $2.6 billion of this sum from a fund Iran had established in 2000 (primarily to protect its economy from an oil price collapse). Yet Iran imports about a third of its gasoline, meaning that such an increase in expenditures on gasoline subsidies will have an even greater economic impact. Legislators were opposed to spending any additional funds to pay for gasoline imports in 2007. Moreover, Zanganeh has reported that Iran's oil fields have a natural decline rate of 200,000 to 300,000 barrels per day and need to be upgraded and modernized.

Iran's other economic problems include a drain on its budget caused by a rapidly growing population (population has more than doubled in 20 years), a sizable rate of inflation (15 percent), a bleak unemployment rate, a high poverty level, inefficient and expensive governing bodies, and inefficient state monopolies for the key sectors of Iran's economy.

RELATIONS WITH THE UNITED STATES

Relations between the United States and Iran have been strained. One tactic that Iran has used to cushion its oil-dependent economy is to accept euro dollars as payment for the oil it exports to EU and Asian countries. Since this measure has the potential to destabilize the value of the U.S. dollar, now the standard currency for trading in the global oil market, some have claimed that the true reason for U.S. opposition to Iran is not Iran's support of terrorism or possible nuclear weapons program, but its challenging of the supremacy of the U.S. dollar in the international oil market. Yet strained U.S.-Iranian relations have a long history—and have been due to a series of energy-related conflicts, rather than to the nuclear issue or the euro issue alone. For example, when Mohammad Mossadegh was prime minister of Iran (1951–53), he nationalized Iran's oil industry, which was originally under British control. In response Britain enforced an economic blockade against Iran and, working with the U.S. Central Intelligence Agency (CIA), helped to instigate Mossadegh's overthrow by Shah Mohammad Reza Pahlavi. Mossadegh was sent to prison, where he died in 1967. The shah opened Iran's oil reserves to Western companies. Whereas Mossadegh's oil policies were antagonistic to the United States and Britain, Pahlavi was an American-friendly leader who opened Iranian oil reserves to Western companies. However, Islamic fundamentalists who viewed the shah as a puppet of Western powers drove him out in 1979 during the Iranian Revolution. A new Islamic republic emerged, based on fundamentalist Islamic principles and led by the national spiritual leader Ayatollah Khomeini (1979–89).

Khomeini nationalized Iran's oil industry. Under his leadership Iran became the first modern state ruled by fundamentalist Islamic principles. When the shah entered the United States for medical treatment in 1979, Islamist militants seized the U.S. embassy in Iran and held 52 Americans hostage for 444 days. The hostage crisis, combined with the oil crisis from Arab oil embargoes, seriously challenged the United States during the presidency of Jimmy Carter (1977–81). It was at this time that the United States severed diplomatic ties with Iran, and it has considered Iran a rogue state ever since.

U.S. SANCTIONS

Perhaps the greatest strains on Iran's economy are isolation from the international community and the economic sanctions that have been imposed on the country by the United States. A sanction is a ban on imports from a country, imposed as a penalty on countries that pose a risk to the safety and security of the international community. The United States began imposing sanctions on Iran after the hostage crisis in 1979, including a ban on importing oil from Iran. Although some sanctions were lifted after the release of the hostages in 1981, each successive U.S. administration has imposed some restriction on trade with Iran, including an almost continuous ban on oil imports from Iran. The U.S. president George W. Bush, who has called Iran part of "the axis of evil," extended sanctions originally imposed in 1995 by the U.S. president Bill Clinton (1993–2001), barring U.S. companies and their foreign subsidiaries from doing business with Iran or investing in the development of Iran's petroleum resources. According to the U.S. government, the sanctions are a U.S. response to Iran's support of terrorism, its disruption of peace efforts in the Middle East, and its pursuit of a nuclear weapons program.

NUCLEAR POWER

Yet Iran has claimed that its pursuit of nuclear power is a part of an attempt to diversify its energy assets to bolster its largely oil-dependent economy while making good use of its significant reserves of uranium ore. Iran is a signatory of the Nuclear Non-Proliferation Treaty (NPT), which it ratified in 1970 and which binds member countries to an agreement to develop, research, produce, and use nuclear energy for peaceful purposes only. However, in 2003 the International Atomic Energy Agency (IAEA) reported that Iran had during the preceding 18 years secretly pursued an uranium enrichment program. The NPT permits its members to enrich uranium for peaceful purposes, albeit only with IAEA oversight, as the same technology used for producing fuel for nuclear power can be used for producing fuel for a nuclear weapon. The fact that Iran hid its enrichment program for 18 years has raised international concern that the true purpose of the program might be the development of a nuclear weapon. However, the Iranian leadership has continued to be vocal about efforts to enrich uranium, stating that nuclear power is necessary to sustain the level of energy consumed by Iran's booming population. Iran has a need for an alternative fuel that can curtail excessive imports of gasoline and electricity and unhealthy levels of oil consumption.

CHANGE IN LEADERSHIP

In June 2005, Mahmoud Ahmadinejad, a leader known for his militancy and radicalism, became president of Iran. He replaced Mohamed Khatemi, who had been elected by a landslide in 1997. Khatemi's leadership had been more moderate than that of other Iranian leaders—especially more moderate than that of Ahmadinejad. During Khatemi's presidency the U.S. president, Bill Clinton, even waived sanctions against Russian, French, and Malaysian firms for investing in Iran's oil industry. However, only five days after Ahmadinejad's election, Iran resumed its project to enrich uranium at a facility in Isfahan, Iran, that had been suspended in 2004 to allow negotiations with the IAEA to continue. Ahmadinejad defended Iran's decision to conduct nuclear research and contended that Iran had the right to research peaceful uses of nuclear technology. He maintained that the research was for the purpose of generating nuclear power to help his country, which, he claimed, was running short on energy. In September 2005 Iranian leaders threatened to make the investment climate of Iran unfavorable to countries that attempt to hinder its access to nuclear energy technology. Then, in November 2005, Iran rejected a plan whereby it would abstain from uranium enrichment and accept enriched uranium from Russia instead. However, Iran was reported to the UN Security Council in February 2006 because of charges that its nuclear program violated the Nuclear Non-Proliferation Treaty. Ahmadinejad threatened that Iran would "revise its policies" if the rights of the Iranian people were violated.

In April 2006 Ahmadinejad announced that Iran had successfully enriched uranium. The uranium was enriched by using more than 100 centrifuges, devices that separate substances of different densities. This quantity of centrifuges would make the uranium capable of being used in a nuclear reactor, but manufacturing a nuclear bomb would require several thousands of centrifuges. However, that same month, the Institute for Science and International Security (ISIS) published satellite images that allegedly showed new nuclear sites under construction in Iran, including a new tunnel entrance at a nuclear facility in Isfahan. In the meantime, the United States, Great Britain, France, and other nations pressed for UN sanctions against Iran if it continued to pursue uranium enrichment and attempted to institute a series of negotiations to obtain Iran's cooperation in nuclear nonproliferation.

In June 2006 Iranian officials indicated that they would consider proposals to delay their nuclear power program until UN officials determined it would be used for peaceful purposes only. The proposals, developed by the UN Security Council with Germany, would involve the United States in direct talks with Iranian officials for the first time in more than 25 years. On August 31, 2006, however, Iran failed to meet a UN Security Council deadline to either suspend its nuclear program or receive sanctions. The council voted in late 2006 to impose sanctions on Iran's nuclear materials and technology trade. In response Iran vowed to further increase its uranium enrichment activities. In March 2007, after Iran failed to meet a second deadline for halting uranium enrichment, Germany and five permanent council nations, the United States, Russia, China, Britain, and France, drafted a resolution to embargo arms exports and impose financial sanctions on Iran's Revolutionary Guards and one of its banks. It was hoped that these measures would apply enough pressure to cause Iran to stop enriching uranium.

In early 2007, amid the political insurgency and violence that had been escalating in Iraq for four years since U.S. forces first invaded the country, the Bush administration reported that Iran was backing Shiite extremists inside Iraq, supplying them with weapons with which to launch terror attacks against U.S. troops. Thus, tensions between Iran and the United States heightened even further.

OTHER STRATEGIES TO DIVERSIFY ENERGY SOURCES

Iran's highly controversial nuclear energy program aside, there are other programs that Iran has instituted to lessen its economy's great dependency on oil. For example, Iran has attempted to diversify its economy by investing oil revenues in petrochemicals. In late 2006 and early 2007, Iran also began taking steps to institute a rationing system, whereby low domestic gasoline prices would apply only up to a certain quantity but then convert over to full market prices. In addition, Zaganeh called for increased oil production and improved oil recovery in Iran. Plans included doubling national oil production to more than 7 million barrels per day by around 2015.

With so much contention in the history of U.S.-Iranian relations, and with nuclear power a prominent source of even further political tension, it may be a long time before Iran can heal its caustic energy relationship with the United States, and possibly with other Western countries, as well.

Saudi Arabia

OIL RESERVES

Saudi Arabia leads the world in oil production and exports. Its crude oil production costs are low, and its total crude oil production equals about 10.5 to 11.0 million barrels of oil per day. Saudi Arabia maintains that it is actually capable of producing more than that, up to 15 million barrels per day by the year 2020. However, many industry analysts have noted that there are often discrepancies between Saudi Arabia's actual production capabilities and the reserves claims that it reports. These experts highlight the fact that the prospect of running out of Saudi oil is not a far-off, unrealistic notion.[25]

World oil supply and consumption run about 84 to 85 million barrels per day, and Saudi Arabia fulfills more than 10 percent of this demand in the entire world. Oil export revenues, according to the EIA, account for as much as 90 percent of total Saudi export earnings and around 44 percent of Saudi Arabia's GDP.[26] During 2004 and early 2005 Saudi Arabia pulled in massive oil export revenues as a result of high oil prices and increased exports. The country earned about $116 billion in net oil export revenues in 2004, and oil export revenues were expected to reach up to $154 billion in 2006 because of even higher oil prices.

Saudi Arabia exports the greatest portion of its oil to the United States, Europe, and Asia (which now receives about 60 percent of all Saudi oil exports). In the United States, Saudi Arabia is one of three top exporters of oil, supplying approximately 15 percent of U.S. oil in recent years. Saudi Arabia also holds one fifth of the world's proven oil reserves. With its robust oil supply and production capacity, Saudi Arabia has turned little attention to the development of alternative energy sources—so little attention, in fact, that less than 0.1 percent of Saudi Arabia's total energy consumption is derived from renewable sources.[27] Saudi Arabia's arid environment also renders it unsuitable for renewable forms of energy such as hydroelectric power.

ECONOMIC CHALLENGES

With its economy so driven by revenues from oil exports, Saudi Arabia's economic challenges could become a long-term drain on its oil industry profits. Such challenges include a high unemployment rate and a population boom that increases the need for government spending. Saudi Arabia also has artificially low electric power prices because of government-mandated rates and consumer subsidies. These low prices have encouraged an increase in demand that has placed a strain on electric utilities. Electric power demand has been growing by 7 percent each year, according to the EIA. Moreover, because of Saudi Arabia's growing population, its per capita oil export revenues are about 80 percent below the levels Saudi Arabia once achieved during the 1970s and early 1980s.[28] Although Saudi Arabia's state-run oil company, Aramco, has confidently projected that Saudi Arabia's oil production can easily be increased, oil industry analysts, including representatives of the Association for the Study of Peak Oil, analysts from major financial institutions, and former executives of Aramco itself, have claimed that Saudi oil production was, as of 2005, approaching its peak.[29]

One of Saudi Arabia's key energy-related issues is that it is located in the politically unstable Middle East, a major concern to Saudi oil customers such as the United States. A related issue, one that directly impacts Saudi Arabia's economy, is that Saudi Arabia has needed to increase its security spending because of a number of terrorist attacks in the country since 2003. Of special concern is the threat posed to Saudi oil facilities and the 3,000 Western oil workers in the country. Saudi Arabia reportedly has ramped up security spending by 50 percent since 2004. Saudi Arabia's oil fields, refineries, estimated 10,000 miles of oil pipelines, and other oil facilities fall under the protection of the Saudi National Guard, regular Saudi military forces, and Interior Ministry officers.

COUNTERSTRATEGIES

Diversifying the Economy

Another issue for Saudi Arabia is the prevalence of monopolies, especially in the oil industry. Large state corporations, such as the oil firm Saudi Aramco and the Saudi Basic Industries Corporation, dominate the economy. In an effort to diversify the economy, Saudi Arabia has attempted to attract foreign direct investment to help encourage privatization. The Saudi cabinet approved a reduction in taxes on foreign direct investment in June 2004.

Maximizing Oil Production

Another of Saudi Arabia's responses to its energy-related issues is to maximize its oil production. In 2005 Khalid al-Falih, a Saudi Aramco senior vice president, stated that Saudi Arabia would increase oil production to more than 12 million barrels per day by 2009, up to a total of 15 million barrels per day by 2020. The details of this $18-billion plan were released in 2006. The country has also sought to increase the number of its operating drill rigs to more than double 2004 levels. However, according to another Aramco senior vice president, existing Saudi oil fields sustain from a 5 to a 10 percent decline in available reserves each year.

Maintaining Its Lead in Global Oil Production

Saudi Arabia has also attempted to secure its footing as the world's leading oil producer by stepping up production when foreign oil markets demand it. After the U.S. invasion of Iraq in March 2003, which placed Iraqi oil production on hold, Ali bin Ibrahim al-Naimi, Saudi Arabia's minister of petroleum and mineral resources since 1995, asserted Saudi Arabia's commitment to stepping up its own oil production to fill the void in Middle Eastern oil exports. In 2006 he declared that Saudi Arabia would increase its production while keeping oil prices competitive. That same year, however, he also announced the termination of negotiations with foreign oil companies on a plan that would have opened Saudi Arabia's upstream hydrocarbons sector to foreign investment for the first time since the 1970s. Negotiations reportedly fell apart over the extent of gas reserves to be included in the plan and the rate of return that Saudi Arabia would offer to participating companies.

Saudi Arabia has attempted to stay on top of the world oil market in other ways, too. In 1990 Sheik Ahmed Zaki Yamani, former petroleum and mineral resources minister of Saudi Arabia, founded the Centre for Global Energy Studies, an independent organization that analyzes energy-related developments, especially within the oil and natural gas market. In November 2005 Abdullah bin Abd al-Aziz al-Sa'ud, king of Saudi Arabia since August 2005, called on leading consumer countries to cut taxes on petroleum products to help control oil prices during a time of increasing pressure for OPEC to increase production to meet global oil demand. The king has also promoted privatization of his country's assets.

Nuclear Power

As Iran has, Saudi Arabia has run into problems in the Western world over its potential to manufacture nuclear weapons. In 2003 Saudi leadership had claimed that worsening relations with the United States were driving it to consider developing nuclear weapons.[30] A former Saudi Arabian ambassador to the United Nations, Muhammad Khilewi, defected to the United States in 1994 and revealed documents that hinted of an arrangement between Saudi Arabia and Pakistan by which Saudi Arabia would partially fund Pakistan's development of nuclear weapons in exchange for the Saudis' ability to use such weapons should nuclear aggression be directed at them. However, Saudi Arabia has denied that any nuclear weapons program is being pursued on its soil. Yet in 2005 Saudi Arabia also signed the International Atomic Energy Agency (IAEA) Small Quantities Protocol, which allows countries with a low risk of nuclear weapons proliferation to opt out of extensive inspections for nuclear weapons if they make a disclosure about their nuclear activities. By signing this protocol, Saudi Arabia has raised suspicion in the global community that it is in the process of developing nuclear weapons and is trying to prevent detection of its nuclear activities.

Nevertheless, Saudi Arabia's economic foothold as a supplier of U.S. energy seems likely to be maintained as long as oil remains a dominant energy source, for the U.S.-Saudi Arabia energy relationship has been well established ever since Franklin D. Roosevelt, president of the United States from 1932 to 1945, fostered U.S. ties with the oil-producing country of Saudi Arabia, personally visiting its founder, Abd al-Aziz ibn Saud, in 1945. As long as the United States continues to need to fill a gap in its ability to meet its domestic oil demand, Saudi Arabia will continue to be one of the top candidates for the job.

NIGERIA

Oil and Natural Gas Reserves

Nigeria, a member of OPEC, is the largest oil producer in Africa and the 10th largest in the world. It is also a major supplier of oil to both Western Europe and the United States—the fifth-highest supplier of U.S. oil imports, in fact. According to the EIA, Nigeria's oil production averaged about 2.6 million barrels per day in 2005 and has been increasing each year. The Nigerian government announced plans to increase production to 4 million barrels per day by 2010.[31]

Nigeria also has a substantial supply of oil reserves. According to *Oil and Gas Journal,* Nigeria's estimated proven oil reserves are 35.9 billion barrels, the majority of which are in the coastal Niger River Delta region.[32] The Nigerian government has announced plans to expand Nigeria's proven reserves to 40 billion barrels by 2010. In December 2004 Olusegun Obasanjo, president of Nigeria since 1999, also announced a plan to increase oil exports to the United States. Oil from Nigeria now accounts for about 7 percent of U.S. oil imports; Obasanjo's plan would increase this figure to 15 percent.[33]

Nigeria also has one of the top 10 largest reserves of natural gas in the world and the largest in Africa. According to *Oil and Gas Journal,* there are approximately 185 trillion cubic feet of proven natural gas reserves in Nigeria, and Nigeria estimated in October 2004 that its natural gas reserves could actually be as high as 660 trillion cubic feet.

Economic Problems

POVERTY AND LACK OF INFRASTRUCTURE

But despite having a considerable wealth of oil, Nigeria is one of the world's poorest nations. More than 70 percent of its people live in poverty. A small fraction of Nigeria's population benefits financially from Nigeria's oil wealth, but, in many cases, such benefits are derived from corrupt practices, such as stealing oil and reselling it or accepting bribes from foreign oil companies in exchange for the ability to develop oil fields.[34]

Nigeria's economy is also heavily dependent on oil sector revenues, which account for 80 percent of government revenues. Nigeria's widespread poverty and weak economy have manifested themselves in the energy sector in the form of an inability to establish a sufficient infrastructure to support energy supply and production. This, in turn, has resulted in inconsistent power supply, especially electrical power. Only about 40 percent of Nigeria has access to electricity, and the country experiences frequent power outages. The electricity sector consistently fails to operate up to its full capacity because of poor maintenance and disordered operations. In February 2006 peak electric demand was 7,600 megawatts, but actual generation capability was less than half of that.

In November 2004 electricity generation fell from 3,500 megawatts to 2,566 megawatts, even though Nigeria has 5,900 megawatts of installed electricity generating power. This caused widespread power outages. The state-owned National Electric Power Authority (NEPA), which dominates the Nigerian electric power sector, blamed the power outrages on low water levels that reduced output at hydropower stations. The lack of consistent power supply to support industrial growth has debilitated the economy of Nigeria even further, preventing the country from keeping pace with the industrialization of the nations to which it provides so many of its energy exports.

Energy infrastructure inadequacies have also led to frequent natural gas flaring in Nigeria. Since Nigeria does not possess the necessary technology or systems to collect and process its large supply of natural gas, an estimated 75 percent of natural gas is disposed of by being burned, or flared, each year. Nigeria flares more natural gas than any country in the world—43 percent of total annual natural gas production. The high incidence of natural gas flaring, with its associated pollutant and safety risks, is one of Nigeria's top energy-related problems.

ENERGY SECURITY

Energy security is also a major problem in Nigeria, as a result of insufficient infrastructure and a large, impoverished, discontented population. Considerable political and ethnic strife plague the oil-rich Niger Delta region. There have been frequent attacks on oil and electricity infrastructure in this region. Violence, kidnappings, and oil facility seizure and sabotage disrupt Nigerian oil production in the region, sometimes causing major oil companies to suspend or close down production and lose hundreds of thousands of barrels of oil production each day. The region has been beset by attacks from the Movement for the Emancipation of the Niger Delta (MEND), a group of militants led by Major-General Godswill Tamuno. In early 2006 MEND blew up oil pipelines, held foreign oil workers hostage, and sabotaged major oil fields. Tamuno has stated that the group wants their land's oil wealth released from foreign interests, and all foreign oil companies and their employees out of the region.

From January to September 2004 there were an estimated 581 cases of oil pipeline vandalism in Nigeria. In December 2004 a pipeline explosion caused the death of 26 people. On May 12, 2006, 200 people were killed in Nigeria when a vandalized gasoline pipeline exploded, sending a fireball over Nigerians who were attempting to drain fuel from the pipeline into cans. (Pipeline vandalism in Nigeria is mainly caused by Nigerians who break into pipelines by drilling or other means in order to steal fuel and sell it.)

Counterstrategies

Here are some of the steps that Nigeria has taken toward rectifying some of these serious energy-related issues:

- An antivandalism law that outlines penalties for pipeline vandalism, including life imprisonment.
- The Electric Power Sector Reform (EPSR) Act, signed into law in March 2004, which aims to reform the nation's unreliable and disorganized electricity generation system. The law enables private power companies to participate in electricity generation, transmission, and distribution and separates NEPA into 11 electricity distribution firms, six electricity-generating companies, and an electricity transmission company. The law calls for all of these operations to be privatized. Nigeria's passage of the EPSR Act set in motion the privatization of NEPA and a long-awaited reform of the nation's unreliable and disorganized electricity generation system.
- An effort to reform natural gas infrastructure and end natural gas flaring. Nigeria set a goal of ending natural gas flaring by 2008; it also established an infrastructure to begin collecting natural gas and processing it into liquefied natural gas (LNG). By the end of 2004 the country had reported a 30 percent reduction in natural gas flaring. The government also announced plans to stop subsidizing natural gas by 2007 and to develop its natural gas infrastructure fully by 2010. The primary goal is to raise

revenues from natural gas exports to 50 percent of Nigeria's oil revenues by 2010. The government estimated, however, that $15 billion in private sector investments would be necessary to accomplish this.

- A plan to increase petroleum profit taxes that are paid by multinational oil companies in Nigeria. This plan was designed to provide Nigeria with better resources to fund infrastructure development, including oil exploration and production development, in the hope of reaching increased production targets. In March 2005 Nigeria began offering licensing of inland and offshore land for oil exploration and production in the Gulf of Guinea.

It is hoped that such measures will allow Nigeria to make use of its considerable natural resources base and emerge into a developed nation with a better quality of life for all of its people.

VENEZUELA

Oil Reserves

Venezuela has the largest oil reserves in the Western Hemisphere, about 79.7 billion barrels, according to *Oil and Gas Journal*.[35] According to the EIA, Venezuela was the world's eighth-largest net oil exporter in 2005 and reportedly rakes in about $1 billion per month in profits from oil exports. Oil exports also account for about a third of its GDP. Venezuela is a top crude oil supplier to the United States, providing about 11 percent of U.S. oil.

Most of Venezuela's oil lies only 60 feet below ground, so it has been relatively easy for the country to obtain and produce. Yet Venezuela and oil industry analysts offer different figures for Venezuela's actual crude oil production rate. For example, the Venezuelan government states that the country produces 3.3 million barrels of crude oil per day, while analysts and the EIA estimate the actual figure to be 2.8–2.9 quatrillion billion barrels per day.[36] Hugo Rafael Chávez Frías (Hugo Chávez), president of Venezuela since 1999, has sometimes been accused of falsifying Venezuela's oil production figures to manipulate prices in the world oil market. In any case, stores of oil believed to be lying yet untapped in Venezuela's oil-rich Orinoco River Basin are expected to have enough oil in them to add billions of barrels to Venezuela's reserves figures and make Venezuela the country with the most oil reserves in the entire world.

Founding Member of OPEC

Venezuela is also a founding member of OPEC (the only non-Arab founding member) and has considerable leverage in the world oil market. Juan

Pablo Pérez Alfonzo (Juan Pérez), former Venezuelan oil minister, OPEC cofounder, and the so-called Father of OPEC, laid the groundwork for some of Venezuela's influence in the worldwide oil industry today. He worked to form arrangements between major oil companies and oil-exporting countries that were more favorable to the latter. He also successfully persuaded industry leaders that the prices that oil companies declared for tax purposes were below their actual value. He established contact with Iran and the Soviet Union in the hopes of forming a partnership to limit the major oil companies' control of the international oil market. Pérez was forced into exile in 1949 after a military coup by Marcos Pérez Jiménez, who formed a government with the backing of the United States. Juan Pérez returned to the country and his position as oil minister in 1959 when Jimenez was ousted from power.

Foreign Policy and Oil Production

Venezuela is no stranger to controversial political leaders, and this problem interacts, quite unfavorably, with the stability of the energy industry. For example, Rafael Caldera Rodríguez (Rafael Caldera), president of Venezuela in 1969–74 and 1994–99, began his second term with several challenges to overcome: the collapse of the country's banking sector, falling oil prices, foreign debt, and inflation. Under his presidency the government announced a plan to expand the country's gold and diamond mining operations to reduce its dependency on the oil sector, but many Venezuelans considered this idea impractical.

Since taking office, Hugo Chávez, Venezuela's most recent leader, has been criticized by Venezuela's wealthier citizens for commandeering Venezuela's economy and by the United States and other importers of Venezuelan oil for extracting higher revenues on oil exports by scaling back oil production.[37] In 2001 he signed a controversial energy law known as the Hydrocarbons Law. This law increased the royalties that private companies must pay to the government for producing oil in Venezuela and requires foreign oil investments to be in the form of joint ventures with Venezuela's state-owned oil company, Petroleos de Venezuela, SA (PdVSA). The law is controversial because it guarantees PdVSA a majority share of any new projects, while there are many doubts about PdVSA's ability to fund sufficient investment in expanding crude oil production. The investment climate in Venezuela's oil sector has grown more unfriendly under Chávez; in November 2004 he announced a new royalty rate that was the highest rate allowable under Venezuela's prior hydrocarbons laws.

Early in Chávez's presidency, he also expanded his presidential powers through a new constitution and strengthened Venezuela's ties with Cuba and Middle Eastern oil-producing nations. Chávez is particularly outspoken about his disdain for U.S. foreign policy, often threatening to cut off oil supplies to the United States. U.S. President George W. Bush and Secretary of State Condoleezza Rice have called Chávez's leadership dictatorial and a threat to the international community. But Rafael Ramirez, Venezuela's minister of energy, only reiterated Chávez's threats to cut off oil supplies if the United States shows any signs of aggression toward Venezuela.

Chávez's actions as president have stirred disapproval not only in the United States but in Chávez's own country. In late 2002 Chávez's opponents led a nationwide revolt against him. The country's oil workers went on strike, and all oil operations ceased, hurting Venezuela's economy as well as leading to a spike in U.S. oil prices in early 2003. The Venezuelan economy entered a recession at the end of 2002 because of the strike. Later Chávez dismissed nearly half of the country's oil workforce.

Nonetheless, Chávez's popularity in Latin America as a whole increases as revenues from Venezuela's high-priced oil exports flow into the region. Furthermore, Chávez has redirected some of Venezuela's oil wealth into programs to help its 3 billion residents who live in poverty. A portion of oil profits is used to fund free public health insurance, provide discounted groceries, and give factory jobs to this population. The price of gasoline under Chávez's leadership is also subsidized, costing only 11 cents per gallon—a situation that some consider to be an unnecessary drain on Venezuela's oil profits. In 2005, under the leadership of Chávez, Venezuela also began providing subsidized heating oil to low-income families in selected parts of the United States. In 2006 Chávez more than doubled the number of U.S. households and states included in this program. Some of Chávez's critics have been suspicious of his motives for offering cheaper heating fuel in the United States. However, despite suspicion or opposition that any of his policies may have provoked among his critics, Chávez won reelection in his country in December 2006. Soon after beginning his new term in January 2007, he took steps to nationalize key sectors of the Venezuelan economy, including the electric power sector.

In conclusion, the United States is not the only country (by far) that has faced the issues of energy crises and shortfalls; affordability and economics; foreign oil dependence, political instability, and foreign relations; infrastructure issues and production limitations; and environmental issues and the potential of renewable and alternative sources of energy. Other countries have both faced and found possible solutions to these problems. Some solutions have worked and some have not. As long as oil continues to be the primary source of energy in the United States and other parts of the world international experiences with energy are likely to continue to be marked by both startling successes and, at times, discouraging failures.

---

[1] International Energy Agency. *Challenges: Renewables in Global Energy Supply: An EIA Fact Sheet.* Paris: International Energy Agency, 2007, p. 3.

[2] Ibid.

[3] Ibid.

[4] International Energy Agency. *Oil Crises and Climate Change: 30 Years of Energy Use in IEA Countries.* Paris: International Energy Agency, 2004, p. 35.

[5] International Energy Agency. *Challenges: Renewables in Global Energy Supply: An EIA Fact Sheet.* Paris: International Energy Agency, 2007, p. 3.

[6] *BP Statistical Review of World Energy 2006*, p. 2. Available online. URL: www.bp.com/statisticalreview. Accessed March 15, 2007.

[7] The relevant data can be found in the EIA's *Country Analysis Briefs: China* (August 2006), p. 2. Available online. URL: http://www.eia.doe.gov/emeu/cabs/contents.html.

[8] Ibid.

[9] James Regan. "A Cooling China Prompts Relief." *International Herald Tribune*, Bloomberg News, January 11, 2005. Available online. URL: http://www.iht.com/articles/2005/01/10/bloomberg/sxchiecon.php. Accessed February 17, 2006. See also Mergent Industry Reports, International Annual Reports, *Oil and Gas: Asia Pacific*, Mergent, Inc., January 1, 2005, Country Profiles: China.

[10] Energy Information Administration. *Country Analysis Briefs: China* (August 2006), p. 2.

[11] U.S. Census Bureau. "Census Bureau Frames U.S. in Global Context; Identifies Aging, Fertility Trends." Press release, February 6, 2002. Available online. URL: http://www.census.gov/Press-Release/www/releases/archives/fertility/000318.html. Accessed January 29, 2006.

[12] Energy Information Administration, *Greenhouse Gases, Climate Change, and Energy*, chapter 1. Available online. URL: http://www.eia.doe.gov/oiaf/1605/ggccebro/chapter1.html. Accessed August 15, 2006.

[13] Energy Information Administration. *Country Analysis Briefs: China* (2005), p. 4.

[14] Steven Mufson. "The Yangtze Dam: Feat or Folly?" *Washington Post*, November 9, 1997, p. 101.

[15] Ibid.

[16] See the EIA's *Country Analysis Briefs: Germany* (December 2006), p. 1. Available online. URL: http://www.eia.doe.gov/emeu/cabs/contents.html.

[17] Ibid., pp. 8, 10, and 12.

[18] Ibid., p. 3.

[19] Ibid., p. 2.

[20] Ibid., pp. 10–11.

[21] See the EIA's *Country Analysis Briefs: Iran* (August 2006), p. 1. Available online. URL: http://www.eia.doe.gov/emeu/cabs/contents.html.

[22] Ibid., p. 10.

[23] As reported by Flashpoints: Guide to World Conflicts, in *Country Briefings: Iran.* Available online. URL: http://www.flashpoints.info/countries-conflicts/Iran-web/Iran_briefing.htm. Accessed February 19, 2006.

[24] See the EIA's *Country Analysis Briefs: Iran* (2005), p. 1.

[25] For the relevant data and discussion, see the EIA's *Country Analysis Briefs: Saudi Arabia* (February 2007), pp. 1–3. Available online. URL: http://www.eia.doe.gov/emeu/cabs/contents.html.

[26] Ibid., p. 1.

[27] EIA, *Environmental Issues in Saudi Arabia,* "Renewable Energy." Available online. URL: http://www.eia.doe.gov/emeu/cabs/saudenv.html. Accessed August 15, 2006.

[28] See the EIA's *Country Analysis Briefs: Saudi Arabia* (2005), p. 2.

[29] See the April, June, July, August, and September 2005 entries in the timeline of "peak oil" news, as posted by the energy expert Jeremy Legget on *The Guardian* online "Comments" portal. Available online. URL: http://commentisfree.guardian.co.uk/jeremy_leggett/2006/04/two_clocks_ticking_ever_loudly.html. Accessed August 16, 2006.

[30] Ewen MacAskill and Ian Traynor. "Saudis Consider Nuclear Bomb." *The Guardian,* September 18, 2003. Available online. URL: http://www.guardian.co.uk/print/0,3858,4755775103681,00.html. Accessed August 16, 2006.

[31] See the EIA's *Country Analysis Briefs: Nigeria* (March 2006), pp. 1–2. Available online. URL: http://www.eia.doe.gov/emeu/cabs/contents.html.

[32] These *Oil and Gas Journal* statistics are as quoted by another source, the EIA's *Country Analysis Briefs: Nigeria* (March 2006), p. 1.

[33] Ibid., p. 3.

[34] Dudley Althaus. "An African Tale of Looted Oil Money, Vanished Ship." *Houston Chronicle,* December 17, 2004. Available online. URL: http://www.chron.com/disp/story.mpl/world/2934700.html. Accessed August 16, 2006.

[35] These *Oil and Gas Journal* statistics are as quoted by another source, the EIA's *Country Analysis Briefs: Venezuela* (September 2006), p. 2. Available online. URL: http://www.eia.doe.gov/emeu/cabs/contents.html.

[36] Ibid., p. 5.

[37] For an analysis of these criticisms, see Kevin Sullivan's "Embattled Chavez Taps Oil Cash in a Social, Political Experiment." *Washington Post,* June 18, 2004, p. A19.

PART II

Primary Sources

4

United States Documents

Many of the energy-related policies, programs, laws, and statements issued by the United States attempt to tackle several energy-related concerns at once. The most recent major U.S. energy law is the Energy Policy Act of 2005, excerpts of which are reproduced here. In its original form this law is more than 1,700 pages long and contains regulations whose stated purpose is to take on many, if not all of, the key U.S. energy-related issues. However, for ease of use, this law and the other U.S. documents included here have been grouped according to the predominant issue or issues addressed by the excerpts chosen for inclusion, into the following sections:

Shortages in Energy Supply, Crises in the Energy Sector, and Foreign Oil Dependence

Affordability of Energy Resources and Energy's Impact on the Economy

Energy's Environmental Impact

Renewable and Alternative Energy Technologies

Nuclear Energy

SHORTAGES IN ENERGY SUPPLY, CRISES IN THE ENERGY SECTOR, AND FOREIGN OIL DEPENDENCE

Energy Policy Act of 2005

The Energy Policy Act of 2005 was signed into U.S. law on August 8, 2005. More than 1,700 pages long, the Energy Policy Act of 2005 is the United States' most recent major energy policy. Its supporters claim that it is a comprehensive energy policy that tackles shortages in energy resources through such measures as developing alternative and renewable energy technologies to supplement the *use of fossil fuels for energy. Among its provisions are an authorization to fill the Strategic Petroleum Reserve to capacity, new reliability standards for electricity utilities for the purpose of modernizing the electrical grid, incentives for expanding the use of nuclear energy for electricity, and a commitment to invest $2 billion over 10 years to support research into environmentally "cleaner" ways of using coal to generate electricity. It also included a measure aimed at tripling, by 2012, the current required amount of biomass-derived fuel (or biofuel), such as ethanol, that must be mixed with gasoline sold in the United States. In addition, it provided for a tax credit of up to $3,400 for owners of hybrid vehicles. Also included in the Energy Policy Act of 2005 was the Set America Free Act of 2005, which calls for the creation of a "United States Commission on North American Energy Freedom" to study the particulars of U.S. foreign oil dependence and to "make recommendations for a coordinated, comprehensive, and long-range national policy to achieve North American energy freedom by 2025." The Energy Policy Act of 2005 also extended daylight savings time in the United States by four weeks on the presumption that Americans will use less electricity if daylight is available for a longer period. The act also authorized subsidies for producers of wind energy and other alternative energies, $50 million in an annual biomass grant, and tax breaks for energy conservation improvements made to homes. However, authorizations of funding are not the same as actual appropriations of funds; authorizations are not guarantees that the funds will actually be directed where stipulated. There is a great deal of uncertainty about whether this law will actually have a real impact on U.S. energy supply.*

Energy Policy Act of 2005
Public Law 109-58
109th Congress
An Act

To ensure jobs for our future with secure, affordable, and reliable energy.
August 8, 2005 H.R. 6

Be it enacted by the Senate and House of Representatives of the United States of America in Congress assembled.

* *** *

TITLE III—OIL AND GAS

Subtitle A—Petroleum Reserve and Home Heating Oil

SEC. 301. PERMANENT AUTHORITY TO OPERATE THE STRATEGIC PETROLEUM RESERVE AND OTHER ENERGY PROGRAMS.

* *** *

(e) FILL STRATEGIC PETROLEUM RESERVE TO CAPACITY.—

(1) IN GENERAL.—The Secretary shall, as expeditiously as practicable, without incurring excessive cost or appreciably affecting the price of petroleum products to consumers, acquire petroleum in quantities sufficient to fill the Strategic Petroleum Reserve to the 1,000,000,000-barrel capacity authorized under section 154(a) of the Energy Policy and Conservation Act (42 U.S.C. 6234(a)), in accordance with the sections 159 and 160 of that Act (42 U.S.C. 6239,6240).

(2) PROCEDURES.—
(A) AMENDMENT.—Section 160 of the Energy Policy and Conservation Act (42 U.S.C. 6240) is amended by inserting after subsection (b) the following new subsection:

"(c) PROCEDURES.—The Secretary shall develop, with public notice and opportunity for comment, procedures consistent with the objectives of this section to acquire petroleum for the Reserve. Such procedures shall take into account the need to—

"(1) maximize overall domestic supply of crude oil (including quantities stored in private sector inventories);

"(2) avoid incurring excessive cost or appreciably affecting the price of petroleum products to consumers;

"(3) minimize the costs to the Department of the Interior and the Department of Energy in acquiring such petroleum products (including foregone revenues to the Treasury when petroleum products for the Reserve are obtained through the royalty-in-kind program);

"(4) protect national security;

"(5) avoid adversely affecting current and futures prices, supplies, and inventories of oil; and

"(6) address other factors that the Secretary determines to be appropriate."

* *** *

TITLE VI—NUCLEAR MATTERS [...]

Subtitle D—Nuclear Security

* *** *

SEC. 244. PARTNERSHIP PROGRAM WITH INSTITUTIONS OF HIGHER EDUCATION.

"(3)(A) The task force, in consultation with Federal, State, and local agencies, the Conference of Radiation Control Program Directors, and the Organization of Agreement States, and after public notice and an opportunity for comment, shall evaluate, and provide recommendations relating to, the security of radiation sources in the United States from potential terrorist threats, including acts of sabotage, theft, or use of a radiation source in a radiological dispersal device.

"(B) Not later than 1 year after the date of enactment of this section, and not less than once every 4 years thereafter, the task force shall submit to Congress and the President a report, in unclassified form with a classified annex if necessary, providing recommendations, including recommendations for appropriate regulatory and legislative changes, for—

"(i) a list of additional radiation sources that should be required to be secured under this Act, based on the potential attractiveness of the sources to terrorists and the extent of the threat to public health and safety of the sources, taking into consideration—

"(I) radiation source radioactivity levels;

"(II) radioactive half-life of a radiation source;

"(III) dispersability;

"(IV) chemical and material form;

"(V) for radioactive materials with a medical use, the availability of the sources to physicians and patients for medical treatment; and

"(VI) any other factor that the Chairperson of the Commission determines to be appropriate;

"(ii) the establishment of, or modifications to, a national system for recovery of lost or stolen radiation sources;

"(iii) the storage of radiation sources that are not used in a safe and secure manner as of the date on which the report is submitted;

"(iv) modifications to the national tracking system for radiation sources;

"(v) the establishment of, or modifications to, a national system (including user fees and other methods) to provide for the proper disposal of radiation sources secured under this Act;

"(vi) modifications to export controls on radiation sources to ensure that foreign recipients of radiation sources are able and willing to adequately control radiation sources from the United States;

"(vii)(I) any alternative technologies available as of the date on which the report is submitted that may perform some or all of the functions performed by devices or processes that employ radiation sources; and

"(II) the establishment of appropriate regulations and incentives for the replacement of the devices and processes described in subclause (I)—

"(aa) with alternative technologies in order to reduce the number of radiation sources in the United States; or

"(bb) with radiation sources that would pose a lower risk to public health and safety in the event of an accident or attack involving the radiation source; and

"(viii) the creation of, or modifications to, procedures for improving the security of use, transportation, and storage of radiation sources, including—

"(I) periodic audits or inspections by the Commission to ensure that radiation sources are properly secured and can be fully accounted for;

"(II) evaluation of the security measures by the Commission;

"(III) increased fines for violations of Commission regulations relating to security and safety measures applicable to licensees that possess radiation sources;

"(IV) criminal and security background checks for certain individuals with access to radiation sources (including individuals involved with transporting radiation sources);

"(V) requirements for effective and timely exchanges of information relating to the results of criminal and security background checks between the Commission and any State with which the Commission has entered into an agreement under section 274 b.;

"(VI) assurances of the physical security of facilities that contain radiation sources (including facilities used to temporarily store radiation sources being transported); and

"(VII) the screening of shipments to facilities that the Commission determines to be particularly at risk for sabotage of radiation sources to ensure that the shipments do not contain explosives.

* *** *

TITLE VII—VEHICLES AND FUELS

Subtitle B—Hybrid Vehicles, Advanced Vehicles, and Fuel Cell Buses

PART 1—HYBRID VEHICLES

SEC. 711. HYBRID VEHICLES.

The Secretary shall accelerate efforts directed toward the improvement of batteries and other rechargeable energy storage systems, power electronics, hybrid systems integration, and other technologies for use in hybrid vehicles.

SEC. 712. EFFICIENT HYBRID AND ADVANCED DIESEL VEHICLES.

(a) PROGRAM.—The Secretary shall establish a program to encourage domestic production and sales of efficient hybrid and advanced diesel vehicles. The program shall include grants to automobile manufacturers to encourage domestic production of efficient hybrid and advanced diesel vehicles.

(b) AUTHORIZATION OF APPROPRIATIONS.—There are authorized to be appropriated to the Secretary for carrying out this section such sums as may be necessary for each of the fiscal years 2006 through 2015.

PART 2—ADVANCED VEHICLES

SEC. 721. PILOT PROGRAM.

(a) ESTABLISHMENT.—The Secretary in consultation with the Secretary of Transportation, shall establish a competitive grant pilot program (referred to in this part as the "pilot program"), to be administered through the Clean Cities Program of the Department, to provide not more than 30 geographically dispersed project grants to State governments, local governments, or metropolitan transportation authorities to carry out a project or projects for the purposes described in subsection (b).

(b) GRANT PURPOSES.—A grant under this section may be used for the following purposes:

(1) The acquisition of alternative fueled vehicles or fuel cell vehicles, including—

(A) passenger vehicles (including neighborhood electric vehicles); and

(B) motorized 2-wheel bicycles or other vehicles for use by law enforcement personnel or other State or local government or metropolitan transportation authority employees.

(2) The acquisition of alternative fueled vehicles, hybrid vehicles, or fuel cell vehicles, including—

(A) buses used for public transportation or transportation to and from schools;

(B) delivery vehicles for goods or services; and

(C) ground support vehicles at public airports (including vehicles to carry baggage or push or pull airplanes toward or away from terminal gates).

(3) The acquisition of ultra-low sulfur diesel vehicles.

(4) Installation or acquisition of infrastructure necessary to directly support an alternative fueled vehicle, fuel cell vehicle, or hybrid vehicle project funded by the grant, including fueling and other support equipment.

(5) Operation and maintenance of vehicles, infrastructure, and equipment acquired as part of a project funded by the grant.

* *** *

PART 3—FUEL CELL BUSES

SEC. 731. FUEL CELL TRANSIT BUS DEMONSTRATION.

(a) IN GENERAL—The Secretary, in consultation with the Secretary of Transportation, shall establish a transit bus demonstration program to make competitive, merit-based awards for 5-year projects to demonstrate not more than 25 fuel cell transit buses (and necessary infrastructure) in 5 geographically dispersed localities.

(b) PREFERENCE.—In selecting projects under this section, the Secretary shall give preference to projects that are most likely to mitigate congestion and improve air quality.

(c) AUTHORIZATION OF APPROPRIATIONS.—There are authorized to be appropriated to the Secretary to carry out this section $10,000,000 for each of fiscal years 2006 through 2010.

Subtitle C—Clean School Buses

SEC. 741 CLEAN SCHOOL BUS PROGRAM.

(a) DEFINITIONS.—In this section:

(1) ADMINISTRATOR.—The term "Administrator" means the Administrator of the Environmental Protection Agency.

(2) ALTERNATIVE FUEL.—The term "alternative fuel" means—

(A) liquefied natural gas, compressed natural gas, liquefied petroleum gas, hydrogen, or propane;

(B) methanol or ethanol at no less than 85 percent by volume; or

(C) biodiesel conforming with standards published by the American Society for Testing and Materials as of the date of enactment of this Act.

(3) CLEAN SCHOOL BUS.—The term "clean school bus" means a school bus with a gross vehicle weight of greater than 14,000 pounds that—

(A) is powered by a heavy duty engine; and

(B) is operated solely on an alternative fuel or ultra-low sulfur diesel fuel.

(4) ELIGIBLE RECIPIENT.—

(A) IN GENERAL.—Subject to subparagraph (B), the term "eligible recipient" means—

(i) 1 or more local or State governmental entities responsible for—

(I) providing school bus service to 1 or more public school systems; or

(II) the purchase of school buses;

(ii) 1 or more contracting entities that provide school bus service to 1 or more public school systems; or

(iii) a nonprofit school transportation association.

(B) SPECIAL REQUIREMENTS.—In the case of eligible recipients identified under clauses (ii) and (iii), the Administrator shall establish timely and appropriate requirements for notice and may establish timely and appropriate requirements for approval by the public school systems that would be served by buses purchased or retrofit using grant funds made available under this section.

(5) RETROFIT TECHNOLOGY.—The term "retrofit technology" means a particulate filter or other emissions control equipment that is verified or certified by the Administrator or the California Air Resources Board as an effective emission reduction technology when installed on an existing school bus.

(6) ULTRA-LOW SULFUR DIESEL FUEL—The term "ultra-low sulfur diesel fuel" means diesel fuel that contains sulfur at not more than 15 parts per million.

(b) PROGRAM FOR RETROFIT OR REPLACEMENT OF CERTAIN EXISTING SCHOOL BUSES WITH CLEAN SCHOOL BUSES.—

(1) ESTABLISHMENT.—

(A) IN GENERAL.—The Administrator, in consultation with the Secretary and other appropriate Federal departments and agencies, shall establish a program for awarding grants on a competitive basis to eligible recipients for the replacement, or retrofit (including repowering, aftertreatment, and remanufactured engines) of, certain existing school buses.

(B) BALANCING.—In awarding grants under this section, the Administrator shall, to the maximum extent practicable, achieve an appropriate balance between awarding grants—

(i) to replace school buses; and

(ii) to install retrofit technologies.

(2) PRIORITY OF GRANT APPLICATIONS.—

(A) REPLACEMENT.—In the case of grant applications to replace school buses, the Administrator shall give priority to applicants that propose to replace school buses manufactured before model year 1977.

(B) RETROFITTING.—In the case of grant applications to retrofit school buses, the Administrator shall give priority to applicants that propose to retrofit school buses manufactured in or after model year 1991.

(3) USE OF SCHOOL BUS FLEET.—

(A) IN GENERAL—All school buses acquired or retrofitted with funds provided under this section shall be operated as part of the school bus fleet for which the grant was made for not less than 5 years.

(B) MAINTENANCE, OPERATION, AND FUELING.—New school buses and retrofit technology shall be maintained, operated, and fueled according to manufacturer recommendations or State requirements.

(4) RETROFIT GRANTS.—The Administrator may award grants for up to 100 percent of the retrofit technologies and installation costs.

(5) REPLACEMENT GRANTS.—

(A) ELIGIBILITY FOR 50 PERCENT GRANTS.—The Administrator may award grants for replacement of school buses in the amount of up to one-half of the acquisition costs (including fueling infrastructure) for—

(i) clean school buses with engines manufactured in model year 2005 or 2006 that emit not more than—

(I) 1.8 grams per brake horsepower-hour of non-methane hydrocarbons and oxides of nitrogen; and

(II) .01 grams per brake horsepower-hour of particulate matter, or

(ii) clean school buses with engines manufactured in model year 2007, 2008, or 2009 that satisfy regulatory requirements established by the Administrator for emissions of oxides of nitrogen and particulate matter to be applicable for school buses manufactured in model year 2010.

(B) ELIGIBILITY FOR 25 PERCENT GRANTS.—The Administrator may award grants for replacement of school buses in the amount of up to one-fourth of the acquisition costs (including fueling infrastructure) for—

(i) clean school buses with engines manufactured in model year 2005 or 2006 that emit not more than—

(I) 2.5 grams per brake horsepower-hour of non-methane hydrocarbons and oxides of nitrogen; and

(II) .01 grams per brake horsepower-hour of particulate matter, or

(ii) clean school buses with engines manufactured in model year 2007 or thereafter that satisfy regulatory requirements established by the Administrator for emissions of oxides of nitrogen and particulate matter from school buses manufactured in that model year.

(6) ULTRA-LOW SULFUR DIESEL FUEL.—

(A) IN GENERAL.—In the case of a grant recipient receiving a grant for the acquisition of ultra-low sulfur diesel fuel school buses with engines manufactured in model year 2005 or 2006, the grant recipient shall provide, to the satisfaction of the Administrator—

(i) documentation that diesel fuel containing sulfur at not more than 15 parts per million is available for carrying out the purposes of the grant; and

(ii) a commitment by the applicant to use that fuel in carrying out the purposes of the grant.

...

(d) AUTHORIZATION OF APPROPRIATIONS.—There are authorized to be appropriated to the Administrator to carry out this section, to remain available until expended—

(1) $55,000,000 for each of fiscal years 2006 and 2007; and

(2) such sums as are necessary for each of fiscal years 2008, 2009, and 2010.

* *** *

TITLE XIII—ENERGY POLICY TAX INCENTIVES

* *** *

Subtitle C—Conservation and Energy Efficiency Provisions

* *** *

SEC. 1337. BUSINESS SOLAR INVESTMENT TAX CREDIT.

(a) INCREASE IN ENERGY PERCENTAGE.—Section 48(a)(2)(A) (relating to energy percentage), as amended by this Act, is amended to read as follows:

"(A) IN GENERAL—The energy percentage is—

"(i) 30 percent in the case of—

"(I) qualified fuel cell property,

"(II) energy property described in paragraph (3)(A)(i) but only with respect to periods ending before January 1, 2008, and

"(III) energy property described in paragraph (3)(A)(ii), and

"(ii) in the case of any energy property to which clause (i) does not apply, 10 percent.".

(b) HYBRID SOLAR LIGHTING SYSTEMS.—Subparagraph (A) of section 48(a)(3) is amended by striking "or" at the end of clause (i), by redesignating clause (ii) as clause (iii), and by inserting after clause (i) the following new clause:

"(ii) equipment which uses solar energy to illuminate the inside of a structure using fiber-optic distributed sunlight but only with respect to periods ending before January 1, 2008, or".

(c) LIMITATION ON USE OF SOLAR ENERGY TO HEAT SWIMMING POOLS.—Clause (i) of section 48(a)(3)(A) is amended by inserting "excepting property used to generate energy for the purposes of heating a swimming pool," after "solar process heat,".

(d) EFFECTIVE DATE.—The amendments made by this section shall apply to periods after December 31, 2005, in taxable years ending after such date, under rules similar to the rules of section 48(m) of the Internal Revenue Code of 1986 (as in effect on the day before the date of the enactment of the Revenue Reconciliation Act of 1990).

Subtitle D—Alternative Motor Vehicles and Fuels Incentives

SEC. 1341. ALTERNATIVE MOTOR VEHICLE CREDIT.

(a) IN GENERAL—Subpart B of part IV of subchapter A of chapter 1 (relating to foreign tax credit, etc.) is amended by adding at the end the following new section:

"SEC. 30B. ALTERNATIVE MOTOR VEHICLE CREDIT.

"(a) ALLOWANCE OF CREDIT.—There shall be allowed as a credit against the tax imposed by this chapter for the taxable year an amount equal to the sum of—

"(1) the new qualified fuel cell motor vehicle credit determined under subsection (b),

"(2) the new advanced lean burn technology motor vehicle credit determined under subsection (c),

"(3) the new qualified hybrid motor vehicle credit determined under subsection (d), and

"(4) the new qualified alternative fuel motor vehicle credit determined under subsection (e).

"(b) NEW QUALIFIED FUEL, CELL MOTOR VEHICLE CREDIT.— "(1) IN GENERAL.—For purposes of subsection (a), the new qualified fuel cell motor vehicle credit determined under this subsection with respect to a new qualified fuel cell motor vehicle placed in service by the taxpayer during the taxable year is—

"(A) $8,000 ($4,000 in the case of a vehicle placed in service after December 31, 2009), if such vehicle has a gross vehicle weight rating of not more than 8,500 pounds,

"(B) $10,000, if such vehicle has a gross vehicle weight rating of more than 8,500 pounds but not more than 14,000 pounds,

"(C) $20,000, if such vehicle has a gross vehicle weight rating of more than 14,000 pounds but not more than 26,000 pounds, and

"(D) $40,000, if such vehicle has a gross vehicle weight rating of more than 26,000 pounds.

"(2) INCREASE FOR FUEL EFFICIENCY.—

"(A) IN GENERAL.—The amount determined under paragraph (1)(A) with respect to a new qualified fuel cell motor vehicle which is a passenger automobile or light truck shall be increased by—

"(i) $1,000, if such vehicle achieves at least 150 percent but less than 175 percent of the 2002 model year city fuel economy,

"(ii) $1,500, if such vehicle achieves at least 175 percent but less than 200 percent of the 2002 model year city fuel economy, "(iii) $2,000, if such vehicle achieves at least 200 percent but less than 225 percent of the 2002 model year city fuel economy,

"(iv) $2,500, if such vehicle achieves at least 225 percent but less than 250 percent of the 2002 model year city fuel economy,

"(v) $3,000, if such vehicle achieves at least 250 percent but less than 275 percent of the 2002 model year city fuel economy,

"(vi) $3,500, if such vehicle achieves at least 275 percent but less than 300 percent of the 2002 model year city fuel economy, and

"(vii) $4,000, if such vehicle achieves at least 300 percent of the 2002 model year city fuel economy.

"(B) 2002 MODEL YEAR CITY FUEL ECONOMY.—For purposes of subparagraph (A), the 2002 model year city fuel economy with respect to a vehicle shall be determined in accordance with the following tables:

"(i) In the case of a passenger automobile:

"If vehicle inertia weight class is:	**The 2002 model year city fuel economy is:**
1,500 or 1,750 lbs	45.2 mpg
2,000 lbs	39.6 mpg
2,250 lbs	35.2 mpg
2,500 lbs	31.7 mpg
2,750 lb	28.8 mpg
3,000 lbs	26.4 mpg
3,500 lbs	22.6 mpg
4,000 lbs	19.8 mpg
4,500 lbs	17.6 mpg
5,000 lbs	15.9 mpg
5,500 lbs	14.4 mpg
6,000 lbs	13.2 mpg
6,500 lbs	12.2 mpg
7,000 to 8,500 lbs	11.3 mpg.

"(ii) In the case of a light truck:

"If vehicle inertia weight class is:	**The 2002 model year city fuel economy is:**
1,500 or 1,750 lbs	39.4 mpg
2,000 lbs	35.2 mpg
2,250 lbs	31.8 mpg
2,500 lbs	29.0 mpg

2,750 lb	26.8 mpg
3,000 lbs	24.9 mpg
3,500 lbs	21.8 mpg
4,000 lbs	19.4 mpg
4,500 lbs	17.6 mpg
5,000 lbs	16.1 mpg
5,500 lbs	14.8 mpg
6,000 lbs	13.7 mpg
6,500 lbs	12.8 mpg
7,000 to 8,500 lbs	12.1 mpg.

* *** *

"(c) NEW ADVANCED LEAN BURN TECHNOLOGY MOTOR VEHICLE CREDIT.—

"(1) IN GENERAL.—For purposes of subsection (a), the new advanced lean burn technology motor vehicle credit determined under this subsection for the taxable year is the credit amount determined under paragraph (2) with respect to a new advanced lean burn technology motor vehicle placed in service by the taxpayer during the taxable year.

"(2) CREDIT AMOUNT.—

"(A) FUEL ECONOMY.—

"(i) IN GENERAL.—The credit amount determined under this paragraph shall be determined in accordance with the following table:

"In the case of a vehicle which achieves a fuel economy (expressed as a percentage of the 2002 model year city fuel economy) of—	**The credit amount is—**
At least 125 percent but less than 150 percent	$400
At least 150 percent but less than 175 percent	$800
At least 175 percent but less than 200 percent	$1,200
At least 200 percent but less than 225 percent	$1,600
At least 225 percent but less than 250 percent	$2,000
At least 250 percent	$2,400.

"(ii) 2002 MODEL YEAR CITY FUEL ECONOMY.—For purposes of clause (i), the 2002 model year city fuel economy with respect to a vehicle shall be determined on a gasoline gallon equivalent basis as determined by the Administrator of the Environmental Protection Agency using the tables provided in subsection (b)(2)(B) with respect to such vehicle.

"(B) CONSERVATION CREDIT.—The amount determined under subparagraph (A) with respect to a new advanced lean burn technology motor vehicle shall be increased by the conservation credit amount determined in accordance with the following table:

"In the case of a vehicle which achieve a lifetime fuel savings (expressed in gallons of gasoline) of—	**The conservation credit amount is—**
At least 1,200 but less than 1,800	$250
At least 1,800 but less than 2,400	$500
At least 2,400 but less than 3,000	$750
At least 3,000	$1,000

* *** *

TITLE XIV—MISCELLANEOUS

* *** *

Subtitle B—Set America Free

SEC. 1421. SHORT TITLE.

This subtitle may be cited as the "Set America Free Act of 2005" or the "SAFE Act".

SEC. 1422. PURPOSE.

The purpose of this subtitle is to establish a United States commission to make recommendations for a coordinated and comprehensive North American energy policy that will achieve energy self-sufficiency by 2025 within the three contiguous North American nation area of Canada, Mexico, and the United States.

SEC. 1423. UNITED STATES COMMISSION ON NORTH AMERICAN ENERGY FREEDOM.

(a) ESTABLISHMENT.—There is hereby established the United States Commission on North American Energy Freedom (in this subtitle referred to as the "Commission"). The Federal Advisory Committee Act (5 U.S.C. App.), except sections 3, 7, and 12, does not apply to the Commission.

(b) MEMBERSHIP.—

(1) APPOINTMENT.—The Commission shall be composed of 16 members appointed by the President from among individuals described in paragraph (2) who are knowledgeable on energy issues, including oil and gas exploration and production, crude oil refining, oil and gas pipelines, electricity production and transmission, coal, unconventional hydrocarbon resources, fuel cells, motor vehicle power systems, nuclear energy,

renewable energy, biofuels, energy efficiency, and energy conservation. The membership of the Commission shall be balanced by area of expertise to the extent consistent with maintaining the highest level of expertise on the Commission. Members of the Commission may be citizens of Canada, Mexico, or the United States, and the President shall ensure that citizens of all three nations are appointed to the Commission.

* *** *

(f) REPORT.—Within 12 months alter the effective date of this AR, the Commission shall submit to Congress and the President a final report of its findings and recommendations regarding North American energy freedom.

SEC. 1424. NORTH AMERICAN ENERGY FREEDOM POLICY.

* *** *

Within 90 days after receiving and considering the report and recommendations of the Commission under section 1423, the President shall submit to Congress a statement of proposals to implement or respond to the Commission's recommendations for a coordinated, comprehensive, and long-range national policy to achieve North American energy freedom by 2025.

TITLE XV—ETHANOL AND MOTOR FUELS

Subtitle A—General Provisions

SEC. 1501. RENEWABLE CONTENT OF GASOLINE.

(a) IN GENERAL.—Section 211 of the Clean Air Act (42 U.S.C. 7545) is amended—

(1) by redesignating subsection (o) as subsection (r); and

(2) by inserting after subsection (n)thc following:

"(o) RENEWABLE FUEL PROGRAM.—

"(1) DEFINITIONS.—In this section:

"(A) CELLULOSIC BIOMASS ETHANOL.—The term 'cellulosic biomass ethanol' means ethanol derived from any lignocellulosic or hemicellulosic matter that is available on a renewable or recurring basis, including—

"(i) dedicated energy crops and trees;

"(ii) wood and wood residues;

"(iii) plants;

"(iv) grasses;

"(v) agricultural residues;

"(vi) fibers;

"(vii) animal wastes and other waste materials; and

"(viii) municipal solid waste.

The term also includes any ethanol produced in facilities where animal wastes or other waste materials are digested or otherwise used to displace 90 percent or more of the fossil fuel normally used in the production of ethanol.

* *** *

SEC. 1511. RENEWABLE FUEL.

The Clean Air Act is amended by inserting after section 211 (42 U.S.C. 7411) the following:

* *** *

"SEC. 212. RENEWABLE FUEL.

"(a) DEFINITIONS.—In this section:

"(1) MUNICIPAL SOLID WASTE.—The term 'municipal solid waste' has the meaning given the term 'solid waste' in section 3004 of the Solid Waste Disposal Act (42 U.S.C. 6903).

"(2) RFG STATE.—The term 'RFG State' means a State in which is located one or more covered areas (as defined in section 211(k)(10)(D)).

"(3) SECRETARY.—The term 'Secretary' means the Secretary of Energy.

"(b) CELLULOSIC BIOMASS ETHANOL AND MUNICIPAL SOLID WASTE LOAN GUARANTEE PROGRAM.—

"(1) IN GENERAL.—Funds may be provided for the cost (as defined in the Federal Credit Reform Act of 1990 (2 U.S.C. 661 et seq.)) of loan guarantees issued under title XIV of the Energy Policy Act to carry out commercial demonstration projects for cellulosic biomass and sucrose-derived ethanol.

"(2) DEMONSTRATION PROJECTS.—

"(A) IN GENERAL.—The Secretary shall issue loan guarantees under this section to carry out not more than 4 projects to commercially demonstrate the feasibility and viability of producing cellulosic biomass ethanol or sucrose-derived ethanol, including at least 1 project that uses cereal straw as a feedstock and 1 project that uses municipal solid waste as a feedstock.

"(B) DESIGN CAPACITY.—Each project shall have a design capacity to produce at least 30,000,000 gallons of cellulosic biomass ethanol each year.

"(3) APPLICANT ASSURANCES.—An applicant for a loan guarantee under this section shall provide assurances, satisfactory to the Secretary, that—

"(A) the project design has been validated through the operation of a continuous process facility with a cumulative output of at least 50,000 gallons of ethanol;

"(B) the project has been subject to a full technical review;

"(D) the project, with the loan guarantee, is economically viable; and

"(E) there is a reasonable assurance of repayment of the guaranteed loan.

* *** *

"(c) AUTHORIZATION OF APPROPRIATIONS FOR RESOURCE CENTER.—There is authorized to be appropriated, for a resource center to further develop bioconversion technology using low-cost biomass for the production of ethanol at the Center for Biomass-Based Energy at the Mississippi State University and the Oklahoma State University, $4,000,000 for each of fiscal years 2005 through 2007.

"(d) RENEWABLE FUEL PRODUCTION RESEARCH AND DEVELOPMENT GRANTS.—

"(1) IN GENERAL—The Administrator shall provide grants for the research into, and development and implementation of, renewable fuel production technologies in RFG States with low rates of ethanol production, including low rates of production of cellulosic biomass ethanol.

"(2) ELIGIBILITY.—

"(A) IN GENERAL.—The entities eligible to receive a grant under this subsection are academic institutions in RFG States, and consortia made up of combinations of academic institutions, industry, State government agencies, or local government agencies in RFG States, that have proven experience and capabilities with relevant technologies.

(B) APPLICATION.—To be eligible to receive a grant under this subsection, an eligible entity shall submit to the Administrator an application in such manner and form, and accompanied by such information, as the Administrator may specify.

"(3) AUTHORIZATION OF APPROPRLATIONS.—There is authorized to be appropriated to carry out this subsection $25,000,000 for each of fiscal years 2006 through 2010.

"(e) CELLULOSIC BIOMASS ETHANOL CONVERSION ASSISTANCE.—

"(1) IN GENERAL.—The Secretary may provide grants to merchant producers of cellulosic biomass ethanol in the United States to assist the producers in building eligible production facilities described in paragraph (2) for the production of cellulosic biomass ethanol.

"(2) ELIGIBLE PRODUCTION FACILITIES.—A production facility shall be eligible to receive a grant under this subsection if the production facility—

"(A) is located in the United States; and

"(B) uses cellulosic biomass feedstocks derived from agricultural residues or municipal solid waste.

"(3) AUTHORIZATION OF APPROPRIATIONS.—There is authorized to be appropriated to carry out this subsection—

"(A) $250,000,000 for fiscal year 2006; and

"(B) $400,000,000 for fiscal year 2007."

SEC. 1512. CONVERSION ASSISTANCE FOR CELLULOSIC BIOMASS, WASTE-DERIVED ETHANOL, APPROVED RENEWABLE FUELS.

Section 211 of the Clean Air Act (42 U.S.C. 7545) is amended by adding at the end the following:

"(r) CONVERSION ASSISTANCE FOR CELLULOSIC BIOMASS, WASTE-DERIVED ETHANOL, APPROVED RENEWABLE FUELS.—

"(1) IN GENERAL.—The Secretary of Energy may provide grants to merchant producers of cellulosic biomass ethanol, waste-derived ethanol, and approved renewable fuels in the United States to assist the producers in building eligible production facilities described in paragraph (2) for the production of ethanol or approved renewable fuels.

"(2) ELIGIBLE PRODUCTION FACILITIES.—A production facility shall be eligible to receive a grant under this subsection if the production facility—

"(A) is located in the United States; and

"(B) uses cellulosic or renewable biomass or waste-derived feedstocks derived from agricultural residues, wood residues, municipal solid waste, or agricultural byproducts.

"(3) AUTHORIZATION OF APPROPRIATIONS.—There are authorized to be appropriated the following amounts to carry out this subsection:

"(A) $100,000,000 for fiscal year 2006.

"(B) $250,000,000 for fiscal year 2007.

"(C) $400,000,000 for fiscal year 2008.

"(4) DEFINITIONS.—For the purposes of this subsection:

"(A) The term 'approved renewable fuels' are fuels and components of fuels that have been approved by the Department of Energy, as defined in section 301 of the Energy Policy Act of 1992 (42 U.S.C. 13211), which have been made from renewable biomass.

"(B) The term 'renewable biomass' is, as defined in Presidential Executive Order 13134, published in the Federal Register on August 16, 1999, any organic matter that is available on a renewable or recurring basis (excluding old-growth timber), including dedicated energy crops and trees, agricultural food and feed crop residues, aquatic plants, animal wastes, wood and wood residues, paper and paper residues, and other vegetative waste materials. Old-growth timber means timber of a forest from the late successional stage of forest development."

SEC. 1514. ADVANCED BIOFUEL TECHNOLOGIES PROGRAM.

(a) IN GENERAL.—Subject to the availability of appropriations under subsection (d), the Administrator of the Environmental Protection Agency shall, in consultation with the Secretary of Agriculture and the Biomass Research and Development Technical Advisory Committee established under section 306 of the Biomass Research and Development Act of 2000 (Public Law 106—224; 7 U.S.C. 8101 note), establish a program, to be known as the "Advanced Biofuel Technologies Program", to demonstrate advanced technologies for the production of alternative transportation fuels.

PRIORITY.—In carrying out the program under subsection (a), the Administrator shall give priority to projects that enhance the geographical diversity of alternative fuels production and utilize feedstocks that represent 10 percent or less of ethanol or biodiesel fuel production in the United States during the previous fiscal year.

(c) DEMONSTRATION PROJECTS.—

(1) IN GENERAL.—As part of the program under subsection (a), the Administrator shall fund demonstration projects—

(A) to develop not less than 4 different conversion technologies for producing cellulosic biomass ethanol; and

(B) to develop not less than 5 technologies for coproducing value-added bioproducts (such as fertilizers, herbicides, and pesticides) resulting from the production of biodiesel fuel.

(2) ADMINISTRATION.—Demonstration projects under this subsection shall be—

(A) conducted based on a merit-reviewed, competitive process; and

(B) subject to the cost-sharing requirements of section 988.

(d) AUTHORIZATION OF APPROPRIATIONS.—There are authorized to be appropriated to carry out this section $110,000,000 for each of fiscal years 2005 through 2009.

Source: Excerpted from the Web site of the U.S. Government Printing Office at http://frwebgate.access.gpo.gov/cgi-bin/getdoc.cgi?dbname=109_cong_public_laws&docid=f:publ058.109.

Energy Policy and Conservation Act of 1975

The Energy Policy and Conservation Act of 1975 is a U.S. law enacted in response to the 1973–74 energy crisis caused by the Arab oil embargo. Its purpose was to reduce U.S. dependency on high-priced oil imported from politically unstable countries, to prepare the United States for energy shortage conditions, and to improve energy efficiency and conservation. One of its measures for protecting the country from future oil shortages was the establishment of the SPR. It authorized the stockpiling of up to one billion barrels of petroleum in the SPR to tap into during energy emergencies. The law also gave the president of the United States the authority to withdraw crude oil from the SPR in response to an energy emergency and distribute it to oil companies by competitive sale. In addition, it established incentives for increasing domestic oil production, including off-shore drilling in the United States, and called for increased production of other energy sources. Another policy the act established is the Corporate Average Fuel Economy (CAFE) program, which sets minimum requirements for the number of miles per gallon of gas that cars and light trucks should be able to achieve. Under the CAFE program automobile manufacturers are required to average at least a minimum specified miles per gallon for each category of cars they make.

Energy Policy and Conservation Act of 1975
Public Law 94-163-94th Congress

* *** *

Sec. 6234. — Strategic Petroleum Reserve

(a) Establishment

(1) A Strategic Petroleum Reserve for the storage of up to 1 billion barrels of petroleum products shall be created pursuant to this part. By the end of the 3-year period which begins on December 22, 1975, the Strategic Petroleum Reserve (or the Early Storage Reserve authorized by section 6235 of this title, if no Strategic Petroleum Reserve Plan has become effective pursuant to the provisions of section 6239(a) of this title) shall contain not less than 150 million barrels of petroleum products.

(2) 'Beginning on October 24, 1992, the President shall take actions to enlarge the Strategic Petroleum Reserve to 1,000,000,000 barrels as rapidly as possible. Such actions may include —

(A) petroleum acquisition, transportation, and injection activities at the highest practicable fill rate achievable, subject to the availability of appropriated funds;

(B) contracting for petroleum product not owned by the United States as specified in part C of this subchapter;

(C) contracting for petroleum product for storage in facilities not owned by the United States, except that no such product may be stored in such facilities unless petroleum product stored in facilities owned by the United States on the date such product is delivered for storage is at least 750,000,000 barrels;

(D) carrying out the activities described in section 6240(h) of this title;

(E) the transferring of oil from the Naval Petroleum Reserve; and

(F) other activities specified in this subchapter.

(b) Strategic Petroleum Reserve Plan

The Secretary, not later than December 15, 1976, shall prepare and transmit to the Congress, in accordance with section 6421 of this title, a Strategic Petroleum Reserve Plan. Such Plan shall comply with the provisions of this section and shall detail the Secretary's proposals for designing, constructing, and filling the storage and related facilities of the Reserve.

(c) Levels of crude oil to be stored

(1) To the maximum extent practicable and except to the extent that any change in the storage schedule is justified pursuant to subsection (e)(6) of this section, the Strategic Petroleum Reserve Plan shall provide that:

(A) within 7 years after December 22, 1975, the volume of crude oil stored in the Reserve shall equal the total volume of crude oil which was imported into the United States during the base period specified in paragraph (2);

(B) within 18 months after December 22, 1975, the volume of crude oil stored in the Reserve shall equal not less than 10 percent of the goal specified in subparagraph (A);

(C) within 3 years after December 22, 1975, the volume of crude oil stored in the Reserve shall equal not less than 25 percent of the goal specified in subparagraph (A); and

(D) within 5 years after December 22, 1975, the volume of crude oil stored in the Reserve shall equal not less than 65 percent of the goal specified in subparagraph (A).

Volumes of crude oil initially stored in the Early Storage Reserve and volumes of crude oil stored in the Industrial Petroleum Reserve, and the Regional Petroleum Reserve shall be credited toward attainment of the storage goals specified in this subsection.

(2) The base period shall be the period of the 3 consecutive months, during the 24-month period preceding December 22, 1975, in which average monthly import levels were the highest.

(d) Plan objectives

The Strategic Petroleum Reserve Plan shall be designed to assure, to the maximum extent practicable, that the Reserve will minimize the impact of any interruption or reduction in imports of refined petroleum products and residual fuel oil in any region which the Secretary determines is, or is likely to become, dependent upon such imports for a substantial portion of the total energy requirements of such region. The Strategic Petroleum Reserve Plan shall be designed to assure, to the maximum extent practicable, that each noncontiguous area of the United States which does not have overland access to domestic crude oil production has its component of the Strategic Petroleum Reserve within its respective territory.

(e) Plan provisions

The Strategic Petroleum Reserve Plan shall include:

(1) a comprehensive environmental assessment;

(2) a description of the type and proposed location of each storage facility (other than storage facilities of the Industrial Petroleum Reserve) proposed to be included in the Reserve;

(3) a statement as to the proximity of each such storage facility to related facilities;

(4) an estimate of the volumes and types of petroleum products proposed to be stored in each such storage facility;

(5) a projection as to the aggregate size of the Reserve, including a statement as to the most economically-efficient storage levels for each such storage facility;

(6) a justification for any changes, with respect to volumes or dates, proposed in the storage schedule specified in subsection (c) of this section, and a program schedule for overall development and completion of the Reserve (taking into account all relevant factors, including cost effectiveness, the need to construct related facilities, and the ability to obtain sufficient quantities of petroleum products to fill the storage facilities to the proposed storage levels);

(7) an estimate of the direct cost of the Reserve, including —

(A) the cost of storage facilities;

(B) the cost of the petroleum products to be stored;

(C) the cost of related facilities; and

(D) management and operation costs;

(8) an evaluation of the impact of developing the Reserve, taking into account —

(A) the availability and the price of supplies and equipment and the effect, if any, upon domestic production of acquiring such supplies and equipment for the Reserve;

(B) any fluctuations in world, and domestic, market prices for petroleum products which may result from the acquisition of substantial quantities of petroleum products for the Reserve;

(C) the extent to which such acquisition may support otherwise declining market prices for such products; and

(D) the extent to which such acquisition will affect competition in the petroleum industry;

(9) an identification of the ownership of each storage and related facility proposed to be induded in the Reserve (other than storage and related facilities of the Industrial Petroleum Reserve);

(10) an identification of the ownership of the petroleum products to be stored in the Reserve in any case where such products are not owned by the United States;

(11) a statement of the manner in which the provisions of this part relating to the establishment of the Industrial Petroleum Reserve and the Regional Petroleum Reserve will be implemented; and

(12) a Distribution Plan setting forth the method of drawdown and distribution of the Reserve.

(f) Purpose of drawdown and distribution; requests for funds for storage

(1) The drawdown and distribution of petroleum products from the Strategic Petroleum Reserve is authorized only under section 6241 of this title, and drawdown and distribution of petroleum products for purposes other than those described in section 6241 of this title shall be prohibited.

(2) In the Secretary's annual budget submission, the Secretary shall request funds for acquisition, transportation, and injection of petroleum products for storage in the Reserve. If no requests for funds are made, the Secretary shall provide a written explanation of the reason therefore

Sec. 6241. — Drawdown and distribution of Reserve

(a) Power of Secretary

The Secretary may drawdown and distribute the Reserve only in accordance with the provisions of this section.

(b) Drawdown and distribution of Reserve in accordance with Distribution Plan contained in Strategic Petroleum Reserve Plan

Distribution Plan contained in Strategic Petroleum Reserve Plan

Except as provided in subsections (c), (f), and (g) of this section, no drawdown and distribution of the Reserve may be made except in accordance with the provisions of the Distribution Plan contained in the Strategic Petroleum Reserve Plan which has taken effect pursuant to section 6239(a) of this title.

(c) Drawdown and distribution of Early Storage Reserve in accordance with Distribution Plan contained in Early Storage Reserve Plan

Reserve Plan

Drawdown and distribution of the Early Storage Reserve may be made in accordance with the provisions of the Distribution Plan contained in the Early Storage Reserve Plan until the Strategic Petroleum Reserve Plan has taken effect pursuant to section 6239(a) of this title.

(d) Presidential finding prerequisite to implementation of Distribution Plan

(1) Neither the Distribution Plan contained in the Strategic Petroleum Reserve Plan nor the Distribution Plan contained in the Early Storage Reserve Plan may be implemented, and no drawdown and distribution of

the Reserve or the Early Storage Reserve may be made, unless the President has found that implementation of either such Distribution Plan is required by a severe energy supply interruption or by obligations of the United States under the international energy program.

(2) For purposes of this section, in addition to the circumstances set forth in section 6202(8) of this title, a severe energy supply interruption shall be deemed to exist if the President determines that —

(A) an emergency situation exists and there is a significant reduction in supply which is of significant scope and duration;

(B) a severe increase in the price of petroleum products has resulted from such emergency situation; and

(C) such price increase is likely to cause a major adverse impact on the national economy.

(e) Price levels and allocation procedures

The Secretary may, by rule, provide for the allocation of any petroleum product withdrawn from the Strategic Petroleum Reserve in amounts specified in (or determined in a manner prescribed by) and at prices specified in (or determined in a manner prescribed by) such rules. Such price levels and allocation procedures shall be consistent with the attainment, to the maximum extent practicable, of the objectives specified in section 753(b)(1) of title 15.

(f) Removal or disposal of products in Industrial Petroleum Reserve

The Secretary may permit any importer or refiner who owns any petroleum products stored in the Industrial Petroleum Reserve pursuant to section 6236 of this title to remove or otherwise dispose of such products upon such terms and conditions as the Secretary may prescribe.

(g) Directive to carry out test drawdown

(1) The Secretary shall conduct a continuing evaluation of the Distribution Plan. In the conduct of such evaluation, the Secretary is authorized to carry out test drawdown and distribution of crude oil from the Reserve. If any such test drawdown includes the sale or exchange of crude oil, then the aggregate quantity of crude oil withdrawn from the Reserve may not exceed 5,000,000 barrels during any such test drawdown or distribution.

(2) The Secretary shall carry out such drawdown and distribution in accordance with the Distribution Plan and implementing regulations and contract provisions, modified as the Secretary considers appropriate taking into consideration the artificialities of a test and the absence of a severe energy supply interruption. To meet the requirements of subsections (d) and (e) of section 6239 of this title, the Secretary shall transmit any such modification

of the Plan, along with explanatory and supporting material, to both Houses of the Congress no later than 15 calendar days prior to the offering of any crude oil for sale under this subsection.

(3) At least part of the crude oil that is sold or exchanged under this subsection shall be sold or exchanged to or with entities that are not part of the Federal Government.

(4) The Secretary may not sell any crude oil under this subsection at a price less than that which the Secretary determines appropriate and, in no event, at a price less than 90 percent of the sales price, as estimated by the Secretary, of comparable crude oil being sold in the same area at the time the Secretary is offering crude oil for sale in such area under this subsection.

(5) The Secretary may cancel any offer to sell or exchange crude oil as part of any drawdown and distribution under this subsection if the Secretary determines that there are insufficient acceptable offers to obtain such crude oil.

(6) (A) The minimum required fill rate in effect for any fiscal year shall be reduced by the amount of any crude oil drawdown from the Reserve under this subsection during such fiscal year.

(B) In the case of a sale of any crude oil under this subsection, the Secretary shall, to the extent funds are available in the SPR Petroleum Account as a result of such sale, acquire crude oil for the Reserve within the 12-month period beginning after the completion of the sale. Such acquisition shall be in addition to any acquisition of crude oil for the Reserve required as part of a fill rate established by any other provision of law.

(7) Rules, regulations, or orders issued in order to carry out this subsection which have the applicability and effect of a rule as defined in section 551(4) of title 5 shall not be subject to the requirements of subchapter II of chapter 5 of such title or to section 6393 of this title.

(8) The Secretary shall transmit to both Houses of the Congress a detailed explanation of the drawdown and distribution carried out under this subsection. Such explanation may be a part of any report made to the President and the Congress under section 6245 of this title.

(h) Prevention or reduction of adverse impact of severe domestic energy supply interruptions

(1) If the President finds that —

(A) a circumstance, other than those described in subsection (d) of this section, exists that constitutes, or is likely to become, a domestic or international energy supply shortage of significant scope or duration; and

(B) action taken under this subsection would assist directly and significantly in preventing or reducing the adverse impact of such shortage,

then the Secretary may, subject to the limitations of paragraph (2), draw down and distribute the Strategic Petroleum Reserve.

(2) In no case may the Reserve be drawn down under this subsection —

(A) in excess of an aggregate of 30,000,000 barrels with respect to each such shortage;

(B) for more than 60 days with respect to each such shortage;

(C) if there are fewer than 500,000,000 barrels of petroleum product stored in the Reserve; or

(D) below the level of an aggregate of 500,000,000 barrels of petroleum product stored in the Reserve.

(3) During any period in which there is a drawdown and distribution of the Reserve in effect under this subsection, the Secretary shall transmit a monthly report to the Congress containing an account of the drawdown and distribution of petroleum products under this subsection and an assessment of its effect.

(4) In no case may the drawdown under this subsection be extended beyond 60 days with respect to any domestic energy supply shortage.

(i) Exchange of withdrawn products

Notwithstanding any other law, the President may permit any petroleum products withdrawn from the Strategic Petroleum Reserve in accordance with this section to be sold and delivered for refining or exchange outside of the United States, in connection with an arrangement for the delivery of refined petroleum products to the United States.

(j) Purchases from Strategic Petroleum Reserve by entities in insular areas of United States and Freely Associated States

(1) Definitions

In this subsection:

(A) Binding offer

The term "binding offer" means a bid submitted by the State of Hawaii for an assured award of a specific quantity of petroleum product, with a price to be calculated pursuant to paragraph (2) of this subsection, that obligates the offeror to take title to the petroleum product without further negotiation or recourse to withdraw the offer.

(B) Category of petroleum product

The term "category of petroleum product" means a master line item within a notice of sale.

(C) Eligible entity

The term "eligible entity" means an entity that owns or controls a refinery that is located within the State of Hawaii.

(D) Full tanker load

The term "full tanker load" means a tanker of approximately 700,000 barrels of capacity, or such lesser tanker capacity as may be designated by the State of Hawaii.

(E) Insular area

The term "insular area" means the Commonwealth of Puerto Rico, the Commonwealth of the Northern Mariana Islands, the United States Virgin Islands, Guam, American Samoa, the Freely Associated States of the Republic of the Marshall Islands, the Federated States of Micronesia, and the Republic of Palau.

(F) Offering

The term "offering" means a solicitation for bids for a quantity or quantities of petroleum product from the Strategic Petroleum Reserve as specified in the notice of sale.

(G) Notice of sale

The term "notice of sale" means the document that announces —

(i) the sale of Strategic Petroleum Reserve products;

(ii) the quantity, characteristics, and location of the petroleum product being sold;

(iii) the delivery period for the sale; and

(iv) the procedures for submitting offers.

(2) In general

In the case of an offering of a quantity of petroleum product during a drawdown of the Strategic Petroleum Reserve —

(A) the State of Hawaii, in addition to having the opportunity to submit a competitive bid, may —

(i) submit a binding offer, and shall on submission of the offer, be entitled to purchase a category of a petroleum product specified in a notice of sale at a price equal to the volumetrically weighted average of the successful bids made for the remaining quantity of the petroleum product within the category that is the subject of the offering, and

(ii) submit one or more alternative offers, for other categories of the petroleum product, that will be binding if no price competitive contract is awarded for the category of petroleum product on which a binding offer is submitted under clause (i); and

(B) at the request of the Governor of the State of Hawaii, a petroleum product purchased by the State of Hawaii at a competitive sale or through a binding offer shall have first preference in scheduling for lifting.

(3) Limitation on quantity

(A) In general

In administering this subsection, in the case of each offering, the Secretary may impose the limitation described in subparagraph (B) or (C) that results in the purchase of the lesser quantity of petroleum product.

(B)Portion of quantity of previous imports

The Secretary may limit the quantity of a petroleum product that the State of Hawaii may purchase through a binding offer at any offering to 1/12 of the total quantity of imports of the petroleum product brought into the State during the previous year (or other period determined by the Secretary to be representative).

(C) Percentage of offering

The Secretary may limit the quantity that may be purchased through binding offers at any offering to 3 percent of the offering.

(4) Adjustments

(A) In general

Notwithstanding any limitation imposed under paragraph (3), in administering this subsection, in the case of each offering, the Secretary shall, at the request of the Governor of the State of Hawaii, or an eligible entity certified under paragraph (7), adjust the quantity to be sold to the State of Hawaii in accordance with this paragraph.

(B) Upward adjustment

The Secretary shall adjust upward to the next whole number increment of a full tanker load if the quantity to be sold is —

(i) less than 1 full tanker load; or

(ii) greater than or equal to 50 percent of a full tanker load more than a whole number increment of a full tanker load.

(C) Downward adjustment

The Secretary shall adjust downward to the next whole number increment of a full tanker load if the quantity to be sold is less than 50 percent of a full tanker load more than a whole number increment of a full tanker load.

(5) Delivery to other locations

The State of Hawaii may enter into an exchange or a processing agreement that requires delivery to other locations, if a petroleum product of similar value or quantity is delivered to the State of Hawaii.

(6) Standard sales provisions

Except as otherwise provided in this chapter, the Secretary may require the State of Hawaii to comply with the standard sales provisions applicable to purchasers of petroleum products at competitive sales.

(7) Eligible entities

(A) In general

Subject to subparagraphs (B) and (C) and notwithstanding any other provision of this paragraph, if the Governor of the State of Hawaii certifies to the Secretary that the State has entered into an agreement with an eligible entity to carry out this chapter, the eligible entity may act on behalf of the State of Hawaii to carry out this subsection.

(B) Limitation

The Governor of the State of Hawaii shall not certify more than one eligible entity under this paragraph for each notice of sale.

(C) Barred company

If the Secretary has notified the Governor of the State of Hawaii that a company has been barred from bidding (either prior to, or at the time that a notice of sale is issued), the Governor shall not certify the company under this paragraph.

(8) Supplies of petroleum products

At the request of the Governor of an insular area, the Secretary shall, for a period not to exceed 180 days following a drawdown of the Strategic Petroleum Reserve, assist the insular area or the President of a Freely Associated State in its efforts to maintain adequate supplies of petroleum products from traditional and nontraditional suppliers

Source: Excerpted from http://caselaw.lp.findlaw.com/casecode/uscodes/42/chapters/77/toc.html.

AFFORDABILITY OF ENERGY RESOURCES AND ENERGY'S IMPACT ON THE ECONOMY

International Energy Agency (IEA) Report: *Energy Prices and Taxes, 1st Quarter 2004:* "Thirty Years of Energy Prices and Savings"

The IEA uses "energy intensity" as an index to measure energy efficiency, an "intensity" being the energy used per unit of activity—for example, the gasoline used per mile driven by a car. The IEA has reported that, as a result of declining energy intensities, significant energy savings began around the world in 1973, when the first oil "price shock" hit, and lasted through the mid- to late 1980s. The IEA concluded that the changes caused by the 1970s oil crises, including the revisions in energy policy that came out of the crises, had a substantially greater impact on reducing energy consumption than energy conservation and efficiency policies implemented in the 1990s. Declining energy intensities, according to the IEA, led to significantly reduced energy costs since 1973, but the reductions in energy intensities have been much more modest since the late 1980s, so the rate of energy savings has slowed since then. According to the IEA energy use for cars is much greater in countries with low fuel prices and there is a correlation between higher fuel prices and lower vehicle fuel intensity and travel per capita. See below for a full report on the IEA findings.

International Energy Agency (IEA) Report:
Energy Prices and Taxes, 1st Quarter 2004:
"Thirty Years of Energy Prices and Savings"[1]

Fridtjof Unander, Senior Energy Analyst
Energy Technology Policy Division

INTRODUCTION

The IEA publication that this article is based on examines how energy efficiency and other factors such as economic structure, income, fuel mix and prices have shaped developments in energy use and CO_2 emissions in IEA countries since the IEA was founded 30 years ago. The analysis is based on a newly developed database with detailed information on energy use in the manufacturing, households, service and transport sectors. This information is used to develop disaggregated energy intensities within each sector. For example, the analysis includes energy intensities for five uses and six major electric appliances in the household sector; for six sub-sectors in manufacturing; and four different modes of passenger travel. The energy intensities are defined as energy use per unit of activity, e.g. space heating per unit of dwelling area, energy use per unit value-added of production of ferrous metals, and energy use per kilometre driven by cars.

The results of this study show that declining energy intensities across most sectors and countries led to significant energy savings from 1973 through the mid to late 1980s. However, since then, declines in energy intensities have been much more modest with the result that the rate of energy savings has slowed. This implies that the changes caused by the oil price shocks in the 1970s and the resulting energy policies did considerably more to control growth in energy demand and reduce CO_2 emissions than the energy efficiency and climate policies implemented in the 1990s.

[1] This article is based on the IEA publication *30 Years of Energy Use in IEA Countries,* which was published in March 2004.

This article discusses absolute levels and the evolution over time of selected energy intensities related to energy price development and also how changes in prices, fuel mix and energy intensities have affected energy costs for manufacturing, households and passenger travel.

MANUFACTURING

Energy Prices

The end of the high fossil fuel price era in 1986 offers an important explanation for the slow-down in energy intensity declines.

Oil prices shot up in the wake of the embargo in 1973–4 and were further exacerbated by supply disruptions induced by the Iran-Iraq war in 1979. The manufacturing sector faced significant price hikes: in the late 1970s and early 1980s industrial oil prices skyrocketed by 200–300% before falling back almost to the 1973 level by 1987. Since prices dropped in the mid-1980s, there have been only relatively small price fluctuations until the late 1990s. Natural gas and coal went through similar developments in most IEA countries.

It is hard not to attribute part of the decline in manufacturing energy intensities observed before 1986 to higher energy prices. This stimulated the application of energy-saving technologies. Other factors also affected the rate of energy intensity decline. The rate of growth in manufacturing output influences the rate of investment in new technology and the utilisation of production factors, including energy. Thus both prices and the rate of economic growth are important determinants for how manufacturing energy intensities change over time.

Shifts in the production processes themselves have also lowered manufacturing energy intensities. Examples include increased use of recycled feedstocks (e.g., scrap metal, recycled paper) and shifts from primary to secondary production, i.e., shifts away from raw steel and aluminium production to secondary and from raw paper to recycled paper production. These shifts represent structural changes within sub-sectors. To isolate the effects of these "micro" structural changes requires more disaggregated data than what are generally available on an internationally consistent basis.

Energy Costs as Share of Total Production Cost

The share of energy in total production costs varies significantly across countries and sub-sectors, but has fallen everywhere.

The cost of energy for manufacturing depends on the energy intensity of the products produced, the mix of fuels used and the price of those fuels. For energy-intensive industries, energy costs constitute a significant share of total production costs. . . . Yet there are relatively large differences in this share among the countries included in this figure. Some of these differences are due to variations in the level of sub-sector energy intensities. Differences in fuel prices also play a role, although higher intensities, especially in energy-intensive industries, tend to be related to lower prices. Access to cheap energy is often a stimulant for the production of energy-intensive materials. For example, in Australia and Norway, where energy has been relatively inexpensive, the production of aluminium—a very energy intensive process—constitutes an important share of the production of primary metals and thus drives up the average intensity for this sector.

In France, the United Kingdom and United States, the energy cost share fell significantly between 1982 and 1998 in all sectors. Most sectors in Japan also saw a reduction in this share, but less than in the three other countries, even though Japanese industries reduced energy intensities at a faster rate than most other countries through the 1980s. A closer examination of the data for intermediate production cost and value-added, shows that value-added increased relative to product costs in Japanese industries. This indicates that the use of other production factors also became more efficient in parallel with the energy intensity reductions. Compared to other countries, the somewhat higher share of energy expenditures in Japan in 1998 could thus be related to that the use of other production factors is more efficient, though quantification of this is beyond the scope of this study.

Decomposition of Changes in Energy Costs

Declines in energy costs have slowed with slowing energy intensity reductions.

.... Energy cost relative to value-added in this sector fell significantly before 1990 in all countries.... In Japan energy expenditures relative to value-added fell more than relative to intermediate product.... The decline in Japan was due to the combination of rapidly falling energy intensities and significant declines in fuel-weighted energy prices that led to energy costs falling by almost 8% per year on average relative to value-added between 1982 and 1990. These two factors also reduced energy costs per value-added in the United States, United Kingdom and France, but to a more modest degree. Examining data for other sub-sectors show to a large extent the same picture, expenditures fell as intensities and fuel prices declined.

The impact from changes in the fuel mix in primary metals production was modest. In other sub-sectors where more significant fuel switching took place, energy costs generally increased where electricity and gas took shares from coal, while costs fell where more expensive oil (per energy unit) was replaced by natural gas.

After 1990 energy costs relative to value-added in the primary metals sector only declined significantly in the United Kingdom, where energy expenditures fell as prices dropped. For all countries the lack of a considerable decline in energy intensities limited further reductions in energy costs. The same tendency can be observed in other manufacturing sub-sectors.

It is thus tempting to conclude that today the lower energy costs—which results from both successful energy efficiency improvements and lower energy prices—has made investments in energy efficiency less attractive than investing in ways to reduce other production costs compared to a couple of decades ago.

HOUSEHOLDS

Residential Electricity Prices

Only moderate changes in real electricity prices since 1973, but significant differences among countries.

In the majority of IEA countries residential electricity prices in real terms have undergone less dramatic changes than oil prices, and to some extent less than coal and gas prices. Fossil fuel prices increased significantly in the aftermath of the oil price shocks in 1973–1974 and 1979, and fell again with the crash in crude oil prices in 1986. For the IEA-11 countries, electricity prices between 1973 and 1986 increased moderately in most and declined in a few. After 1986, electricity prices fluctuated somewhat, but with a general downward trend, especially over the last few years. In fact, only in Denmark where electricity taxes increased during this period, were real prices in 2000 higher than in 1986.

There is a striking variation in residential electricity price levels among countries. In 2000, there was a 3.5 factor difference between Denmark with the highest price in this group and Norway with the lowest price. Since both these countries belong to the common Nordic electricity market, the price differential is mostly due to differences in taxation. This divergence in policy between the two neighbouring countries reflects very different resource endowments; Norwegian consumers' access to inexpensive hydropower has been considered a public good, while in energy import-dependent Denmark, reducing fuel inputs for power generation through taxation of electricity has been more generally practiced.

Electricity and Useful Heat Prices

The price paid for useful space heating varies much less than electricity prices.

In countries such as Norway, where the share of electricity in space heating is high, the price of electricity is a good indicator of what consumers pay to heat their homes. Most countries, however, rely primarily on other energy carriers for heating. To compare what households actually pay for space heating, it is thus more revealing to investigate fuel-consumption-weighted prices.... This weighted price, calculated by multiplying the consumption of each heating fuel by its price and dividing by total useful space heating, is compared against electricity prices in 1998.

As the figure indicates the difference between the electricity price and the useful space-heating price is smallest in Norway followed by Sweden, the two countries that use the most electricity for space heating. For the other countries, the useful space-heating price is 50% or less of the electricity price. In Denmark, for example, consumers pay about three times more for electricity than they pay per unit of useful heat provided to their homes. The differences would have been even larger if electricity prices were compared against fuel-weighted price per final energy for space heating instead of per unit of useful energy.

Consequently, the variations among countries in space heating prices are much less than for electricity prices. This indicates that non-price factors such as building codes, climate (homes in cold climates tend to be better insulated) and cultural aspects (e.g. preference for indoor temperature) are important to explain differences in the level of space heating intensities. However, the strong variations in electricity prices indicate differences in the economic incentives to undertake energy efficiency improvements of electricity-specific uses such as lighting and appliances.

Energy Expenditures by End Use

Appliances are taking over from space heating as the most expensive residential end use in countries with warm to moderate climates.

With space heating being the most significant residential energy use in most countries, it is not surprising that it constitutes the largest share of household energy expenditures.... Prices for space heating are normally much lower than for electricity. This means that the expenditure share for space heating is generally lower, and the expenditure share for electricity-specific applications is generally higher, than their shares of residential energy consumption would imply.

The low variation in space heating prices from country to country indicates that variations in consumption levels are important to explain differences in expenditures for space heating. While for electricity dominated end uses, differences in expenditures are clearly a result of both of price and consumption level divergences.

Besides Japan, all [IEA] countries... have seen a reduction in the expenditure share for space heating between 1973 and 1998, driven mostly by reduced intensities. In 1998 only consumers in Sweden and Norway spent 50% or more of their energy budgets to heat their homes. On the other hand, only in Japan did electric appliances account for more than half of energy expenditures, although the share of this end use has increased in all countries since 1973.

Energy Expenditures as a Share of Personal Income

IEA households today spend roughly the same share of their incomes on energy as in 1973.

The share of disposable incomes that IEA households pay for energy has varied significantly over the last three decades. In the mid 1970s it was between 2.5 and 4% in a selected group of IEA countries. The share increased to between 4 and 5.5% by the mid 1980s and then started to decline in most countries. By 1998, the shares were back into the same band as in the mid 1970s.

What percentage of total income is spent on household energy depends on the price of purchased fuel, the mix of fuels used and the level of residential energy demand per unit of income. To better understand the development of the shares in these countries, it is interesting to look at how each of these components has evolved.

This can be done by decomposing changes in the expenditure share into changes in residential energy demand per unit of income, real fuel prices and fuel mix through holding all but one factor constant at the 1990-level. In this manner, the impact from changes in prices can be calculated as the change in real prices for oil, natural gas, coal and electricity, weighted at the 1990 fuel mix and consumption levels. Conversely, the change in expenditure share resulting from changes in the fuel mix is calculated holding relative fuel prices and energy demand per unit of income constant at the 1990-level. Then the impact from changes in energy demand per unit of income is calculated with prices and fuel mix at 1990-levels.

Factors Affecting the Development of Energy Expenditures

Strong growth in energy expenditures due to higher prices reversed after 1982.

.... The share of energy expenditures in disposable income increased rapidly between 1973 and 1982 in all the countries shown, indicating that energy was a much bigger burden on household budgets. The average annual growth ranged from 2.9% in the United States to 6.8% in France. In the United States, the expenditure share grew even though energy demand per unit of income fell by more than 3% per year on average. This growth was primarily driven by increased real prices for fuels and electricity; although an increased share of electricity in the fuel mix also contributed (electricity is generally more expensive than other energy carriers). Increased prices also drove up expenditures in the other countries, as did the higher share of electricity, except in the United Kingdom, where the electricity share actually fell between 1973 and 1982.

After 1982 the picture changed: the share of energy expenditures fell in the above countries. The main reason for this turn-around was a strong decline in energy prices in all countries except Norway. On the other hand, increased shares of electricity in the fuel mix driven by more appliances continued to induce a moderate upward pressure on energy expenditures in most countries. The fail in prices was augmented by a decline in energy demand per unit of income in all countries except Japan, where residential energy demand grew slightly faster than income. Taken over the whole period, lower growth in energy consumption relative to income helped reduce the importance of energy in the household budgets in all IEA-11 countries, except Japan. Clearly, without the energy savings achieved between 1973 and 1998, IEA households would have seen much more of their incomes being spent on energy.

PASSENGER TRANSPORT

Gasoline Prices

Gasoline prices have varied considerably both over time and across IEA countries.

Fuel prices for passenger transport are an important river of travel demand, mode choice and fuel intensity. Fuel prices have varied considerably across IEA countries and continue to do so, particularly when fuel axes are taken into account. In 2000, prices ranged from a low of about 0.35 cents per litre in the United States to $1.20 per litre in Italy (adjusted using a PPP-weighted exchange rate).

Fuel taxes have also varied considerably over time. Oil prices peaked in the early 1980s, and generally have been declining in real terms for the past 20 years. Price trends for countries such as Japan and the United States reflect this, since these countries have not substantially increased fuel taxes, and real gasoline prices in 2000 were comparable or lower than in the mid 1970s. On the other hand, most European countries offset much of the oil price decline by raising fuel taxes. In some cases, such as Germany and Finland, taxes have been increased at a greater rate than the oil price decline, yielding a steady rise in retail fuel prices since the 1980s and mid 1990s.

Between 1998 and 2000, fuel prices increased significantly in many countries, particularly in Europe. This was due both to increases in world oil prices and, in a few countries, increases in fuel taxes, such as in the United Kingdom and Germany. While these represent relatively minor changes compared to historical price fluctuations, for a number of European countries the recent increases have put fuel prices, in real dollar terms, near all-time highs, and led to protests in Europe during 1999.

Fuel Costs Per Kilometre Driven

Car fuel costs per kilometre generally have fallen since the early 1980s. By combining fuel price information with data or stock-average fuel intensity of cars, the average cost per kilometre of driving is derived. The cost of driving shows a somewhat similar pattern of variation by country and over time as for gasoline prices, with peak costs occurring in the early to mid-1980s. However, there are two noticeable effects related to vehicle fuel intensity. First, fuel costs ($ per car-km) have declined more during the late 1980s and 1990s than fuel prices, reflecting the impact of reduced fuel intensity on lowering travel costs.

Second, apart from the United States, the range between countries in fuel cost per kilometre narrowed considerably during the 1990s, with a slight "convergence" toward $0.06-$0.08 per kilometre driven. This may reflect that over time, drivers have responded to fuel price changes by buying types of vehicles that bring their driving costs into an acceptable range that does not vary too much by country. However, the United States has moved away from this emerging convergence, with a cost per kilometre declining steadily through the 1990s and below $0.04 by 1998. Thus despite the relatively high fuel intensity of U~S. cars, fuel prices in the late 1990s were low enough for driving to be far cheaper than in other countries. On the other hand, without the significant drop in car intensities during the 1970s and 1980s the cost of driving in the United States today would have been more comparable to other IEA countries.

Fuel Use per Capita versus Fuel Prices

Energy use for cars is much higher in countries with low fuel prices. .... The United States is by itself, with almost twice the car fuel use per capita and half the average fuel price of any other country. Japan and the European countries, with fuel prices two to three times higher, use around two-thirds less fuel per capita than the United States. Canada and Australia fall into a middle group, both in terms of fuel price and fuel use per capita.

Looking at the Japan and Europe grouping, there is no clear relationship between fuel prices and fuel use. Other factors explain most of the variation: country size, income, vehicle prices, taxes, ownership levels, and availability and extent of mass transit. For example, Japan, with the lowest average fuel price and lowest energy use of the countries in the Europe/Japan group, is also one of the smaller countries and the one with the highest rail transit availability and use.

The United States, Canada and Australia are geographically the three largest countries, which partially explain why their energy use per capita is higher. But size does not explain the difference between the United States and Canada/Australia. This difference is clearly influenced by lower energy prices in the United States, and also higher incomes.

Vehicle Travel and Intensities versus Fuel Prices

Higher fuel prices correlate with lower vehicle fuel intensity and lower travel per capita, though the travel effect is fairly weak.

. . . . The European countries have relatively similar levels of car travel per capita, despite that real fuel prices vary significantly. In Japan travel rates are the lowest of all countries, even if the fuel price is relatively low. This underscores the impact of Japan's small geographical size and well developed mass-transit system on car fuel use. . . .

The bottom part of Figure 8 shows a relatively clear correlation between higher fuel prices and lower car fuel intensity, also within the group of European countries. Interestingly, in this case Japan is no longer an outlier of the general trend: its relative high fuel intensity and low fuel price bring it into the same range as Australia and Canada. Thus the low car fuel use per capita relative to fuel price in Japan is a result of modest car use, not low fuel intensity. Similarly, Australia and Canada, which represent mid-case countries for per capita car fuel use, are close to the United States in fuel intensity, but closer to European countries in terms of car travel per capita.

Source: International Energy Agency (IEA) Report: *Energy Prices and Taxes, 1st Quarter 2004:* "Thirty Years of Energy Prices and Savings," pp. 1–2, which is available online at http://www.iea.org/Textbase/Papers/2005/cost.pdf.

Capitol Hill Hearing Testimony: Statement of Brian Castelli, Executive Vice President and COO, Alliance to Save Energy, House Committee on Energy and Commerce Subcommittee on Energy and Air Quality (November 2, 2005)

In addition to members of Congress, political activists have also pressured the U.S. government to do something about rising energy costs—and something more than the provisions laid out in the Energy Policy Act of 2005. For example, a leader of the Alliance to Save Energy, a nonprofit coalition of business, government, environmental, and consumer groups that advocates energy efficiency programs and reduced energy use, addressed the U.S. House Energy and Subcommittee on Energy and Air Quality in November 2005. Brian Castelli, the Alliance's chief operating officer and executive vice president, criticized what he viewed as loopholes in the Energy Policy Act of 2005 that would only result in higher oil prices. Castelli's testimony highlights certain energy-price-related concerns that other Americans have also expressed, such as a need to revise CAFE standards according to realistic emissions testing and to classify SUVs and minivans as passenger vehicles instead of light trucks. Castelli also lamented the "startling and immediate effects of Hurricane Katrina and Rita" on energy prices, which, he says, show that we need to bring energy demand and supplies into balance.

CAPITOL HILL HEARING TESTIMONY: Statement of Brian Castelli, Executive Vice President and COO, Alliance to Save Energy, Before the Committee on House Energy Commerce Subcommittee on Energy Air Quality, November 2, 2005

The Time Is Now for Energy Efficiency Measures

The recently enacted energy bill is really an important to-do list rather than a completed product. Existing federal programs also have a tremendous potential for cost-effective energy savings. Yet the fiscal year 2006 budget request for energy efficiency is down 14 percent after inflation just since 2002. Here are some key implementation and funding needs:

Consumer education: The fastest way to address an energy supply shortage, and probably the only way to have a significant impact on prices this winter, is consumer education.

Tax incentives: To make the new energy-efficiency incentives effective, we need the implementing rules out as soon as possible, we need to move up the effective dates, and, we need to extend the incentives beyond the two years most are scheduled to last.

Appliance standards: We remain very concerned that DOE is years behind deadlines in setting about 20 standards required under law. This program requires effective oversight and increased funding.

Building codes assistance: More funding is needed for the Gulf states and for a new program to encourage states to adopt the latest codes and then assist them in achieving high rates of compliance.

State and utility energy-efficiency programs: Funding is needed for an innovative new pilot program to assist several states in the design and implementation of energy efficiency resource programs.

Federal energy management: More funding is needed for DOE's Federal Energy Management Program to ramp up ESPC use and to undertake other activities required in the energy bill.

The Time Is Now to Move Beyond the Energy Bill

There is a gaping hole in the energy bill where transportation policies should be. The Alliance estimates that the energy bill will save virtually no oil. Two policies that could be as effective as a straightforward increase in Corporate Average Fuel Economy (CAFE) standards are:

Close CAFE loopholes: Several reforms are needed to close major loopholes and bring actual fuel economies closer to current standards: base

CAFE on realistic testing of fuel economy, treat SUVs and minivans like the passenger vehicles they are, include heavier SUVs under CAFE, and treat "dual-fuel" vehicles based on actual alternative fuel use.

Vehicle fuel use "feebate": A new, innovative approach to efficiency of cars and light trucks is a national feebate system. Such a system would impose a national security surcharge, or "fee" on inefficient vehicles, and then use the funds collected to provide a "rebate" to fuel efficient vehicles.

There also are a number of other policies to save natural gas that should be reconsidered in light of sharply higher natural gas prices. The Alliance recommends two sets of policies:

Federal building codes: The energy bill passed over several opportunities to improve standards for buildings regulated by or paid for by the federal government. They include standards for manufactured housing, new public housing and new housing with federally subsidized mortgages, subsidized rebuilding after disasters, and privatized military housing.

State and utility energy-efficiency programs: Energy efficiency resource standards (EERS) in several states now require electricity and natural gas utilities to meet customer needs in part through demand-side management programs rather than by constructing new facilities and purchasing energy. A requirement that state PUCs consider this and other energy efficiency policies would save natural gas.

Energy efficiency is our largest energy resource, and it should be our first energy priority. We hope you will both work to ensure the fine energy-efficiency provisions of the last energy bill are fully funded and implemented, and use the increasing pressure for action to fill the gaping holes in that bill.

Introduction

My name is Brian T. Castelli and I serve as the Executive Vice President and Chief Operating Officer of the Alliance to Save Energy, a bipartisan, nonprofit coalition of more than 90 business, government, environmental and consumer leaders. The Alliance's mission is to promote energy efficiency worldwide to achieve a healthier economy, a cleaner environment, and greater energy security. . . . The United States has only 2 percent of the world's known oil reserves, and 5 percent of the world's people, but uses 25 percent of the world's oil. And now the same pattern is being repeated with natural gas.

Although measures to increase energy supplies are necessary, we must not fool ourselves into believing that we can produce our way out of the problem. U.S. production of oil and of natural gas is lower than it was in 1970, while our energy consumption has steadily risen. Even the National Petroleum Council has concluded that natural gas supplies from traditional North American production will not be able to meet projected demand, and that "greater energy efficiency and conservation are vital near-term and long-term mechanisms for moderating price levels and reducing volatility." It is time to turn serious attention to the demand side of the equation, to reducing our energy use.

Energy Efficiency Is America's Greatest Energy Resource

Energy efficiency is the nation's greatest energy resource—efficiency now contributes more than any other single energy resource to meeting our nation's energy needs, including oil, natural gas, coal, or nuclear power. The Alliance to Save Energy estimates that without the energy efficiency gains since 1973 we would now be using at least 39 quadrillion Btu more energy each year, or 40% of our actual energy use. Much of these savings resulted from federal energy policies and programs like appliance and motor vehicle standards, research and development, and the Energy Star program.

Energy efficiency is the quickest, cheapest, and cleanest way not only to tackle our current energy cost issues, but also to meet the anticipated future growth in energy demand in the U.S. The enormous contribution of energy efficiency to meeting our energy needs is achieved with little or no negative impact on our wilderness areas, our air quality, or the global climate. Energy efficiency enhances our national and energy security by lessening requirements for foreign energy sources. Further, energy efficiency is invulnerable to supply disruptions; is rarely subject to siting disputes; is available in all areas in large or small quantities; and generally costs much less than it would to buy additional energy.

Energy-efficiency and conservation measures have a proven track record of balancing demand and supply much faster than drilling, constructing power plants, or new import facilities. When a series of rolling blackouts and electricity price spikes hit California in 2000–2001, the state undertook a massive electricity efficiency outreach campaign that reduced peak summer power demand by 10 percent and reduced overall electricity use by 7 percent in less than a year, thus helping avoid further shortages. The cost was just 3 cents per kWh.

The American Council for an Energy-Efficient Economy estimates that a small decrease in natural gas demand (2–4 percent) could result in a decrease in wholesale natural gas prices of as much as 25 percent over the next few years, with vast savings for consumers and energy-intensive industries.

The Time Is Now to Make the Energy Bill Real

Many of the policies needed to increase use of energy efficiency as a major energy source are enacted, and many of the programs are in place. But these policies must be carried out, and the programs must be funded, or they will do no good. In particular, the recently enacted energy bill (the Energy Policy Act of 2005, P.L. 109-58) is really an important "to-do list," rather than a completed product. The Alliance to Save Energy estimates that the new energy law could save 5 percent of all U.S. energy use by 2020, and a higher percentage of natural gas—if it is fully implemented and funded. If the law is widely ignored, the savings will be a fraction of that amount.

Existing federal programs also have a tremendous potential for energy savings. A 2001 National Research Council report found that every dollar invested in 17 Department of Energy (DOE) energy efficiency research and development (R&D) programs returned nearly $20 to the U.S. economy in the form of new products, new jobs, and energy cost savings to American homes and businesses. Environmental benefits were estimated to be of a similar magnitude. DOE itself estimates that its efficiency and renewables programs will result in major savings, including $134 billion in energy bills, 157 GW of avoided new conventional power plants, 1.9 quads of natural gas, and 213 MMTC of greenhouse gas emissions in 2025. Yet the fiscal year 2006 budget request for energy efficiency is down 14 percent after inflation just since 2002, and core research and development funding (excluding grants and the fuel cell FreedomCar program) is down 31 percent in those four years.

Following are some key implementation and funding needs for programs that have the potential to save large quantities of natural gas and, in some cases, oil. Note that one of the most effective ways of reducing natural gas consumption is to reduce electric demand, as most peaking power plants and most new power plants are fueled by natural gas. Similarly, any reduction in the consumption of oil products (gasoline, jet fuel, etc) helps reduce the stress on heating oil supplies.

Consumer education: As was demonstrated in California, the fastest way to address an energy supply shortage, and probably the only way to have a significant impact on prices this winter, is consumer education and associated incentive programs. In particular, there is an immediate need for funding for the energy efficiency public information campaign authorized in the energy bill section 134. This important program was authorized by the Congress at $90 million per year, from fiscal year 2006 through fiscal year 2010. It is intended to provide consumers with energy saving tips like maintaining and repairing heating and cooling ducts and equipment, insulating and weatherizing homes and buildings, properly maintaining tires and cars, and purchasing energy-efficient products and equipment. Importantly, the program could ensure that consumers and businesses are made aware of the important energy-efficiency tax incentives included in the energy bill (see below). It also could be coordinated with, and could support, other programs, including the appliance rebate program and state demand-side management programs. Coupling such efforts would optimize use of federal funding and ensure the greatest impact in terms of changing consumer behavior.

Additional funding is equally important for the Energy Star program. Energy Star is a successful voluntary deployment program at EPA and DOE that has made it easy for consumers to find and buy many energy-efficient products. Energy Star is the best demonstration of how effective government consumer education can be. For every federal dollar spent, Energy Star produces average energy bill savings of $75 and sparks $15 in investment of new technology. Last year alone, Americans, with the help of Energy Star, prevented 30 million metric tons of greenhouse gas emissions—equivalent to the annual emissions from 20 million vehicles, and saved about $10 billion on their utility bills.

Tax incentives: The energy bill included important tax incentives for highly energy-efficient new homes, improvements to existing homes, commercial buildings, heating and cooling equipment, appliances, and hybrid vehicles. These incentives for consumers and businesses have the potential to help transform markets to embrace energy-efficient technologies and thus to help the best buildings, vehicles, and equipment become mainstream.

However, there are several immediate needs to make these incentives effective. First, IRS, with the assistance of DOE, must get the implementing rules out as soon as possible. Many important details and interpretations were left to the agencies. We also hope these rules will make determining eligibility and applying for these incentives as simple as possible. Without clear rules and procedures, only "free-riders"—those who were going to buy the products anyway—likely will take a chance on the incentives, and the opportunity for meaningful and sustainable market transformation will be lost.

Second, we need to move up the effective date of the incentives from January 1, 2006. Under current law consumers that want to put in a better furnace or new windows need to wait until next year, well into the winter heating season, if they want to take advantage of the incentives. We already have begun consumer education programs aimed at winter heating; however, we are reluctant to inform consumers of the "soon-to-be available" incentives, as purchases that are important to managing energy use and costs this winter may be delayed until the current effective date of the incentives. This creates a conundrum, as the incentives are an important tool to change consumer behavior, but represent a potential barrier to immediate action, which is what we are seeking to encourage.

Third, we need to extend the incentives beyond the two years most are scheduled to last. It is almost impossible to plan and build a commercial building in two years, so large segments of the market are effectively excluded from the incentive by the short time horizon. For the best-selling hybrid vehicles, the tax incentives may have an even shorter horizon, as the law includes a per-manufacturer phase-out triggered by the sale of 60,000 eligible vehicles. The incentives were mostly planned to last four to five years, and their effectiveness will be multiplied if the eligibility is extended and the manufacturer vehicle volume cap is removed or increased.

Appliance standards: National appliance and equipment efficiency standards provide an efficiency baseline that American consumers can trust, provide uniform national rules for manufacturers, and slash wasteful energy consumption with one broad and effective stroke. The federal appliance standards program has been among the most effective of all efficiency measures. The program already has saved an estimated 2.5 percent of all U.S. electricity use and saved consumers billions of dollars in energy bills.

The energy bill includes a package of fifteen new energy-efficiency standards that were negotiated between energy-efficiency advocates, product manufacturers, and states. DOE is required to set standards for additional products, as well as to update many of the standards set in law.

The Alliance is pleased that DOE recently codified the legislated standards in rules. However, we remain very concerned that DOE is years behind the statutory deadline for setting about 20 standards required under previous laws. For example, an updated standard for residential furnaces and boilers was due in 1994. This is one of DOE's highest priorities as it is one of the most effective ways to save natural gas. The most recent delay, announced last December, means that DOE will not set this standard until late 2007 at the earliest, and that the standards will not go into effect until at least 2010. According to the American Council for an Energy-Efficient Economy, each year of delay in just three of these national standards—residential furnaces and boilers, distribution transformers, and commercial air conditioners and heat pumps—has locked in $7.1 billion in higher energy costs for consumers and businesses.

Largely due to the delays in the DOE program, a number of states are setting state standards on products not regulated by the federal government in order to reduce the cost, reduce the environmental impact, and increase the reliability of their energy systems. In addition, the work on state standards has been a key incentive to reaching agreements on federal standards.

The new energy bill adds additional standards to DOE's list of responsibilities. Even the legislated standards require test procedures that were not included in DOE's recent rulemaking. This program requires effective and vigilant oversight. In addition, as establishing standards requires a rigorous, time consuming, and costly rulemaking process, increased funding to the

DOE standards program is critical to ensuring that the enormous potential of this program is achieved.

. . .

Building codes assistance: While residential and commercial building codes are implemented at the state level, the states rely on DOE for technical specifications, training, and implementation assistance. Full adoption of and compliance with up-to-date building codes could save almost as much energy as appliance standards. The energy bill includes an authorization of $25 million per year for building codes assistance to states. Part of the funding increase would be for a new program to encourage states to adopt the latest codes and then assist them in achieving high rates of compliance. Such assistance is especially critical in the Gulf states to ensure that the massive rebuilding in the wake of the hurricanes is performed at least to minimally acceptable efficiency standards. We urge funding for this program.

In addition, we are concerned that DOE is significantly behind in providing information and guidance to the states on both residential and commercial building energy codes. DOE is required within one year of a residential or commercial model energy code update to make a determination on whether that update save energy; however, DOE still has not made the required determinations on the 2003 residential IECC, the 2004 Supplement, the newly adopted 2006 IECC, the 2001 ASHRAE commercial standard, or the 2004 ASHRAE standard. DOE must applv the necessary human and financial resources to ensure timely determinations on the codes.

State and utility energy-efficiency programs: Over the last two decades, states worked with regulated utilities using "Integrated Resource Planning" and demand-side management programs to avoid the need for about 100 300-Megawatt (MW) power plants. However, utility spending on public benefit programs nationwide has been cut significantly since the mid-1990's. In recent years some states have adopted innovative policies to rebuild these programs, including public benefits funds, energy efficiency performance standards, incentive rate structures, and priority in infrastructure planning. But the benefits of these programs have not spread to many other states.

The energy bill requires a study by DOE along with the National Association of Regulatory Utility Commissioners (NARUC) and the National Association of State Energy Officials (NASEO) of "best practices" among the states in demand side management (DSM) and other energy efficiency resource programs (Sec. 139). In addition, it authorizes $5 million per year for an innovative new pilot program to provide funding assistance to several states (3 to 7) to assist in the design and implementation of energy efficiency resource programs designed to lower electricity and natural gas demand by 0.75% a year. Again, funding is needed for this program.

Federal energy management: America's largest, single energy consumer is the federal government. According to the 1998 Alliance to Save Energy report, Leading by Example: Improving Energy Productivity in Federal Government Facilities, the federal government wastes $1 billion in taxpayer dollars each year on its buildings that use energy inefficiently.

DOE's Federal Energy Management Program (FEMP) is a rare example of a program that actually saves the government money. At an average cost of $20 million per year, FEMP has helped cut federal building energy waste by nearly 21 percent from 1985-1999—a reduction that now saves federal taxpayers roughly $1 billion each year in reduced energy costs. However, much more can be done, and the added targets, standards, and authorities in the energy bill will help.

We are especially pleased that the energy bill extended authority for Energy Savings Performance Contracts (ESPC5) through FY 2016. This unique program allows federal agencies to contract with the private sector to upgrade the efficiency of federal buildings. The contractors put up the money for the improvements and are paid back out of the utility bill savings. By law the payments can be no more than the savings. The agency saves energy at no additional cost, the companies build their business and create jobs, and the government saves money and pollution. Unfortunately, the ESPC program is still trying to get back on its feet after its authorization lapsed for a year in 2004.

The advice and assistance of FEMP is critical to the success of this program. FEMP support also is necessary for successful implementation of other federal energy management provisions in the energy bill—to provide guidance on building metering ("You can't manage what you can't measure"), help agencies comply with the product procurement rules, and help agencies meet the overall energy reduction targets. Without FEMP's support, the federal energy management title probably is not worth the paper it is printed on. More funding is needed to ramp up ESPC use and to undertake these other activities.

The Time Is Now to Move Beyond the Energy Bill

Although the energy bill includes a number of programs with the potential to save natural gas, of course many other effective policies were not included. And even though the major authors of the bill all cited high gasoline prices as a key rationale for the bill, there is a gaping hole where transportation policies should be. The Alliance estimates that the energy bill will save virtually no oil-small savings from the hybrid tax credit and other policies will be roughly canceled out by the extension of the fuel economy standard loophole for "dual-fueled" vehicles. This hole was noted by virtually every major editorial page in the country, and even noted by the authors of the bill as gasoline prices jumped even higher in the wake of Hurricane Katrina.

We cannot afford to wait thirteen years for another energy bill to fill in this hole, or another thirty years for an effective transportation policy. The economic, environmental, and national security costs of our insatiable oil demand are too high. While the Alliance recognizes that politically this is one of the most difficult areas to address, we must act now to bring our oil use under control.

Additional Opportunities for Oil Savings

The Alliance recommends consideration of two policies that could be as effective as a straightforward increase in Corporate Average Fuel Economy (CAFE) standards, and we hope will be more politically palatable: Close CAFE loopholes: CAFE standards passed by Congress in 1975 led to a 70 percent increase in America's gas mileage over the subsequent decade. However, CAFE standards have remained static for almost two decades due to political gridlock. The current standards of 27.5 miles per gallon for automobiles and 21 mpg for light trucks are roughly the same as in the mid-1980s. Furthermore, real on-road fuel economies are much lower than those numbers would suggest—the average fuel economy of cars and light trucks is only around 20 mpg. And as the sales of SUVs have exploded, average vehicle fuel economy has actually declined since 1988. Even if the political will to raise CAFE standard numbers does not exist, there are several reforms that could close major loopholes and thereby bring actual fuel economies closer to standards already required under existing law:

—"Truth-in-testing" loophole: By law, CAFE is based on the fuel economy tests that were used for model year 1975. EPA recognized that those tests are inaccurate, and in 1984 started reducing reported fuel economies by about 15%. Because driving patterns have changed, real gas mileage is likely 20–25% below CAFE numbers. Testing procedures for CAFE need to be updated to reflect increased congestion, higher speed limits, use of air conditioning, more powerful vehicles, and other changes.

—"SUV" loophole: When light trucks were given a lower standard, pickup trucks and vans were used primarily for businesses and farming, and represented only about 20% of vehicles sold. Today, about half of all light-duty vehicles sold in America qualify as "light trucks" for CAFE. Most of those are SUVs and minivans, most are used as passenger or family vehicles, and they average roughly 40% more fuel for each mile driven than the average passenger car. SUVs and minivans should be reclassified as what they are: passenger vehicles.

—"Hummer" loophole: CAFE standards only apply to vehicles under 8,500 pounds (gross vehicle weight). In fact, EPA does not even test or report the fuel economy of larger vehicles, yet their mileage is generally much lower. Manufacturers are selling more and more of these super-large SUVs and pickup trucks, such as GM Hummers and Ford Excursions. CAFE standards should cover these heavier vehicles.

—"Dual fuel" loophole: Automakers that produce vehicles that can run either on gasoline or on an alternative fuel, usually ethanol, can claim CAFE credit as if the vehicles ran on the alternative fuel one-half of the time. Unfortunately, dual fueled vehicles today run on gasoline 99% of the time. With only a few hundred ethanol fueling stations, the infrastructure does not exist to supply these vehicles with ethanol. This credit has allowed manufacturers to put more gas guzzlers on the road, and thus increases gasoline use. It should be modified to require actual use of the alternative fuel.

Vehicle fuel use "feebate": A new, innovative approach to efficiency of cars and light trucks is a national "feebate" system. Such a system would impose a national security surcharge, or "fee" on inefficient vehicles, and then use the funds collected to provide a "rebate" to fuel efficient vehicles.

How would a national feebate work? In one approach, a fee or rebate would apply to manufacturers of all new light-duty passenger vehicles-including SUVs and minivans. The amount would be based on 25 cents per gallon of gasoline estimated to be used over the lifetime of the vehicle. The fee or rebate would then be determined relative to a mid-point fuel economy. This dividing line between fees and rebates would be set each year such that the total fees would just pay for all the rebates, so there would be no net revenue or cost to the government. A feebate would create an incentive for manufacturers to use fuel-efficient technologies in the vehicles they produce, and hence should increase the availability of efficient vehicles, as well as creating an incentive for consumers to purchase more efficient vehicles. As fuel economies increased, the mid-point fuel economy would be ratcheted up, creating an incentive for continual improvement, but never out of line with the existing market.

Additional Opportunities for Natural Gas Savings

As I mentioned, there also are a number of other policies to save natural gas that should be reconsidered in light of sharply higher natural gas prices. The Alliance recommends consideration of two additional sets of policies.

Federal building codes: Although the energy bill requires new, stricter standards for energy efficiency in buildings owned by the federal government, it passed over several opportunities to improve standards for buildings regulated by or paid for by the federal government. These standards could help transform the housing market and make the federal government into a market leader rather than a market laggard. They include:

—Manufactured housing: Even before the hurricanes, "mobile homes" accounted for 131,000 buildings last year, about one in twelve new homes. Because they are manufactured in central factories, they are regulated not by the states but by the federal Department of Housing and Urban Development (HUD). Like many states, HUD is years behind in adopting up-to-date building codes-their standard has not been modified since 1996. These buildings are used like site-built homes. There is no reason they should not meet the same current model energy code. Setting this floor would reduce the energy bills of mobile home owners, many of whom are low income and many of whom rely on federal LIHEAP assistance, by 9 percent.—Federally subsidized housing: New public housing and new housing with federally assisted mortgages also must meet a federal standard, currently the 1992 Model Energy Code as set in the Energy Policy Act of 1992. This standard should be updated to the most recent model codes. Rebuilding with federal subsidies in the wake of the recent hurricanes and other natural disasters also should be subject to a federal standard to ensure recipients receive high-quality homes and that neither the victims nor the federal government pay for unnecessarily high energy bills.

To ensure cost-effective energy savings based on criteria with which local builders and manufacturers are already familiar, both manufactured and site-built homes built with federal disaster assistance should qualify for the Energy Star label.

—Privatized military housing: About 37,000 units of housing are being built each year with federal assistance in order to move service members out of the barracks and into newer private housing. The federal government indirectly pays the energy bills through an energy allowance. We should require that this housing too qualify for the Energy Star label.

State and utility energy-efficiency programs: As described above, a number of states are implementing innovating energy-efficiency policies and funding mechanisms. Several states have recently passed an energy efficiency resource standard (EERS), requiring electricity utilities to meet customer needs in part through demand-side management (energy efficiency and load reduction) programs rather than by constructing new facilities and purchasing energy. An EERS sets a specific target for demand or use reduction due to DSM programs, and requires monitoring and verification of the program savings. The programs have generally been found to save electricity much more cheaply than it could be generated and delivered. Several of the states are now implementing an EERS as part of or alongside renewable electricity generation standards.

. . .

Conclusion

. . .

With respect to energy efficiency provisions, which must be a cornerstone of any such energy policy, we hope you will both work to ensure the fine provisions of the last energy bill are fully funded and implemented, and use the increasing pressure for action to fill the gaping holes in that bill. Additional administrative and legislative action to save natural gas and oil is the only way to assure that we give the American people immediate, cost-effective and sustainable assistance in addressing spiraling energy costs and an ever-less secure energy future.

Source: *Federal Document Clearing House Congressional Testimony*. Washington, D.C.: Congressional Quarterly, Inc. November 2, 2005.

ENERGY'S ENVIRONMENTAL IMPACT

Clean Air Act of 1990

The Clean Air Act is *the chief U.S. law instituted to improve air quality in the nation. It established regulations whose purpose was to minimize the impact of harmful pollutants in the United States, including those emitted by electric power plants and automobiles. First passed as the Air Pollution Control Act in 1955, the law was reinstituted in 1970 as the Clean Air Act with the essential features that it has today. It was also revised in 1977, 1990, and 1997. Under the act, the U.S. Environmental Protection Agency (EPA) sets limits on the amount of emissions allowable anywhere in the United States from pollutants considered to be primary causes of ozone depletion and harm to the environment. These pollutants include sulfur dioxide, ground-level ozone, nitrogen oxides, carbon monoxide, particulates, and volatile organic compounds. The law gives each state the authority to develop its own guidelines regarding how that state will clean up polluted areas and improve air quality, basing such regulations on the needs of that particular state's industries, geography, population patterns, and so on—but states' regulations are prohibited from being any weaker than the original requirements of the act. The 1970 version of the act mandated inspections of new or remodeled coal-burning power plants to ensure that they were utilizing the cleanest-burning technologies possible to limit emission of sulfur dioxide. This requirement, known as the new source review, did not apply to older, unremodeled plants. Many people felt that this exemption encouraged the power industry to keep old, heavily polluting plants in operation. President George W. Bush waived the new source review requirement for remodeled plants* *altogether in 2003. Environmentalists viewed this decision as one that would further increase pollution from coal. In addition, the 1990 version of the act set limits on the amount of sulfur allowable in diesel fuel (later, this requirement was extended to gasoline). It also contained a provision to reduce carbon monoxide in certain states in the winter, calling for refiners to add oxygenating (oxygen-containing) compounds to gasoline in those states during that season. The 1990 amendments also established a new emissions-reduction program, with a goal of reducing sulfur dioxide emissions by 10 million tons and annual nitrogen oxide emissions by 2 million tons from 1980 levels for all man-made sources.*

CLEAN AIR ACT OF 1990

Public Law 10-549

101st Congress

TITLE I – AIR POLLUTION PREVENTION AND CONTROL

Part A – Air Quality and Emission Limitations

FINDINGS AND PURPOSES

Sec. 101. (a) The Congress finds—

(1) that the predominant part of the Nation's population is located in its rapidly expanding metropolitan and other urban areas, which generally cross the boundary lines of local jurisdictions and often extend into two or more States;

(2) that the growth in the amount and complexity of air pollution brought about by urbanization, industrial development, and the increasing use of motor vehicles, has resulted in mounting dangers to the public health and welfare, including injury to agricultural crops and livestock, damage to and the deterioration of property, and hazards to air and ground transportation;

(3) that air pollution prevention (that is, the reduction or elimination, through any measures, of the amount of pollutants produced or created at the source) and air pollution control at its source is the primary responsibility of States and local governments; and

(4) that Federal financial assistance and leadership is essential for the development of cooperative Federal, State, regional, and local programs to prevent and control air pollution.

(b) The purposes of this title are—

(1) to protect and enhance the quality of the Nation's air resources so as to promote the public health and welfare and the productive capacity of its population;

(2) to initiate and accelerate a national research and development program to achieve the prevention and control of air pollution;

(3) to provide technical and financial assistance to State and local governments in connection with the development and execution of their air pollution prevention and control programs; and

(4) to encourage and assist the development and operation of regional air pollution prevention and control programs.

(c) Pollution Prevention.—A primary goal of this Act is to encourage or otherwise promote reasonable Federal, State, and local governmental actions, consistent with the provisions of this Act, for pollution prevention.

....

RESEARCH RELATING TO FUELS AND VEHICLES

Sec. 104. (a) The Administrator shall give special emphasis to research and development into new and improved methods, having industry wide application, for the prevention and control of air pollution resulting from the combustion of fuels. In furtherance of such research and development he shall—

(1) conduct and accelerate research programs directed toward development of improved, cost-effective techniques for—

(A) control of combustion byproducts of fuels,

(B) removal of potential air pollutants from fuels prior to combustion,

(C) control of emissions from the evaporation of fuels,

(D) improving the efficiency of fuels combustion so as to decrease atmospheric emissions, and

(E) producing synthetic or new fuels which, when used, result in decreased atmospheric emissions.

(2) provide for Federal grants to public or nonprofit agencies, institutions, and organizations and to individuals, and contracts with public or private agencies, institutions or persons, for payment of (A) part of the cost of acquiring, constructing, or otherwise securing for research and development purposes, new or improved devices or methods having industry wide application of preventing or controlling discharges into the air of various types of pollutants; (B) part of the cost of programs to develop low emission alternatives to the present internal combustion engine; (C) the cost to purchase vehicles and vehicle engines, or portions thereof, for research, development, and testing purposes; and (D) carrying out the other provisions of this section, without regard to sections 3648 and 3709 of the Revised Statutes (31 U.S.C. 529; 41 U.S.C. 5): Provided, That research or demonstration contracts awarded pursuant to this subsection or demonstration contracts awarded pursuant to this subsection (including contracts for construction) may be made in accordance with, and subject to the limitations provided with respect to research contracts of the

military departments in, section 2353 of title 10, United States Code, except that the determination, approval, and certification required thereby shall be made by the Administrator: Provided further, That no grant may be made under this paragraph in excess of $1,500,000;

(3) determine, by laboratory and pilot plant testing, the results of air pollution research and studies in order to develop new or improved processes and plant designs to the point where they can be demonstrated on a large and practical scale;

(4) construct, operate, and maintain, or assist in meeting the cost of the construction, operation, and maintenance of new or improved demonstration plants or processes which have promise of accomplishing the purposes of this Act;

(5) study new or improved methods for the recovery and marketing of commercially valuable byproducts resulting from the removal of pollutants.

(b) In carrying out the provisions of this section, the Administrator may -

(1) conduct and accelerate research and development of cost-effective instrumentation techniques to facilitate determination of quantity and quality of air pollutant emissions, including, but not limited to, automotive emissions;

(2) utilize, on a reimbursable basis, the facilities of existing Federal scientific laboratories;

(3) establish and operate necessary facilities and test sites at which to carry on the research, testing, development, and programming necessary to effectuate the purposes of this section;

(4) acquire secret processes, technical data, inventions, patent applications, patents, licenses, and an interest in lands, plants, and facilities, and other property or rights by purchase, license, lease, or donation; and

(5) cause on-site inspections to be made of promising domestic and foreign projects, and cooperate and participate in their development in instances in which the purposes of the Act will be served thereby.

(c) Clean Alternative Fuels.—The Administrator shall conduct a research program to identify, characterize, and predict air emissions related to the production, distribution, storage, and use of clean alternative fuels to determine the risks and benefits to human health and the environment relative to those from using conventional gasoline and diesel fuels. The Administrator shall consult with other Federal agencies to ensure coordination and to avoid duplication of activities authorized under this subsection.

....

AIR QUALITY CONTROL REGIONS

Sec. 107. (a) Each State shall have the primary responsibility for assuring air quality within the entire geographic area comprising such State by submitting an implementation plan for such State which will specify the manner in which national primary and secondary ambient air quality standards will be achieved and maintained within each air quality control region in such State.

(b) For purposes of developing and carrying out implementation plans under section 110—

(1) an air quality control region designated under this section before the date of enactment of the Clean Air Amendments of 1970, or a region designated after such date under subsection (c), shall be an air quality control region; and

(2) the portion of such State which is not part of any such designated region shall be an air quality control region, but such portion may be subdivided by the State into two or more air quality control regions with the approval of the Administrator.

(c) The Administrator shall, within 90 days after the date of enactment of the Clean Air Amendments of 1970, after consultation with appropriate State and local authorities, designate as an air quality control region any interstate area or major intrastate area which he deems necessary or appropriate for the attainment and maintenance of ambient air quality standards. The Administrator shall immediately notify the Governors of the affected States of any designation made under this subsection.

* *** *

Source: Excerpted from the 1990 Clean Air Act as amended in 1990 is available on the EPA's Web site at http://www.thecre.com/fedlaw/legal14air/contents.htm2.

President Bush Discusses Global Climate Change (June 11, 2001)

President George W. Bush did not ratify the Kyoto Protocol, which would make it mandatory for the United States to abide by. Bush has objected to its exclusion of developing countries from the reduction, which is based on the fact that industrialization was considered vital to their economies. The president has said that the reduction in emissions would seriously harm the country's economy and that the protocol is too lenient on developing countries whose consumption of these fuels is on the rise. See below for President Bush's speech on global climate change, given on June 11, 2001.

President Bush Discusses Global Climate Change, June 11, 2001

THE PRESIDENT: Good morning. I've just met with senior members of my administration who are working to develop an effective and science-based approach to addressing the important issues of global climate change.

This is an issue that I know is very important to the nations of Europe, which I will be visiting for the first time as President. The earth's well-being is also an issue important to America. And it's an issue that should be important to every nation in every part of our world.

The issue of climate change respects no border. Its effects cannot be reined in by an army nor advanced by any ideology. Climate change, with its potential to impact every corner of the world, is an issue that must be addressed by the world.

The Kyoto Protocol was fatally flawed in fundamental ways. But the process used to bring nations together to discuss our joint response to climate change is an important one. That is why I am today committing the United States of America to work within the United Nations framework and elsewhere to develop with our friends and allies and nations throughout the world an effective and science-based response to the issue of global warming.

My Cabinet-level working group has met regularly for the last 10 weeks to review the most recent, most accurate, and most comprehensive science. They have heard from scientists offering a wide spectrum of views. They have reviewed the facts, and they have listened to many theories and suppositions. The working group asked the highly-respected National Academy of Sciences to provide us the most up-to-date information about what is known and about what is not known on the science of climate change.

First, we know the surface temperature of the earth is warming. It has risen by .6 degrees Celsius over the past 100 years. There was a warming trend from the 1890s to the 1940s. Cooling from the 1940s to the 1970s. And then sharply rising temperatures from the 1970s to today.

There is a natural greenhouse effect that contributes to warming. Greenhouse gases trap heat, and thus warm the earth because they prevent a significant proportion of infrared radiation from escaping into space. Concentration of greenhouse gases, especially C02, have increased substantially since the beginning of the industrial revolution. And the National Academy of Sciences indicate that the increase is due in large part to human activity.

Yet, the Academy's report tells us that we do not know how much effect natural fluctuations in climate may have had on warming. We do not know how much our climate could, or will change in the future. We do not know how fast change will occur, or even how some of our actions could impact it.

For example, our useful efforts to reduce sulfur emissions may have actually increased warming, because sulfate particles reflect sunlight, bouncing it back into space. And, finally, no one can say with any certainty what constitutes a dangerous level of warming, and therefore what level must be avoided.

The policy challenge is to act in a serious and sensible way, given the limits of our knowledge. While scientific uncertainties remain, we can begin now to address the factors that contribute to climate change.

There are only two ways to stabilize concentration of greenhouse gases. One is to avoid emitting them in the first place; the other is to try to capture them after they're created. And there are problems with both approaches. We're making great progress through technology, but have not yet developed cost-effective ways to capture carbon emissions at their source; although there is some promising work that is being done.

And a growing population requires more energy to heat and cool our homes, more gas to drive our cars. Even though we're making progress on conservation and energy efficiency and have significantly reduced the amount of carbon emissions per unit of GDP.

Our country, the United States is the world's largest emitter of man-made greenhouse gases. We account for almost 20 percent of the world's man-made greenhouse emissions. We also account for about one-quarter of the world's economic output. We recognize the responsibility to reduce our emissions. We also recognize the other part of the story—that the rest of the world emits 80 percent of all greenhouse gases. And many of those emissions come from developing countries.

This is a challenge that requires a 100 percent effort; ours, and the rest of the world's. The world's second-largest emitter of greenhouse gases is China. Yet, China was entirely exempted from the requirements of the Kyoto Protocol.

India and Germany are among the top emitters. Yet, India was also exempt from Kyoto. These and other developing countries that are experiencing rapid growth face challenges in reducing their emissions without harming their economies. We want to work cooperatively with these countries in their efforts to reduce greenhouse emissions and maintain economic growth.

Kyoto also failed to address two major pollutants that have an impact on warming: black soot and tropospheric ozone. Both are proven health hazards. Reducing both would not only address climate change, but also dramatically improve people's health.

Kyoto is, in many ways, unrealistic. Many countries cannot meet their Kyoto targets. The targets themselves were arbitrary and not based upon science. For America, complying with those mandates would have a negative economic impact, with layoffs of workers and price increases for consumers. And when you evaluate all these flaws, most reasonable people will understand that it's not sound public policy.

That's why 95 members of the United States Senate expressed a reluctance to endorse such an approach. Yet, America's unwillingness to embrace a flawed treaty should not be read by our friends and allies as any abdication of responsibility. To the contrary, my administration is committed to a leadership role on the issue of climate change.

We recognize our responsibility and will meet it—at home, in our hemisphere, and in the world. My Cabinet-level working group on climate change is recommending a number of initial steps, and will continue to work on additional ideas. The working group proposes the United States help lead the way by advancing the science on climate change, advancing the technology to monitor and reduce greenhouse gases, and creating partnerships within our hemisphere and beyond to monitor and measure and mitigate emissions.

I also call on Congress to work with my administration to achieve the significant emission reductions made possible by implementing the clean energy technologies proposed in our energy plan. Our working group study has made it clear that we need to know a lot more.

The U.N. Framework Convention on Climate Change commences to stabilizing concentrations at a level that will prevent dangerous human interference with the climate; but no one knows what that level is. The United States has spent $18 billion on climate research since 1990—three times as much as any other country, and more than Japan and all 15 nations of the EU combined.

Today, I make our investment in science even greater. My administration will establish the U.S. Climate Change Research Initiative to study areas of uncertainty and identify priority areas where investments can make a difference.

I'm directing my Secretary of Commerce, working with other agencies, to set priorities for additional investments in climate change research, review such investments, and to improve coordination amongst federal agencies. We will fully fund high-priority areas for climate change science over the next five years. We'll also provide resources to build climate observation systems in developing countries and encourage other developed nations to match our American commitment.

And we propose a joint venture with the EU, Japan and others to develop state-of-the-art climate modeling that will help us better understand the causes and impacts of climate change. America's the leader in technology and innovation. We all believe technology offers great promise to significantly reduce emissions—especially carbon capture, storage and sequestration technologies.

So we're creating the National Climate Change Technology Initiative to strengthen research at universities and national labs, to enhance partnerships in applied research, to develop improved technology for measuring and monitoring gross and net greenhouse gas emissions, and to fund demonstration projects for cutting-edge technologies, such as bioreactors and fuel cells.

Even with the best science, even with the best technology, we all know the United States cannot solve this global problem alone. We're building partnerships within the Western Hemisphere and with other like-minded countries. Last week, Secretary Powell signed a new CONCAUSA Declaration with the countries of Central America, calling for cooperative efforts on science research, monitoring and measuring of emissions, technology development, and investment in forest conservation.

We will work with the Inter-American Institute for Global Change Research and other institutions to better understand regional impacts of climate change. We will establish a partnership to monitor and mitigate emissions. And at home, I call on Congress to work with my administration on the initiatives to enhance conservation and energy efficiency outlined in my energy plan, to implement the increased use of renewables, natural gas and hydropower that are outlined in the plan, and to increase the generation of safe and clean nuclear power.

By increasing conservation and energy efficiency and aggressively using these clean energy technologies, we can reduce our greenhouse gas emissions by significant amounts in the coming years. We can make great progress in reducing emissions, and we will. Yet, even that isn't enough.

I've asked my advisors to consider approaches to reduce greenhouse gas emissions, including those that tap the power of markets, help realize the promise of technology and ensure the widest-possible global participation. As we analyze the possibilities, we will be guided by several basic principles. Our approach must be consistent with the long-term goal of stabilizing greenhouse gas concentrations in the atmosphere. Our actions should be measured as we learn more from science and build on it.

Our approach must be flexible to adjust to new information and take advantage of new technology. We must always act to ensure continued economic growth and prosperity for our citizens and for citizens throughout the world. We should pursue market-based incentives and spur technological innovation.

And, finally, our approach must be based on global participation, including that of developing countries whose net greenhouse gas emissions now exceed those in the developed countries.

....

I look forward to continued discussions with our friends and allies about this important issue.

Thank you for coming.

Source: Excerpted from Web site of the White House (www.whitehouse.gov/news/releases/2001/06/20010611-2.html).

Asia-Pacific Partnership on Clean Development and Climate, Inaugural Ministerial Meeting, Sydney: Charter for the Asia-Pacific Partnership on Clean Development and Climate (January 12, 2006)

The Charter of the Asia-Pacific Partnership on Clean Development and Climate (AP6) was ratified at the inaugural ministerial meeting of the Asia-Pacific Partnership on Clean Development and Climate in Sydney, Australia, on January 12, 2006. The AP6 is a nontreaty agreement between the United States, Australia, India, Japan, China, and South Korea to cooperate on developing and implementing technology to reduce greenhouse gas emissions. Member countries, which are responsible for about 50 percent of global greenhouse gas emissions, first announced their partnership on July 28, 2005, at an Association of South East Asian Nations (ASEAN) Regional Forum meeting. On January 12, 2006, the member countries agreed to a formal charter, communiqué, and a work plan for addressing climate change, energy security, and air pollution. The AP6 agreement allows countries to set their own goals for reducing emissions, and there is no mechanism for mandating compliance. Because AP6, unlike the Kyoto Protocol, imposes no mandatory limits on greenhouse gas emissions, environmentalists and nations that have ratified the Kyoto Protocol have called AP6 meaningless.

Asia-Pacific Partnership on Clean Development and Climate, Inaugural Ministerial Meeting, Sydney, January 12, 2006: Charter for the Asia-Pacific Partnership on Clean Development and Climate

We, the representatives of the national governments of Australia, China, India, Japan, the Republic of Korea, and the United States of America (collectively referred to as the "Partners"), meeting in Sydney, Australia on 12 January 2006:

Guided by our Vision Statement for a New Asia-Pacific Partnership on Clean Development and Climate of 28 July 2005 (Annex I), which is an integral part of this Charter;

Bearing in mind that the purposes of the partnership are consistent with the principles of the United Nations Framework Convention on Climate Change and other relevant international instruments, and are intended to complement but not replace the Kyoto Protocol;

Decide to create the Asia-Pacific Partnership on Clean Development and Climate (referred to as the "Partnership") and set forth the following non-legally binding Charter for the Partnership. This Partnership will serve as a framework for supporting agile, constructive, and productive international cooperation among the Partners to meet our development, energy, environment, and climate change objectives.

1 Shared Vision

1.1 The Partners have come together voluntarily to advance clean development and climate objectives, recognizing that development and poverty eradication are urgent and overriding goals internationally. By building on the foundation of existing bilateral and multilateral initiatives, the Partners will enhance cooperation to meet both our increased energy needs and associated challenges, including those related to air pollution, energy security, and greenhouse gas intensities, in accordance with national circumstances. The Partners recognize that national efforts will also be important in meeting the Partnership's shared vision.

2 Purposes

2.1 The purposes of the Partnership are to:

2.1.1 Create a voluntary, non-legally binding framework for international cooperation to facilitate the development, diffusion, deployment, and transfer of existing, emerging and longer term cost-effective, cleaner, more efficient technologies and practices among the Partners through concrete and substantial cooperation so as to achieve practical results;

2.1.2 Promote and create enabling environments to assist in such efforts;

2.1.3 Facilitate attainment of our respective national pollution reduction, energy security and climate change objectives; and

2.1.4 Provide a forum for exploring the Partners' respective policy approaches relevant to addressing interlinked development, energy, environment, and climate change issues within the context of clean development goals, and for sharing experiences in developing and implementing respective national development and energy strategies.

3 Functions

3.1 Through this Partnership, the Partners are to cooperate to:

3.1.1 Exchange information on Partners' respective policy approaches relevant to addressing interlinked development, energy, environment, and climate change issues within the context of clean development, including any gaps and overlaps in national policy approaches, as well as other areas of mutual interest;

3.1.2 Share experiences and exchange information about developing and implementing national clean development strategies and efforts to reduce greenhouse gas intensities;

3.1.3 Identify, assess, and address barriers to the promotion and creation of an enabling environment for development, diffusion, deployment, and transfer of existing, emerging and longer term cost-effective, cleaner, more efficient, and transformational technologies and practices in accordance with the Partners' priorities;

3.1.4 Identify and implement bilateral and multilateral cooperative activities among Partners for the development, deployment, diffusion, and transfer of existing, emerging and longer term cost-effective, cleaner, more efficient, and transformational technologies, in accordance with the Partners' priorities;

3.1.5 Facilitate collaboration among existing bilateral and multilateral initiatives and promote information-sharing on climate-related technologies of respective Partners;

3.1.6 Incorporate human and institutional capacity-building elements, as appropriate, into activities as a means to strengthen cooperative efforts;

3.1.7 Engage the private sector as an integral part of the cooperative activities of the Partnership, as well as development banks, research institutions, and other relevant governmental, intergovernmental, and non-governmental organizations, as appropriate;

3.1.8 Develop and implement work programs decided by the Partners; and

3.1.9 Assess regularly the progress of the Partnership to ensure its effectiveness.

3.2 Each Partner will undertake activities contemplated by this Charter in accordance with the laws, regulations, and policies under which it operates and applicable international instruments to which it is a party.

4 Organization

4.1 A Policy and Implementation Committee and an Administrative Support Group will be formed to facilitate implementation of the Partnership.

4.2 The Policy and Implementation Committee will govern the overall framework, policies, and procedures of the Partnership, periodically review progress of collaboration, and provide direction to the Administrative Support Group. It will be responsible for management of the implementation of the cooperative activities of the Partnership, and for engaging representatives of the private sector, as well as representatives of development banks, research institutions, and other relevant governmental, intergovernmental, and non-governmental organizations, as appropriate. It will undertake activities in the promotion and creation of enabling environments within Partners and in support of Partners' efforts to meet relevant national-level clean development objectives. The Policy and Implementation Committee may form appropriate task forces and other subgroups to assist it in its work. The Policy and Implementation Committee should meet as often as is determined necessary by its members to accomplish its work, and may focus its agenda on policy issues or technical issues, or both, as appropriate. Policy and Implementation Committee decisions are to be made by consensus of the Partners on the Committee.

4.3 The Administrative Support Group, which serves as the principal coordinator of the Partnership's communications and activities, will be responsible for: (1) organizing meetings of the Partnership; (2) arranging special activities, such as teleconferences and workshops; (3) coordinating and communicating information regarding actions of the Partnership; (4) serving as a clearinghouse of information regarding the Partnership; (5) maintaining procedures and responsibilities for key functions that are approved by the Policy and Implementation Committee; and (6) performing such other tasks as the Policy and Implementation Committee directs. The Administrative Support Group's function will be administrative in nature, and will not include matters of substance except as specifically instructed by the Policy and Implementation Committee.

4.4 The Policy and Implementation Committee comprises representatives from Partners. Each Partner included in Annex II may designate up to three representatives to meetings of the Policy and Implementation Committee.

4.5 The Policy and Implementation Committee may, at its discretion, permit other experts to attend its meetings.

4.6 The United States Government is to serve initially as the Partnership's Administrative Support Group. This arrangement will be reviewed at two year intervals and may be changed by decision of the Policy and Implementation Committee. Each Partner will designate an administrative liaison to serve as its principal point of contact for the Administrative Support Group.

4.7 The Administrative Support Group may, as required, utilize the services of personnel employed by the Partners and made available to the Administrative Support Group. Unless otherwise determined by the Partners, such personnel are to be remunerated by their respective employers and remain subject to their employers' conditions of employment.

4.8 Each Partner will individually determine the nature of its participation in Partnership activities.

5 Funding

5.1 Participation in the Partnership is on a voluntary basis. Each Partner may, at its discretion, contribute funds, personnel, and other resources to the Partnership subject to the laws, regulations, and policies of the Partner. Any costs arising from the activities contemplated in this Charter are to be borne by the Partner that incurs them, unless other arrangements are made.

6 Intellectual Property

6.1 All matters related to intellectual property and the treatment thereof arising from cooperative activities of the Partnership are to be addressed on a case-by-case basis within the specific context in which they appear, bearing in mind the purposes of the Partnership.

7 Amendments

7.1 The Policy and Implementation Committee may amend this Charter and its Annex II at any time by consensus of the Partners on the Committee.

8 Term of Charter

8.1 Cooperation under this Charter will commence on 12 January 2006. Any Partner may terminate its membership upon written notice 90 days prior to the anticipated termination.

We will develop a non binding compact in which the elements of this shared vision, as well as the ways and means to implement it, will be further defined. In particular, we will consider establishing a framework for the partnership, including institutional and financial arrangements and ways to include other interested and like-minded countries.

The partnership will also help the partners build human and institutional capacity to strengthen cooperative efforts, and will seek opportunities to engage the private sector. We will review the partnership on a regular basis to ensure its effectiveness.

The partnership will be consistent with and contribute to our efforts under the UNFCCC and will complement, but not replace, the Kyoto Protocol.

Annex I

Vision Statement of Australia, China, India, Japan, the Republic of Korea, and the United States of America for a New Asia-Pacific Partnership on Clean Development and Climate

28 July 2005

Development and poverty eradication are urgent and overriding goals internationally. The World Summit on Sustainable Development made clear the need for increased access to affordable, reliable and cleaner energy and the international community agreed in the Delhi Declaration on Climate Change and Sustainable Development on the importance of the development agenda in considering any climate change approach.

We each have different natural resource endowments, and sustainable development and energy strategies, but we are already working together and will continue to work to achieve common goals. By building on the foundation of existing bilateral and multilateral initiatives, we will enhance cooperation to meet both our increased energy needs and associated challenges, including those related to air pollution, energy security, and greenhouse gas intensities.

To this end, we will work together, in accordance with our respective national circumstances, to create a new partnership to develop, deploy and transfer cleaner, more efficient technologies and to meet national pollution reduction, energy security and climate change concerns, consistent with the principles of the U.N. Framework Convention on Climate Change (UNFCCC).

The partnership will collaborate to promote and create an enabling environment for the development, diffusion, deployment and transfer of existing and emerging cost-effective, cleaner technologies and practices, through concrete and substantial cooperation so as to achieve practical results. Areas for collaboration may include, but not be limited to: energy efficiency, clean coal, integrated gasification combined cycle, liquefied natural gas, carbon capture and storage, combined heat and power, methane capture and use, civilian nuclear power, geothermal, rural/village energy systems, advanced transportation, building and home construction and operation, bioenergy, agriculture and forestry, hydropower, wind power, solar power, and other renewables.

The partnership will also cooperate on the development, diffusion, deployment and transfer of longer-term transformational energy technologies that will promote economic growth while enabling significant reductions in greenhouse gas intensities. Areas for mid- to long-term collaboration may include, but not be limited to: hydrogen, nanotechnologies, advanced biotechnologies, next-generation nuclear fission, and fusion energy.

The partnership will share experiences in developing and implementing our national sustainable development and energy strategies, and explore opportunities to reduce the greenhouse gas intensities of our economies.

Annex II

Australia
China
India
Japan

Republic of Korea
United States of America

Source: Excerpted from the Web site of the Australian Department of Foreign Affairs and Trade at http://www.dfat.gov.au/environment/climate/ap6/charter.html.

RENEWABLE AND ALTERNATIVE ENERGY TECHNOLOGIES

Advanced Energy Initiative, National Economic Council (February 2006)

Advanced Energy Initiative is a U.S. energy plan, details of which were released by the National Economic Council in February 2006 and announced by U.S. President George W. Bush during his January 31, 2006, State of the Union address. The Advanced Energy Initiative calls for greater investment in both coal-fired power plants with reduced polluting capacity and advanced nuclear energy designs focused on "clean and safe" technologies. In his State of the Union address President Bush also announced plans to increase research and development funding for the manufacture of batteries for hybrid and electric cars, pollution-free hydrogen-fueled cars, and ethanol fuels derived from corn, wood chips, stalks, or switch grass. (Ethanol is a fuel usually made from biomass; it is an alcohol that can be used in internal combustion engines that have been modified to run on it.)

ADVANCED ENERGY INITIATIVE
NATIONAL ECONOMIC COUNCIL
FEBRUARY 2006

* *** *

Since 2001, the Administration has spent nearly $10 billion to develop cleaner, cheaper, and more reliable alternative energy sources. As a result, America is on the verge of breakthroughs in advanced energy technologies that could transform the way we produce and use energy. To build on this progress, the President's *Advanced Energy Initiative* provides for a 22% increase in funding for clean-energy technology research at the Department of Energy in two vital areas:

1. Changing the way we fuel our vehicles. We can improve our energy security through greater use of technologies that reduce oil use by improving efficiency, expansion of alternative fuels from homegrown biomass, and development of fuel cells that use hydrogen from domestic feedstocks.

2. Changing the way we power our homes and businesses. We can address high costs of natural gas and electricity by generating more electricity from clean coal, advanced nuclear power, and renewable resources such as solar and wind.

Just as our current challenges did not arise overnight, neither will the solutions to these challenges. We must make a sustained commitment to addressing the fundamental causes of high and volatile energy prices, while protecting our national security and the environment. Through the *Advanced Energy Initiative,* we can take new, bold steps towards the goal of a reliable, affordable, and clean energy future for all Americans.

Changing the Way We Fuel Our Vehicles

Crude oil is used to produce a wide array of petroleum products, including gasoline, diesel and jet fuels, heating oil, lubricants, asphalt, plastics, and many other products. Not surprisingly, crude oil markets are monitored closely by consumers, businesses, and governments, because the prices of petroleum-based products depend heavily on the price of crude oil.

The transportation sector receives nearly all of its energy from petroleum products and accounts for two-thirds of U.S. petroleum use, mainly in the form of gasoline and diesel fuel. The Energy Information Administration (EIA) projects that consumption of gasoline and diesel fuel will continue to rise because the expected increase in total miles traveled will outweigh improvements in the efficiency of fuel use per mile traveled. This is projected to lead to a one-third increase in imports of crude oil and a near-doubling in imports of refined products over the next 25 years, such that these imports would account for 62.5% of our total oil use by 2030.

Advanced Energy Initiative Goals—Fueling Our Vehicles

- Develop advanced battery technologies that allow a plug-in hybrid-electric vehicle to have a 40-mile range operating solely on battery charge.
- Foster the breakthrough technologies needed to make cellulosic ethanol cost-competitive with corn-based ethanol by 2012.
- Accelerate progress towards the President's goal of enabling large numbers of Americans to choose hydrogen fuel cell vehicles by 2020.

In 2004, the U.S. consumed 20.7 million barrels of crude oil and refined products per day, approximately 58 percent of which were imported from other countries. About half of these imports come from non-OPEC nations, such as Canada and Mexico, while the other half come from OPEC nations, mainly Saudi Arabia, Venezuela, Nigeria, and Iraq. Oil supply disruptions pose a threat to our economy and national security, and that threat rises the more dependent we are on oil imports, particularly from less stable regions of the world.

...

Crude oil prices, which hovered in the $15-25 per barrel range from the mid-1980s until 2002, have been above $40 since February 2005. Many of the same factors that drove world oil markets during this time, such as low spare world-oil-production capacity and rapid world-oil-demand growth, will continue to affect markets in the near term. Other factors, such as geopolitical instability and weather, are also important but less predictable.

To reduce America's vulnerability to oil supply disruptions, and the associated economic hardship for our Nation's families and businesses, we must reduce our dependence on foreign sources of oil. This means increasing domestic production of oil here at home and expanding capacity to refine crude oil into products that consumers demand. It also means accelerating deployment of efficient hybrid and clean diesel vehicles in the near-term, developing domestic renewable alternatives to gasoline and diesel fuels in the mid-term, and investing in the advanced battery and hydrogen fuel-cell technologies needed for substantial long-term reductions in oil demand.

* *** *

Changing The Way We Power Our Homes And Businesses

Natural gas has numerous uses in homes, industry, commerce, electricity production, and transportation and is a vital component of fertilizer and chemical production. In 2004, the United States consumed 61 billion cubic feet of natural gas per day, primarily for industrial applications, residential use, and electric generation. At present, 85% of U.S. natural gas demand is met through domestic production; the remainder is met largely through transport by pipeline from Canada, with an increasing amount arriving by tanker in the form of liquefied natural gas (LNG) from Trinidad, Algeria, and other countries.

U.S. natural gas consumption is projected to grow to 74 billion cubic feet per day by 2025. Over the past decade, natural gas has been the "fuel of choice" for new natural gas combined-cycle power plants. Compared to coal-fired power plants, natural gas power plants emit less air pollution and cost less to build. As a result, demand for natural gas in the power sector has increased steadily over the past 15 years, even in the face of dramatically higher prices.

Wholesale natural gas prices at Henry Hub on the Louisiana Gulf coast (a common natural gas pricing benchmark) averaged around $2–$3 per thousand cubic feet from 1994 through the middle of 2000. Prices then spiked to a peak of $10.50 per thousand cubic feet in December of 2000 in response to an unusually cold winter and low hydroelectric production in the West before falling back to their previous low levels. Over the past four years, natural gas prices increased substantially from roughly $3 per thousand cubic feet in early 2002 to over $8 per thousand cubic feet recently, with a pronounced price spike due to Hurricanes Katrina and Rita. At present, natural gas prices track high crude oil prices because natural gas is often used as a substitute for oil in power production and heating. Furthermore, the tight balance between supply and demand has led to a more volatile market, which can respond dramatically to weather events and geopolitical developments.

Advanced Energy Initiative Coals—
Powering Our Homes and Businesses

- Complete the President's commitment to $2 billion in clean coal technology research funding, and move the resulting innovations into the marketplace.
- Develop a new Global Nuclear Energy Partnership (GNEP) to address spent nuclear fuel, eliminate proliferation risks, and expand the promise of clean, reliable, and affordable nuclear energy.
- Reduce the cost of solar photovoltaic technologies so that they become cost-competitive by 2015, and expand access to wind energy through technology.

This substantial increase in natural gas prices and volatility has had a negative impact on the U.S. industrial sector. High prices for natural gas translate to increased production costs for U.S. companies, which places them at a disadvantage to their foreign competitors. As a result, many firms have either shut down U.S. production facilities altogether or relocated them to another country where energy costs are more competitive with the global market. According to the National Association of Manufacturers, the chemicals and plastics industries, which rely on natural gas both for energy and as a raw material, have lost 250,000 jobs and $65 billion in business because of rising natural gas prices. High natural gas prices similarly harm the competitiveness of U.S. farm products in global markets, as natural gas is a primary input for fertilizer.

Diversification of our electric power sector will ensure the availability of affordable electricity and ample natural gas supplies. At the same time, increased efficiency will help reduce demand for natural gas. By easing the demand pressure on natural gas, prices will drop and U.S. firms will be more competitive in the global market, keeping jobs here at home.

Source: Excerpted from Web site of the White House at www.whitehouse.gov/stateoftheunion/2006/energy/energy_booklet.pdf.

NUCLEAR ENERGY

"Nuclear power is regenerating interest" by Todd B. Bates, *Asbury Park Press* (February 15, 2006)

Since the passage of the Energy Policy Act of 2005 nine companies have come forward with plans to build 19 new reactors. The EIA is projecting that nuclear energy consumption will grow 65 percent over 2004 consumption levels by 2035. Todd B. Bates's article on nuclear power offers a primer on nuclear energy generation and production, nuclear energy policy, and nuclear safety issues, while discussing recent technological developments and a possible revival of nuclear energy in the United States.

"Nuclear power is regenerating interest"
by Todd B. Bates
Asbury Park Press, February 15, 2006

The future of nuclear power in the United States is bright or dim, depending on your perspective.

No utility has ordered a new nuclear plant since 1978, and no plant ordered after 1973 was built, according to the U.S Energy Information Administration.

But the 2005 federal energy law provides incentives for new plants, and President Bush said in his State of the Union speech last month that more money will be invested in "clean, safe nuclear energy" and other technologies.

Some companies and groups are considering a number of sites and advanced reactor designs, according to the Nuclear Energy Institute, an industry policy group. And the industry anticipates building 12 to 15 new plants by 2015, according to a statement by Skip Bowman, group president.

The future "certainly looks very promising," said Jose Reyes, who heads the Department of Nuclear Engineering and Radiation Health Physics at Oregon State University, "What we see is that many countries are developing these passively safe power plants" that have "a whole new level of safety and reliability."

Jim Riccio, nuclear policy analyst with the environmental group Greenpeace, said nuclear reactors are "so damn expensive" and take a long time to build. New ones won't help control climate change but will create more terrorist targets and produce more radioactive waste, he said.

Renewed interest in nuclear power comes amid widespread concern about global warming and climate change linked, at least in part, to emissions of greenhouse gases.

Unlike power plants that burn fossil fuels such as coal, natural gas and oil, nuclear reactors do not emit greenhouse gases, such as carbon dioxide. The construction of reactors generates such emissions, however.

Nuclear generating capacity in the United States is projected to increase from about 100 gigawatts in 2004 to about 109 gigawatts by 2020 and stay at that level through 2030, according to the U.S. Energy Information Administration Web site.

A typical plant can generate roughly 1 gigawatt, or 1,000 megawatts of electricity. Oyster Creek generates 636 megawatts.

The 9 additional gigawatts that would be added by 2020 includes 6 gigawatts in new plants spurred by the federal energy law and 3.2 gigawatts of increased power capacity at existing plants, according to the Energy Information Administration.

Potential sites for new nuclear reactors include these locations: New Hill, N.C., near Clinton, Ill., in central Virginia, on the Mississippi River in Mississippi and Louisiana, near Augusta, Ga., and in northeastern Alabama.

Risk insurance coverage

Still, nuclear power, which generated 20 percent of U.S. electricity in 2004, would produce only 15 percent in 2030, according to federal protections. Coal power is expected to increase its share from 50 per cent in 2004 to 57 per cent in 2030.

But the nuclear industry expects electricity generated by nuclear plants to rise substantially from 2020 to 2030, according to the Nuclear Energy Institute.

.... Under the federal energy law, risk insurance provisions are aimed at offering financial protection against construction and operational delays beyond the control of a plant's sponsors, according to a November statement by the U.S. Department of Energy.

The provisions are "a critical step in laying the foundation for the renaissance of nuclear power in America," Energy Secretary Samuel W. Bodman said in the statement.

Each of the first two new nuclear plants would receive up to $500 million in risk insurance coverage to offset the cost of any delays, the statement says. Another four plants would receive coverage of up to $250 million apiece.

Riccio, of Greenpeace said the energy bill provides an "amazing amount of largess to the nuclear industry," but it won't be enough.

Spending money on energy efficiency and renewable energy is much more effective in displacing global warming gases than spending it on nuclear power, he said.

"The demise of nuclear power is actually a good thing for global warming because a dollar spent on nuclear doesn't go that far," Riccio said.

Reyes of Oregon State said many companies are looking closely at nuclear plants because they don't emit greenhouse gases.

Greater efficiencies

Because it has been more than 25 years since the 1979 Three Mile Island nuclear power plant accident in Pennsylvania, public confidence has increased in the industry, Reyes said.

Plants are now on line more than 90 percent of the time, a big change from the 1970s, he said.

In the 1970s, nuclear plants produced less than 60 percent of the power they were capable of generating, on average, according to the Energy Information Administration.

More efficient operations and improved maintenance, including dramatically shorter refueling outages have led to the increase since then, according to the World Nuclear Association Web site.

But Peter M. Sandman, a consultant on communicating risks who lives in Princeton Township, said that despite essentially 20 years of bad news about power from polluting fossil fuels, nuclear power is still having trouble making a comeback.

He called it "a stunning inability to earn the public trust."

However, David Schanzer, a utility analyst with Janney Montgomery Scott in Philadelphia, said he thinks the future of nuclear power is "outstanding."

But "even if everything goes as planned right now, it's going to take 10 years before we get a plant up and running," Schanzer said.

Environmentally, nuclear power is the best alternative, he said.

But it takes a while to get a license, for plans to take effect, to get the proper equipment, to find a site and get the necessary cooling water, if that's the source of plant cooling, Schanzer said.

An advanced, 1,000-megawatt nuclear plant built in 2013 would cost about $2 billion in 2004 dollars, not including interest, according to the 2005 Annual Energy Outlook on the Energy Information Administration Web site.

"Gonzo environmentalists"—people who don't want nuclear power because they don't like the idea of it—are one of the biggest obstacles, Schanzer said.

The other major obstacle is finding a proper storage site for spent nuclear fuel, Schanzer said.

Dr. Donald B. Louria, professor in the Department of Preventive Medicine and Community Health at New Jersey Medical School in Newark and board member of the antinuclear Nuclear Policy Research Institute, said he thinks the problems of nuclear power are so great that it should not be encouraged, especially at older plants.

For example, if terrorists strike plants, "awful things could happen," Louria said.

The more the U.S. promotes nuclear power at home, the more other countries will develop plants without the proper safeguards, "so we're going to have more Chernobyls," he said, referring to the world's worst nuclear accident in Ukraine in 1986.

And "nobody knows how to get rid of the waste," Louria said. "These are very, very serious problems."

Source: Excerpted from the *Asbury Park Press,* February 15, 2006, p. A7.

5

International Documents

In this section you will find excerpts from important policies, laws, speeches, and treaties that relate to energy issues around the world. The first two documents address energy issues that are of general importance to the global community. Following these are energy documents that relate specifically to the countries discussed in this book:

China

Germany

Iran

Saudi Arabia

Nigeria

Venezuela

DOCUMENTS OF GENERAL INTERNATIONAL IMPORTANCE

Kyoto Protocol (February 16, 2005)

The Kyoto Protocol is an international treaty that calls for the world's 38 most industrialized countries to reduce fossil fuel emissions by an average of 5 percent below 1990 levels by 2012. The treaty developed out of the work of the United Nations Framework Convention on Climate Change (UNFCCC), which aimed to find ways to minimize greenhouse gas emissions in order to reduce global climate change. As of this writing, more than 165 countries had signed the Kyoto Protocol, agreeing to achieve, by 2012, at least a 5 percent reduction in greenhouse gas emissions below 1990 levels. The agreement went into effect on February 16, 2005.

Kyoto Protocol to the United Nations Framework Convention on Climate Change

...

Article 3

1. The Parties included in Annex I shall, individually or jointly, ensure that their aggregate anthropogenic carbon dioxide equivalent emissions of the greenhouse gases listed in Annex A do not exceed their assigned amounts, calculated pursuant to their quantified emission limitation and reduction commitments inscribed in Annex B and in accordance with the provisions of this Article, with a view to reducing their overall emissions of such gases by at least 5 per cent below 1990 levels in the commitment period 2008 to 2012.

2. Each Party included in Annex I shall, by 2005, have made demonstrable progress in achieving its commitments under this Protocol.

3. The net changes in greenhouse gas emissions by sources and removals by sinks resulting from direct human-induced land-use change and forestry activities, limited to afforestation, reforestation and deforestation since 1990, measured as verifiable changes in carbon stocks in each commitment period, shall be used to meet the commitments under this Article of each Party included in Annex I. The greenhouse gas emissions by sources and removals by sinks associated with those activities shall be reported in a transparent and verifiable manner and reviewed in accordance with Articles 7 and 8.

4. Prior to the first session of the Conference of the Parties serving as the meeting of the Parties to this Protocol, each Party included in Annex I shall provide, for consideration by the Subsidiary Body for Scientific and Technological Advice, data to establish its level of carbon stocks in 1990 and to enable an estimate to be made of its changes in carbon stocks in subsequent years. The Conference of the Parties serving as the meeting of the Parties to this Protocol shall, at its first session or as soon as practicable thereafter, decide upon modalities, rules and guidelines as to how, and which, additional human-induced activities related to changes in greenhouse gas emissions by sources and removals by sinks in the agricultural soils and the land-use change and forestry categories shall be added to, or subtracted from, the assigned amounts for Parties included in Annex I, taking into account uncertainties, transparency in reporting, verifiability, the methodological work of the Intergovernmental Panel on Climate Change, the advice provided by the Subsidiary Body for Scientific and Technological Advice in accordance with Article 5 and the decisions of the Conference of the Parties. Such a decision shall apply in the second and subsequent commitment periods. A Party may choose to apply such a decision on these

additional human-induced activities for its first commitment period, provided that these activities have taken place since 1990.

5. The Parties included in Annex I undergoing the process of transition to a market economy whose base year or period was established pursuant to decision 9/CP.2 of the Conference of the Parties at its second session shall use that base year or period for the implementation of their commitments under this Article. Any other Party included in Annex I undergoing the process of transition to a market economy which has not yet submitted its first national communication under Article 12 of the Convention may also notify the Conference of the Parties serving as the meeting of the Parties to this Protocol that it intends to use an historical base year or period other than 1990 for the implementation of its commitments under this Article. The Conference of the Parties serving as the meeting of the Parties to this Protocol shall decide on the acceptance of such notification.

6. Taking into account Article 4, paragraph 6, of the Convention, in the implementation of their commitments under this Protocol other than those under this Article, a certain degree of flexibility shall be allowed by the Conference of the Parties serving as the meeting of the Parties to this Protocol to the Parties included in Annex I undergoing the process of transition to a market economy.

7. In the first quantified emission limitation and reduction commitment period, from 2008 to 2012, the assigned amount for each Party included in Annex I shall be equal to the percentage inscribed for it in Annex B of its aggregate anthropogenic carbon dioxide equivalent emissions of the greenhouse gases listed in Annex A in 1990, or the base year or period determined in accordance with paragraph 5 above, multiplied by five. Those Parties included in Annex I for whom land-use change and forestry constituted a net source of greenhouse gas emissions in 1990 shall include in their 1990 emissions base year or period the aggregate anthropogenic carbon dioxide equivalent emissions by sources minus removals by sinks in 1990 from land-use change for the purposes of calculating their assigned amount.

8. Any Party included in Annex I may use 1995 as its base year for hydrofluorocarbons, perfluorocarbons and sulphur hexafluoride, for the purposes of the calculation referred to in paragraph 7 above.

9. Commitments for subsequent periods for Parties included in Annex I shall be established in amendments to Annex B to this Protocol, which shall be adopted in accordance with the provisions of Article 21, paragraph 7. The Conference of the Parties serving as the meeting of the Parties to this Protocol

shall initiate the consideration of such commitments at least seven years before the end of the first commitment period referred to in paragraph 1 above.

10. Any emission reduction units, or any part of an assigned amount, which a Party acquires from another Party in accordance with the provisions of Article 6 or of Article 17 shall be added to the assigned amount for the acquiring Party.

11. Any emission reduction units, or any part of an assigned amount, which a Party transfers to another Party in accordance with the provisions of Article 6 or of Article 17 shall be subtracted from the assigned amount for the transferring Party.

12. Any certified emission reductions which a Party acquires from another Party in accordance with the provisions of Article 12 shall be added to the assigned amount for the acquiring Party.

13. If the emissions of a Party included in Annex I in a commitment period are less than its assigned amount under this Article, this difference shall, on request of that Party, be added to the assigned amount for that Party for subsequent commitment periods.

14. Each Party included in Annex I shall strive to implement the commitments mentioned in paragraph 1 above in such a way as to minimize adverse social, environmental and economic impacts on developing country Parties, particularly those identified in Article 4, paragraphs 8 and 9, of the Convention. In line with relevant decisions of the Conference of the Parties on the implementation of those paragraphs, the Conference of the Parties serving as the meeting of the Parties to this Protocol shall, at its first session, consider what actions are necessary to minimize the adverse effects of climate change and/or the impacts of response measures on Parties referred to in those paragraphs. Among the issues to be considered shall be the establishment of funding, insurance and transfer of technology.

Article 4

1. Any Parties included in Annex I that have reached an agreement to fulfil their commitments under Article 3 jointly, shall be deemed to have met those commitments provided that their total combined aggregate anthropogenic carbon dioxide equivalent emissions of the greenhouse gases listed in Annex A do not exceed their assigned amounts calculated pursuant to their quantified emission limitation and reduction commitments inscribed in Annex B and in accordance with the provisions of Article 3. The respective emission level allocated to each of the Parties to the agreement shall be set out in that agreement.

2. The Parties to any such agreement shall notify the secretariat of the terms of the agreement on the date of deposit of their instruments of ratification, acceptance or approval of this Protocol, or accession thereto. The secretariat shall in turn inform the Parties and signatories to the Convention of the terms of the agreement.

3. Any such agreement shall remain in operation for the duration of the commitment period specified in Article 3, paragraph 7.

4. If Parties acting jointly do so in the framework of, and together with, a regional economic integration organization, any alteration in the composition of the organization after adoption of this Protocol shall not affect existing commitments under this Protocol. Any alteration in the composition of the organization shall only apply for the purposes of those commitments under Article 3 that are adopted subsequent to that alteration.

5. In the event of failure by the Parties to such an agreement to achieve their total combined level of emission reductions, each Party to that agreement shall be responsible for its own level of emissions set out in the agreement.

6. If Parties acting jointly do so in the framework of, and together with, a regional economic integration organization which is itself a Party to this Protocol, each member State of that regional economic integration organization individually, and together with the regional economic integration organization acting in accordance with Article 24, shall, in the event of failure to achieve the total combined level of emission reductions, be responsible for its level of emissions as notified in accordance with this Article.

Article 5

1. Each Party included in Annex I shall have in place, no later than one year prior to the start of the first commitment period, a national system for the estimation of anthropogenic emissions by sources and removals by sinks of all greenhouse gases not controlled by the Montreal Protocol. Guidelines for such national systems, which shall incorporate the methodologies specified in paragraph 2 below, shall be decided upon by the Conference of the Parties serving as the meeting of the Parties to this Protocol at its first session.

2. Methodologies for estimating anthropogenic emissions by sources and removals by sinks of all greenhouse gases not controlled by the Montreal Protocol shall be those accepted by the Intergovernmental Panel on Climate Change and agreed upon by the Conference of the Parties at its third session. Where such methodologies are not used, appropriate adjustments shall be applied according to methodologies agreed upon by the Conference of the Parties serving as the meeting of the Parties to this Protocol

at its first session. Based on the work of, *inter alia,* the Intergovernmental Panel on Climate Change and advice provided by the Subsidiary Body for Scientific and Technological Advice, the Conference of the Parties serving as the meeting of the Parties to this Protocol shall regularly review and, as appropriate, revise such methodologies and adjustments, taking fully into account any relevant decisions by the Conference of the Parties. Any revision to methodologies or adjustments shall be used only for the purposes of ascertaining compliance with commitments under Article 3 in respect of any commitment period adopted subsequent to that revision.

3. The global warming potentials used to calculate the carbon dioxide equivalence of anthropogenic emissions by sources and removals by sinks of greenhouse gases listed in Annex A shall be those accepted by the Intergovernmental Panel on Climate Change and agreed upon by the Conference of the Parties at its third session. Based on the work of, *inter alia,* the Intergovernmental Panel on Climate Change and advice provided by the Subsidiary Body for Scientific and Technological Advice, the Conference of the Parties serving as the meeting of the Parties to this Protocol shall regularly review and, as appropriate, revise the global warming potential of each such greenhouse gas, taking fully into account any relevant decisions by the Conference of the Parties. Any revision to a global warming potential shall apply only to commitments under Article 3 in respect of any commitment period adopted subsequent to that revision.

...

Article 24

1. This Protocol shall be open for signature and subject to ratification, acceptance or approval by States and regional economic integration organizations which are Parties to the Convention. It shall be open for signature at United Nations Headquarters in New York from 16 March 1998 to 15 March 1999. This Protocol shall be open for accession from the day after the date on which it is closed for signature. Instruments of ratification, acceptance, approval or accession shall be deposited with the Depositary.

2. Any regional economic integration organization which becomes a Party to this Protocol without any of its member States being a Party shall be bound by all the obligations under this Protocol. In the case of such organizations, one or more of whose member States is a Party to this Protocol, the organization and its member States shall decide on their respective responsibilities for the performance of their obligations under this Protocol. In such cases, the organization and the member States shall not be entitled to exercise rights under this Protocol concurrently.

3. In their instruments of ratification, acceptance, approval or accession, regional economic integration organizations shall declare the extent of their competence with respect to the matters governed by this Protocol. These organizations shall also inform the Depositary, who shall in turn inform the Parties, of any substantial modification in the extent of their competence.

Article 25

1. This Protocol shall enter into force on the ninetieth day after the date on which not less than 55 Parties to the Convention, incorporating Parties included in Annex I which accounted in total for at least 55 per cent of the total carbon dioxide emissions for 1990 of the Parties included in Annex I, have deposited their instruments of ratification, acceptance, approval or accession.

2. For the purposes of this Article, "the total carbon dioxide emissions for 1990 of the Parties included in Annex I" means the amount communicated on or before the date of adoption of this Protocol by the Parties included in Annex I in their first national communications submitted in accordance with Article 12 of the Convention.

3. For each State or regional economic integration organization that ratifies, accepts or approves this Protocol or accedes thereto after the conditions set out in paragraph 1 above for entry into force have been fulfilled, this Protocol shall enter into force on the ninetieth day following the date of deposit of its instrument of ratification, acceptance, approval or accession.

4. For the purposes of this Article, any instrument deposited by a regional economic integration organization shall not be counted as additional to those deposited by States members of the organization.

. . .

Annex A
Greenhouse gases

Carbon dioxide (CO_2)
Methane (CH_4)
Nitrous oxide (N_20)
Hydrofluorocarbons (HFCs)
Perfluorocarbons (PFCs)
Sulphur hexafluoride (SF_6)

Sectors/source categories

Energy
    Fuel combustion
        Energy industries
        Manufacturing industries and construction

Transport
Other sectors
Other
Fugitive emissions from fuels
Solid fuels
Oil and natural gas
Other

Industrial processes
Mineral products
Chemical industry
Metal production
Other production
Production of halocarbons and sulphur hexafluoride
Consumption of halocarbons and sulphur hexafluoride
Other

Solvent and other product use

Agriculture
Enteric fermentation
Manure management
Rice cultivation
Agricultural soils
Prescribed burning of savannas
Field burning of agricultural residues
Other

Waste
Solid waste disposal on land
Wastewater handling
Waste incineration
Other

Annex B

Party	**Quantified emission limitation or reduction commitment (percentage of base year or period)**
Australia	108
Austria	92
Belgium	92
Bulgaria*	92

Canada	94
Croatia*	95
Czech Republic*	92
Denmark	92
Estonia*	92
European Community	92
Finland	92
France	92
Germany	92
Greece	92
Hungary*	94
Iceland	110
Ireland	92
Italy	92
Japan	94
Latvia*	92
Liechtenstein	92
Lithuania*	92
Luxembourg	92
Monaco	92
Netherlands	92
New Zealand	100
Norway	101
Poland*	94
Portugal	92
Romania*	92
Russian Federation*	100
Slovakia*	92
Slovenia*	92
Spain	92
Sweden	92
Switzerland	92
Ukraine*	100
United Kingdom of Great Britain and Northern Ireland	92
United States of America	93

* Countries that are undergoing the process of transition to a market economy.

Source: Excerpted from UNFCCC's Web site at http://unfccc.int/resource/docs/convkp/kpeng.html.

"Crude awakening indeed to possible oil conflicts" by Paul Roberts, *Times Union*, July 2, 2004

In the United States energy security *is also a term that has come to suggest the need to protect oil refineries, pipelines, and nuclear power facilities. For the United States and the entire international community, however, the term now increasingly suggests the need to maintain a level of fuel and electricity production and affordability that will secure a country's economic stability and political power for generations to come. For more information on this topic, consider this article by Paul Roberts, published on July 2, 2004, in the* Times Union.

Paul Roberts

Times Union, July 2, 2004

While some debate whether the war in Iraq was or was not "about oil," another war, this one involving little but oil, has broken out between two of the world's most powerful nations.

For months, China and Japan have been locked in a diplomatic battle over access to the big oil fields in Siberia. Japan, which depends entirely on imported oil, is desperately lobbying Moscow for a 2,300-mile pipeline from Siberia to coastal Japan. But fast-growing China, now the world's second-largest oil user, after the United States, sees Russian oil as vital for its own "energy security" and is pushing for a 1,400-mile pipeline south to Daqing.

The petro-rivalry has become so intense that Japan has offered to finance the $5 billion pipeline, invest $7 billion in development of Siberian oil fields and throw in an additional $2 billion for Russian "social projects"—this despite the certainty that if Japan does win Russia's oil, relations between Tokyo and Beijing may sink to their lowest, potentially most dangerous, levels since World War II.

Asia's undeclared oil war is but the latest reminder that in a global economy dependent largely on a single fuel—oil—"energy security" means far more than hardening refineries and pipelines against terrorist attack. At its most basic level, energy security is the ability to keep the global machine humming—that is, to produce enough fuels and electricity at affordable prices that every nation can keep its economy running, its people fed and its borders defended. A failure of energy security means that the momentum of industrialization and modernity grinds to a halt. And by that measure, we are failing.

In the United States and Europe, new demand for electricity is outpacing the new supply of power and natural gas and raising the specter of more rolling blackouts. In the "emerging" economies, such as Brazil, India and especially China, energy demand is rising so fast it may double by 2020. And this only hints at the energy crisis facing the developing world, where nearly 2 billion people—a third of the world's International Documents population—have almost no access to electricity or liquid fuels and are thus condemned to a medieval existence that breeds despair, resentment and, ultimately, conflict.

In other words, we are on the cusp of a new kind of war—between those who have enough energy and those who do not but are increasingly willing to go out and get it. While nations have always competed for oil, it seems more and more likely that the race for a piece of the last big reserves of oil and natural gas will be the dominant geopolitical theme of the 21st century.

Already we can see the outlines. China and Japan are scrapping over Siberia. In the Caspian Sea region, European, Russian, Chinese and American governments and oil companies are battling for a stake in the big oil fields of Kazakhstan and Azerbaijan. In Africa, the United States is building a network of military bases and diplomatic missions whose main goal is to protect American access to oil fields in volatile places such as Nigeria, Cameroon, Chad and tiny São Tomé—and, as important, to deny that access to China and other thirsty superpowers.

The diplomatic tussles only hint at what we'll see in the Middle East, where most of the world's remaining oil lies. For all the talk of big new oil discoveries in Russia and Africa—and of how this gush of crude will "free" America and other big importers from the machinations of OPEC—the geological facts speak otherwise. Even with the new Russian and African oil, worldwide oil production outside the Middle East is barely keeping pace with demand.

In the run-up to the Iraq war, Russia and France clashed noisily with the United States over whose companies would have access to the oil in post–Saddam Hussein Iraq. Less well known is the way China has sought to build up its own oil alliances in the Middle East—often over Washington's objections. In 2000 Chinese oil officials visited Iran, a country U.S. companies are forbidden to deal with; China also has a major interest in Iraqi oil.

But China's most controversial oil overture has been made to a country America once regarded as its most trusted oil ally: Saudi Arabia. In recent years, Beijing has been lobbying Riyadh for access to Saudi reserves, the largest in the world. In return, the Chinese have offered the Saudis a foothold in what will be the world's biggest energy market—and, as a bonus, have thrown in offers of sophisticated Chinese weaponry, including ballistic missiles and other hardware, that the United States and Europe have refused to sell to the Saudis.

Granted, the United States, with its vast economic and military power, would probably win any direct "hot" war for oil.

The far more worrisome scenario is that an escalating rivalry among other big consumers will spark new conflicts—conflicts that might require U.S. intervention and could easily destabilize the world economy upon which American power ultimately rests.

As demand for oil becomes sharper, as global oil production continues to lag (and as producers such as Saudi Arabia and Nigeria grow more unstable) the struggle to maintain access to adequate energy supplies, always a critical mission for any nation, will become even more challenging and uncertain and take up even more resources and political attention.

This escalation will not only drive up the risk of conflict but will make it harder for governments to focus on long-term energy challenges, such as avoiding climate change, developing alternative fuels and alleviating Third World energy poverty—challenges that are themselves critical to long-term energy security but which, ironically, will be seen as distracting from the current campaign to keep the oil flowing.

Source: Paul Roberts. "Crude awakening indeed to possible oil conflicts." *Times Union* (July 2, 2004): B8.

DOCUMENTS OF CHINA

Renewable Energy Law (2005)

The Renewable Energy Law was signed into law in 2005 in the People's Republic of China. This law has earned praise from some U.S. environmentalists, who say it is a good model for the United States to follow in forming future U.S. energy policies. The law was China's attempt to establish measures to protect the environment, prevent energy shortages, and reduce dependence on energy imports. Effective on January 1, 2006, the law stipulates that electricity power grid operators purchase resources from approved renewable energy producers. The law includes in its definition of renewable energy hydroelectricity, wind power, solar energy, geothermal energy, and marine energy. It also requires that national financial incentives be available to foster state and local development of renewable energy resources, including solar electricity, solar water heating, and renewable energy fuels; that loan and tax discounts be given for renewable energy projects, such as the construction of commercial renewable energy facilities; and that specific penalties be imposed for noncompliance with the law. China projected that the law would boost its capacity to use renewable energy to 10 percent by the year 2020. In addition, China's National Development and Reform Commission (NDRC) unveiled plans to establish specific renewable energy targets for China as the framework for the implementation of the law. In relation to the law, China also announced plans to increase environmental spending from 0.7 percent of GDP in 1996 to 1.7 percent in 2010 and 10 percent in 2020.

Renewable Energy Law, Signed into Law 2005, the People's Republic of China

Chapter 1. General

Article 1—In order to promote the development and utilization of renewable energy, improve the energy structure, diversify energy supplies, safeguard energy security, protect the environment, and realize the sustainable development of the economy and society, this Law is hereby prepared.

Article 2—Renewable energy in this law refers to non-fossil energy of wind energy, solar energy, water energy, biomass energy, geothermal energy, and ocean energy, etc.

Application of this Law in hydropower shall be regulated by energy authorities of the State Council and approved by the State Council.

This Law does not apply to the direct burning of straw, firewood and dejecta, etc. on low-efficiency stove.

Article 3—This Law applies to territory and other sea area of the People's Republic of China.

Article 4—The Government lists the development of utilization of renewable energy as the preferential area for energy development and promotes the construction and development of the renewable energy market by establishing total volume for the development of renewable energy and taking corresponding measures.

The Government encourages economic entities of all ownerships to participate in the development and utilization of renewable energy and protects legal rights and interests of the developers and users of renewable energy on the basis of law.

Article 5—Energy authorities of the State Council implement management for the development and utilization of renewable energy at the national level. Relevant departments of the State Council are responsible for the management of relevant development and utilization of renewable energy within their authorities.

Energy authorities of local people's governments above the county level are responsible for the management of the development and utilization of renewable energy within their own jurisdiction. Relevant departments of local people's governments above the county level are responsible for the management of relevant development and utilization of renewable energy within their authorities.

Chapter 2 Resource Survey and Development Plan

Article 6—Energy authorities of the State Council are responsible for organizing and coordinating national surveys and management of renewable energy resources, and work with related departments to establish technical regulations for resource surveys.

Relevant departments of the State Council, within their respective authorities, are responsible for related renewable energy resource surveys. The survey results will be summarized by the energy authorities in the State Council.

The result of the survey of renewable energy shall be released to the public, with the exception of confidential contents as stipulated by the Government.

Article 7—Energy authorities of the State Council sets middle and long-term target of the total volume for the development and utilization of renewable energy at the national level, which shall be implemented and released to the pubic after being approved by the State Council.

Energy authorities of the State Council shall, on the basis of the target of total volume in the previous paragraph, as well as the economic development and actual situation of renewable energy resources of all provinces, autonomous regions and municipalities, cooperate with people's governments of provinces, autonomous regions and municipalities in establishing middle and long-term target and release it to the public.

Article 8—Energy authorities of the State Council shall, on the basis of the middle and long-term total volume target of renewable energy throughout the country, prepare national renewable energy development and utilization plan, which is to be implemented after being approved by the State Council.

Energy authorities of the people's governments at the level of province, autonomous region and municipality shall, on the basis of the middle and long-term target for the development and utilization of renewable energy, cooperate with relevant authorities of the people's governments at their own level in preparing national renewable energy development and utilization plan for their own administrative regions, which shall be implemented after being approved by people's governments at their own level.

The approved plan shall be released to the public, with the exception of confidential content as stipulated by the government.

In case that the approved plan needs to be modified, approval of the original approving authorities shall be obtained.

Article 9—In preparing the plan for the development and utilization of renewable energy, opinions of relevant units, experts and the public shall be solicited and the scientific reasoning shall be done.

Chapter 3 Industry Guidance and Technology Support

Article 10—Energy authorities in the State Council shall, in accordance with the national renewable energy development plan, prepare and promulgate development guidance catalogs for renewable energy industries.

Article 11—Standardization authorities of the State Council shall set and publicize technical standard for renewable energy electric power and the technical standards for relevant renewable technology and products for which technical requirements need to be standardized at the national level.

For those technical requirements not dealt with in the national standard in the previous paragraph, relevant authorities of the State Council may establish relevant industrial standard, which shall be reported to the standardization authorities of the State Council for filing.

Article 12—The government lists scientific and technical research in the development and utilization of, and the industrialized development of, renewable energy, as the preferential area for hi-tech development and hi-tech industrial development in the national program, and allocates funding for the scientific and technical research, application demonstration and industrialized development of the development and utilization of renewable energy so as to promote technical advancement in the development and utilization of renewable energy, reduce the production cost of renewable energy products and improve the quality of products.

Education authorities of the State Council shall incorporate the knowledge and technology on renewable energy into general and occupational education curricula.

Chapter 4 Promotion and Application.

Article 13—The Government encourages and supports various types of grid-connected renewable power generation.

For the construction of renewable energy power generation projects, administrative permits shall be obtained or filing shall be made in accordance with the law and regulations of the State Council.

In the construction of renewable power generation projects, if there is more than one applicant for project license, the licensee shall be determined through a tender.

Article 14—Grid enterprises shall enter into grid connection agreement with renewable power generation enterprises that have legally obtained administrative license or for which filing has been made, and buy the grid-connected power produced with renewable energy within the coverage of their power grid, and provide grid-connection service for the generation of power with renewable energy.

Article 15—The Government supports the construction of independent renewable power systems in areas not covered by the power grid to provide power service for local production and living.

Article 16—The Government encourages clean and efficient development and utilization of biological fuel and encourages the development of energy crops.

If the gas and heat produced with biological resources conform to urban fuel gas pipeline networks and heat pipeline networks, enterprises operating gas pipeline networks and heat pipeline networks shall accept them into the networks.

The Government encourages the production and utilization of biological liquid fuel. Gas-selling enterprises shall, on the basis of the regulations of energy authorities of the State Council or people's government at the provincial level, include biological liquid fuel conforming to the national standard into its fuel-selling system.

Article 17—The Government encourages workplaces and individuals in the installation and use of solar energy utilization systems of solar energy water-heating system, solar energy heating and cooling system and solar photovoltaic system, etc.

Construction authorities of the State Council shall cooperate with relevant authorities of the State Council in establishing technical economic policies and technical standards with regard to the combination of solar energy utilization system and construction.

Real estate development enterprises shall, on the basis of the technical standards in the previous paragraph, provide necessary conditions for the utilization of solar energy in the design and construction of buildings.

For buildings already built, residents may, on the condition that its quality and safety is not affected, install solar energy utilization system that conform to technical standards and product standards, unless agreement has been otherwise reached between relevant parties.

Article 18—The Government encourages and supports the development and utilization of renewable energy in rural areas.

Energy authorities of local people's governments above the county level shall, on the basis of local economic and social development, ecological protection and health need, etc., prepare renewable energy development plan for the rural area and promote biomass energy like the marsh gas, etc. conversion, household solar energy, small-scale wind energy and small-scale hydraulic energy, etc.

People's government above the county level shall provide financial support for the renewable energy utilization projects in the rural areas.

Chapter 5 Price Management and Fee Sharing

Article 19—Grid power price of renewable energy power generation projects shall be determined by the price authorities of the State Council in the principle of being beneficial to the development and utilization of renewable energy and being economic and reasonable, where timely adjustment shall be made on the basis of the development of technology for the development and utilization of renewable energy. The price for grid-connected power shall be publicized.

For the price of grid-connected power of renewable power generation projects determined through tender as stipulated in the 3rd paragraph of Article 13 hereof, the bid-winning price shall be implemented; however, such a price shall not exceed the level of grid-connected power of similar renewable power generation projects.

Article 20—The excess between the expenses that power grid enterprises purchase renewable power on the basis of the price determined in Article 19 hereof and the expenses incurred in the purchase of average power price generated with conventional energy shall be shared in the selling price. Price authorities of the State Council shall prepare specific methods.

Article 21—Grid connection expenses paid by grid enterprises for the purchase of renewable power and other reasonable expenses may be included into the grid enterprise power transmission cost and retrieved from the selling price.

Article 22—For the selling price of power generated from independent renewable energy power system invested or subsidized by the Government, classified selling price of the same area shall be adopted, and the excess between its reasonable operation, management expenses and the selling price shall be shared on the basis of the method as specified in Article 20 hereof.

Article 23—The price of renewable heat and natural gas that enters the urban pipeline shall be determined on the basis of price management authorities in the principle of being beneficial to the development and utilization of renewable energy and being economic and reasonable.

Chapter 6 Economic Incentives and supervisory measures

Article 24—The Government budget establishes renewable energy development fund to support the following:

1. Scientific and technological research, standard establishment and pilot project for the development and utilization of renewable energy;
2. Construction of renewable energy projects for domestic use in rural and pasturing areas;
3. Construction of independent renewable power systems in remote areas and islands;
4. Surveys, assessments of renewable energy resources, and the construction of relevant information systems;

5. Localized production of the equipment for the development and utilization of renewable energy.

Article 25—Financial institutions may offer preferential loan with financial interest subsidy to renewable energy development and utilization projects that are listed in the national renewable energy industrial development guidance catalogue and conform to the conditions for granting loans.

Article 26—The Government grants tax benefits to projects listed in the renewable energy industrial development guidance catalogue, and specific methods are to be prepared by the State Council.

Article 27—Power enterprises shall authentically and completely record and store relevant materials of renewable energy power generation, and shall accept the inspection and supervision of power supervisory institutions.

Power supervisory institutions shall do the inspection in accordance with stipulated procedures, and shall keep commercial secret and other secret for inspected units.

Chapter 7 Legal Responsibilities

Article 28—If energy authorities of the State Council and the people's governments above the county level as well as other relevant authorities breach this Law and have one of the following, people's government of their own level or relevant authorities of the superior people's governments may order them to make correction, and impose administrative penalty for competent personnel that are liable and other personnel directly liable; in case that such breaches constitute crime, criminal liabilities shall be legally pursued.

1. Failure to make administrative licensing decision in accordance with law;
2. Failure to make an investigation when illegal activities are discovered;
3. Other acts of not legally performing supervision and management responsibilities.

Article 29—If the power grid enterprises breach Article 14 hereof and fail to purchase renewable power in full, which results in economic loss to the renewable power generation enterprises, such power grid enterprises shall be liable for compensation, and the national power supervisory institutions shall order them to make correction within a stipulated period of time; in case of refusal to make correction, a fine of less than the economic loss of the renewable power generation enterprises shall be imposed.

Article 30—In case that enterprises of natural gas pipeline network and heat pipeline network breach paragraph 2 of Article 16 hereof and do not permit the connection of natural gas and heat that conform to the grid connection technical standard into the network, which results in economic loss to the gas and heat production enterprises, relevant enterprises shall be liable for compensation, and energy authorities of the people's government at the provincial level shall order them to make correction within a stipulated period of time; in case of refusal to make correction, a fine of less than said economic loss shall be imposed against them.

Article 31—If gas-selling enterprises breach paragraph 3 of Article 16 hereof and fail to include biological liquid fuel that conforms to the national standard into its fuel-selling system, which results in economic loss to the biological liquid fuel production enterprises, relevant enterprises shall be liable for compensation, and energy authorities of the State Council or people's government at the provincial level shall order them to make correction within a stipulated period of time; in case of refusal to make correction, a fine of less than said economic loss shall be imposed against them.

Chapter 8 Miscellaneous

Article 32—Terms used herein shall have the following meanings:

1. Biomass energy: means energy converted from natural plants, rejecta as well as urban and rural organic waste.
2. Renewable energy independent power system: means independent renewable power system not connected to the power grid.
3. Energy crop: means herbage and wood plants specially planted and used as raw materials of energy.
4. Biological liquid fuels: means methanol, ethanol, bio-diesel and other liquid fuels derived from biomass resources.

Article 33—This Law shall become effective on Jan 1st, 2006.

Source: Excerpted from the Web site of the China Energy Group, a department of China's government (http://china.lbl.gov/publications/re-law-english.pdf).

Speech by Minister Xie Zhenhua on the China-Africa Environmental Cooperation, Nairobi, Kenya, February 21, 2005

There are a number of environmental laws and regulations that China has instituted in order to improve its environmental situation and address its energy issues. For a discussion of some of these laws and regulations, see this speech by Xie Zhenhua, head of China's State Environmental Protection Administration, on the environmental cooperation between China and Africa, given on February 21, 2005, in Nairobi, Kenya.

Distinguished Vice Premier Zeng Peiyan,
Distinguished Mr. Teopfer,
Distinguished Ministers and Ambassadors,
Ladies and Gentlemen,

Good evening! I am very happy to have this opportunity to have dialogue and exchange views with all of you on China-Africa environmental cooperation. A moment ago, Vice Premier Zeng Peiyan, on behalf of the Chinese Government, has put forward three initiatives on China-Africa environmental cooperation. I would like to take this opportunity to give an introduction to China's performances of protecting the environment in the process of development, share with you our experiences and lessons, and put forth suggestions on measures of fulfilling the initiatives proposed by Vice Premier Zeng Peiyan.

China is the biggest developing country in the world with a huge population, comparatively low level of natural resource in per capita terms and fragile environment. China's economy grows at an average rate of 9% in the past 20 years. However, the growth depends largely upon high input, high consumption and heavy pollution. In this period, the GDP in China has quadrupled, and brought about serious pollution and ecological destruction, with the following major aspects: large amount of major pollutants emission and discharge in excess of environmental load, ignorant of natural pattern in the process of resource development, and the worsening trend of ecological destruction has not been curbed. It can be said that the environmental problems that have occurred stage by stage over 100 years in developed countries and resolved gradually have occurred altogether over the past 20 years in China in the process of rapid development. These compressed and compound issues have the character of being closely linked with economic structure, which has caused serious threats to the sustainable development of the economy and society.

The Chinese Government has always attached great importance to environmental protection. It has identified environmental protection as one of the State policies in the 1980s. It identified sustainable development as a national strategy in early 1990s. In the new century, the Chinese Government has put forward to adhere to the people-centered principle, fulfill the scientific development approach, and integrate the coordinated development of the economy, society and the environment. We have clearly stipulated that the environment should be protected in the process of development and active explorations have been conducted . . . China has established a legal system composed of 9 environmental laws, over ten natural resources laws, more than 20 environmental regulations, a series of rules and standards. At the same time, it has kept, strict enforcement of these environmental laws and regulations. . . .

Source: Excerpted from the Web site of the Forum on China-Africa Cooperation (http://www.fmprc.gov.cn/zflt/eng/zyzl/zyjh/+185190.htm).

DOCUMENTS OF GERMANY

Act Revising the Legislation on Renewable Energy Sources in the Electricity Sector (The Renewable Energy Sources Act) (July 21, 2004)

The Renewable Energy Sources Act (Act Revising the Legislation on Renewable Energy Sources in the Electricity Sector) was signed into law in Germany on July 21, 2004. The purpose of the Renewable Energy Sources Act is to increase the percentage of Germany's electricity that is produced through renewable energy sources—from 6.7 percent in 2000 to 12.5 percent by 2010, 20 percent by 2020, and 50 percent by 2050. The law calls for improved financial incentives for the continuing development of not only wind power for electricity, but also hydropower, now the second-largest renewable source of electricity in Germany after wind, in addition to electricity derived from solar, biomass, and geothermal power.

Act Revising the Legislation on Renewable Energy Sources in the Electricity Sector,[1] Germany, July 21, 2004

The Bundestag has adopted the following act:

Section 1 Act on granting priority to renewable energy sources (Renewable Energy Sources Act)

Article 1
Purpose

(1) The purpose of this act is to facilitate a sustainable development of energy supply, particularly for the sake of protecting our climate, nature and the environment, to reduce the costs of energy supply to the national economy, also by incorporating long-term external effects, to protect nature and the environment, to contribute to avoiding conflicts over fossil fuels and to promote the further development of technologies for the generation of electricity from renewable energy sources.

2 (2) This act is further intended to contribute to the increase in the percentage of renewable energy sources in power supply to at least 12.5 per cent by 2010 and to at least 20 per cent by 2020.

Article 2
Scope of application

(1) This act regulates

[1] Act implementing Directive 2001/77/EC of the European Parliament and of the Council of 27 September 2001 on the promotion of electricity produced from renewable energy sources in the internal electricity market (OJ L 283 p. 33), as last amended by the Acts of Accession of 16 April 2003 (OJ L 236 p. 586).

1. priority connections to the grid systems for general electricity supply of plants generating electricity from renewable energy sources and from mine gas within the territory of the Federal Republic of Germany including its exclusive economic zone (territorial application of this act),

2. the priority purchase and transmission of, and payment for, such electricity by the grid system operators and

3. the nation-wide equalisation scheme for the quantity of electricity purchased and paid for.

(1) This act shall not apply to plants of which over 25 per cent are owned by the Federal Republic of Germany or one of its *Länder* and which were commissioned prior to the 1 August 2004.

Article 3
Definitions

(1) Renewable energy sources shall mean hydropower including wave power, tidal power, salt gradient and flow energy, wind energy, solar radiation, geothermal energy, energy from biomass including biogas, landfill gas and sewage treatment plant gas as well as the biodegradable fraction of municipal and industrial waste.

(2) Plant shall mean any independent technical facility generating electricity from renewable energy sources or from mine gas. Several plants generating electricity from equivalent renewable energy sources or from mine gas, if constructed within the territorial application of this act and directly attached to building structures and commonly used installations technically required for operation shall be considered as one plant if Articles 6 to 12 do not provide for otherwise; inverters, access ways, grid connections as well as measuring, administrative and control facilities in particular are not technically required for such operation.

(3) Plant operator shall mean anyone who, notwithstanding the issue of ownership, uses the plant for the purpose of generating electricity from renewable energy sources or from mine gas.

(4) Commissioning shall mean the first time a plant is put into operation, following establishment of operational readiness or its modernisation, if modernisation costs amount to at least 50 per cent of the investment costs required to build a completely new plant including all building structures and installations technically required for its operation.

(5) Capacity of a plant shall mean the effective electrical capacity which the plant may technically produce without time restrictions during regular opera-

tion irrespective of shortterm deviations. When the relevant capacity is determined to calculate the fees, the standby capacity shall not be considered.

(6) Grid system shall mean all the interconnected facilities used for the transmission and distribution of electricity for general supply.

(7) Grid system operators shall mean the operators of all types of voltage systems for general electricity supply. The transmission system operators shall be the responsible grid system operators of high-voltage and extra-high-voltage systems which are used for the supraregional transmission of electricity to downstream systems.

Article 4
Obligation to purchase and transmit electricity

(1) Grid system operators shall immediately and as a priority connect plants generating electricity from renewable energy sources or from mine gas to their systems and guarantee priority purchase and transmission of all electricity from renewable energy sources or from mine gas supplied by such plants.

* *** *

Article 5
Obligation to pay fees

(1) Pursuant to Articles 6 to 12, the grid system operators shall pay fees for electricity generated in plants exclusively using renewable energy sources or mine gas and purchased in accordance with Article 4(l) or (5). The obligation in accordance with the first sentence above shall only apply to plants with a capacity of over 500 kilowatts where the capacity is measured and recorded.

(2) Pursuant to Articles 6 to 12, the upstream transmission system operator shall pay for the quantity of energy which the grid system operator has purchased in accordance with Article 4(6) and paid for in accordance with paragraph (1) above. Any avoided charges for use of the grid system, calculated in accordance with good professional practice, shall be deducted from the fees. Article 4(6) second sentence shall apply *mutatis mutandis.*

Article 6
Fees paid for electricity produced from hydropower

(1) The fees paid for electricity generated in hydroelectric power plants with a capacity up to and including 5 megawatts shall be

1. at least 9.67 cents per kilowatt-hour for plants with a capacity up to and including 500° kilowatts and

2. at least 6.65 cents per kilowatt-hour for plants with a capacity up to and including 5° megawatts.

The first sentence above shall apply to run-of-river power plants with a capacity of up to 500 kilowatts licensed after 31 December 2007 only if they

1. were constructed in the spatial context of an existing barrage weir or dam which wholly or partly existed before or was newly built primarily for purposes other than the generation of electricity from hydropower or

2. without complete weir coverage, and if this has demonstrably brought about a good ecological status or a substantial improvement in relation to the previous status.

(2) Fees for electricity generated in hydroelectric power plants with a capacity ranging from 5 megawatts up to and including 150 megawatts shall only be paid for in accordance with the provisions of this act if

1. the plant was modernised between 1 August 2004 and 31 December 2012,

2. the modernisation has resulted in an increase in the electrical energy of at least 15 per cent and if

3. such modernisation has demonstrably brought about a good ecological status or a substantial improvement in relation to the previous status.

Notwithstanding Article 3(4), hydroelectric power plants with a capacity of over 5 megawatts which meet the requirements of the first sentence above shall be deemed to have been newly commissioned. The first commissioning of a plant in the spatial context of an existing barrage weir or a dam shall also be deemed to represent modernisation within the meaning of the first sentence above. Fees shall only be paid for the additional electricity generated due to modernisation.

Such fees shall be

1. at least 7.67 cents per kilowatt-hour up to and including an increase in capacity of 500 kilowatts,

2. at least 6.65 cents per kilowatt-hour up to and including an increase in capacity of 10 megawatts,

3. at least 6.10 cents per kilowatt-hour up to and including an increase in capacity of 20 megawatts,

4. at least 4.56 cents per kilowatt-hour up to and including an increase in capacity of 50 megawatts and

5. at least 3.70 cents per kilowatt-hour if the increase in capacity exceeds 50 megawatts. If the plant had a capacity of up to and including 5 megawatts prior to 1 August 2004, the 7 quantity of electricity corresponding to this share of capacity shall in addition be paid for in accordance with paragraph (1) above.

(3) Presentation of an official authorisation under water law shall be deemed to be proof of achievement of a good ecological status or of a substantial improvement of the ecological status compared to the previous status in accordance with paragraph (1) second sentence and paragraph (2) first sentence No. 3 above.

(4) As of 1 January 2005, the minimum fees specified in paragraph (2) above shall be reduced for new plants commissioned after that date by one per cent annually of the relevant value for new plants commissioned in the previous year; the amounts payable shall be rounded to two decimals.

(5) Paragraphs (1) to (4) shall not apply to electricity produced from storage power stations.

Article 7

Fees paid for electricity produced from landfill gas, sewage treatment plant gas and mine gas

The fees paid for electricity from landfill gas, sewage treatment plant gas and mine gas shall be

1. at least 7.67 cents per kilowatt-hour up to and including a capacity of 500 kilowatts and
2. at least 6.65 cents per kilowatt-hour up to and including a capacity of 5 megawatts.

The fees paid for electricity from mine gas plants with a capacity of over 5 megawatts shall be 6.65 cents per kilowatt-hour.

Gas withdrawn from a gas network shall be deemed to be landfill gas, sewage treatment plant gas or mine gas if the thermal equivalent of the withdrawn quantity of such gas corresponds to the quantity of landfill gas, sewage treatment plant gas or mine gas fed into the gas network elsewhere within the territorial application of this act.

(2) The minimum fees in accordance with paragraph (1) above shall be increased by 2 cents per kilowatt-hour if the gas fed in pursuant to paragraph (1) third sentence above has been processed to reach the quality of natural gas or if the electricity is produced by fuel cells, gas turbines, steam engines, organic Rankine cycles, multi-fuel plants, especially Kalina cycles, or Stirling engines. For the purpose of adapting this provision to the state of the art, the Federal Ministry for the Environment, Nature Conservation and Nuclear Safety is authorised to issue, in agreement with the Federal Ministry of Consumer Protection, Food and Agriculture and the Federal Ministry of Economics and Labour, an ordinance detailing further processes

or techniques referred to in the first sentence above or exempting some of these processes or techniques from the scope of the first sentence above.

(3) As of 1 January 2005, the minimum fees specified in paragraph (1) above shall be reduced for new plants commissioned after that date by 1.5 per cent annually of the relevant value for new plants commissioned in the previous year; the amounts payable shall be rounded to two decimals.

Article 8

Fees paid for electricity produced from biomass

(1). The fees paid for electricity produced in plants with a capacity of up to and including 20 megawatts using exclusively biomass as defined in an ordinance adopted pursuant to paragraph (7) below shall be

1. at least 11.5 cents per kilowatt-hour up to and including a capacity of 150 kilowatts,
2. at least 9.9 cents per kilowatt-hour up to and including a capacity of 500 kilowatts,
3. at least 8.9 cents per kilowatt-hour up to and including a capacity of 5 megawatts and
4. at least 8.4 cents per kilowatt-hour for a capacity of over 5 megawatts.

Notwithstanding the first sentence above, the fee shall be 3.9 cents per kilowatt-hour if the plant also uses waste wood classified in categories A III and A IV set out in the Waste Wood Ordinance of 15 August 2002 (BGB1. I p. 3302). Gas withdrawn from the gas network shall be deemed to be biomass if the thermal equivalent of the withdrawn quantity of such gas corresponds to the quantity of biogas from biomass fed into the gas network elsewhere within the territorial application of this act.

(2) The minimum fees in accordance with paragraph (1) first sentence Nos. 1 and 2 above shall be increased by 6 cents per kilowatt-hour, and the minimum fees in accordance with paragraph (1) first sentence No. 3 above, by 4 cents per kilowatt-hour, if

1. all electricity was produced

a) from plants or parts of plants which have originated from agricultural, silvicultural or horticultural operations or during landscaping activities and which have not been treated or modified in any way other than for harvesting, conservation or use in the biomass plant,

b) from manure within the meaning of Regulation (EC) No 1774/2002 of the European Parliament and of the Council of 3 October 2002 laying down health rules concerning animal by-products not intended for human consumption (OJ L 273 p. 1), as amended by Commission Regulation (EC)

No 808/2003 of 12 May 2003 (OJ L 117 p.1), or from vinasse generated at an agricultural distillery pursuant to Article 25 of the Spirits Monopoly Act as promulgated in the Federal Law Gazette Part III No. 612-7, last amended by Article 2 of the Act of 23 December 2003 (BGBI. I p. 2924), if that vinasse is not subject to any other recovery requirements pursuant to Article 25(2) No. 3 or paragraph (3) No. 3 of that Article of the Spirits Monopoly Act or

c) from both substance categories;

2. the biomass plant has been licensed exclusively for operation with substances pursuant to No. 1, or, where such a licence is not available, the plant operator provides proof, by keeping a record of the substances used with details and documentation of the type, quantity and origin of the substances used, that no other substances are used and if

3. there are no other biomass plants on the same site which produce electricity from other substances. Notwithstanding the first sentence above, the minimum fees pursuant to paragraph (1) first sentence No. 3 shall be increased by 2.5 cents per kilowatt-hour if the electricity is produced by burning wood. The obligation to pay increased minimum fees in accordance with the first sentence above shall apply as of the date on which the requirements of the first sentence above are fulfilled. The right to payment of increased fees shall finally expire when the requirements of the first sentence above are no longer fulfilled.

(3) The minimum fees in accordance with paragraph (1) first sentence above shall be increased by 2 cents per kilowatt-hour in the case of electricity within the meaning of Article 3(4) of the Combined Heat and Power Generation Act and where proof can be furnished to the grid system operator in accordance with the *Arbeitsblatt FW 308 - Zertifizierung von KWKAnlagen – Ermittlung des KWK-Stromes* of November 2002 (*Bundesanzeiger* No. 218a of 22 November 2002) published by the *Arbeitsgemeinschaft für Wärme und Heizkraftwirtschaft – AGFW e.V.*. For series-produced combined heat and power stations with a capacity of up to and including 2 megawatts, suitable documentation available from the manufacturer, stating the thermal and electrical capacities and the electricity coefficient, may be furnished instead of proof in accordance with the first sentence above.

(4) The minimum fees in accordance with paragraph (1) first sentence Nos. 1 to 3 above shall be increased by another 2 cents per kilowatt-hour if the electricity was produced in plants using combined heat and power generation and if the biomass was converted by thermochemical gasification or dry fermentation and if the gas used for power generation was processed to reach the quality of natural gas or if the electricity is produced by fuel cells, gas turbines,

steam engines, organic Rankine cycles, multi-fuel plants, especially Kalina cycles, 10 or Stirling engines. For the purpose of adapting this provision to the state of the art, the Federal Ministry for the Environment, Nature Conservation and Nuclear Safety is authorised to issue, in agreement with the Federal Ministry of Consumer Protection, Food and Agriculture and the Federal Ministry of Economics and Labour, an ordinance detailing further processes or techniques referred to in the first sentence above or exempting some of these processes or techniques from the scope of the first sentence above.

(5) As of 1 January 2005, the minimum fees specified in paragraph (1) above for new plants commissioned after that date shall be reduced by 1.5 per cent annually of the relevant value for new plants commissioned in the previous year; the amounts payable shall be rounded to two decimals.

(6) The obligation to pay fees shall not apply to electricity produced from plants commissioned after 31 December 2006 where, for the purposes of priming or supporting fuels, biomass within the meaning of the ordinance pursuant to paragraph (7) below or vegetable oil methyl ester is not used exclusively. For plants commissioned prior to 1 January 2007, the share to be attributed to the necessary priming and supporting fuels from fossil fuels shall continue to be deemed to be electricity from biomass after 31 December 2006.

(7) The Federal Ministry for the Environment, Nature Conservation and Nuclear Safety is authorised to issue, in agreement with the Federal Ministry of Consumer Protection, Food and Agriculture and the Federal Ministry of Economics and Labour and with the consent of the *Bundestag,* an ordinance with provisions as to which substances shall be deemed to be biomass within the meaning of this provision, which technical processes may be used to produce electricity and which environmental standards must be complied with.

Article 9

Fees paid for electricity produced from geothermal energy

(1). The fees paid for electricity generated in geothermal energy plants shall amount to

1. at least 15 cents per kilowatt-hour up to and including a capacity of 5 megawatts,

2. at least 14 cents per kilowatt-hour up to and including a capacity of 10 megawatts,

3. at least 8.95 cents per kilowatt-hour up to and including a capacity of 20 megawatts and

4. at least 7.16 cents per kilowatt-hour for a capacity of 20 megawatts and over.

(2) As of 1 January 2010, the minimum fees specified in paragraph (1) above for new plants commissioned after that date shall be reduced by one per cent annually of the relevant value for new plants commissioned in the previous year; the amounts payable shall be rounded to two decimals.

Article 10

Fees paid for electricity produced from wind energy

(1). The fees paid for electricity generated by wind-powered plants shall amount to at least 5 5. cents per kilowatt-hour except as provided in paragraph (3) below. For a period of five years starting from the date of commissioning, the fees shall be increased in accordance with the first sentence above by 3.2 cents per kilowatt-hour for electricity generated in plants which during this period of time achieve 150 per cent of the reference yield calculated for the reference plant as defined in the annex to this act. For any other plants, this period shall be extended by two months for each 0.75 per cent of the reference yield which their yield stays below 150 per cent of the reference yield.

(2) In derogation of paragraph (1) third sentence above, the period stated in paragraph (1) second sentence above shall be extended for electricity generated by plants which

1. replace or modernise existing plants in the same rural district (*Landkreis*), which were commissioned no later than 31 December 1995 and which

2. at least triple the installed capacity (repowering plants) by two months for each 0.6 per cent of the reference yield which their yield stays below 150 per cent of the reference yield.

(3) The fees paid for electricity generated in offshore wind-power plants which arc located at least three nautical miles seawards from the shoreline shall be at least 6.19 cents per kilowatthour.

This shoreline shall be the shoreline as represented on map No. 2920 *"Deutsche Nordseekliste und angrenzende Gewässer"*, 1994 edition, XII, and map No. 2921 *"Deutsche Ostseeküste und angrenzende Gewässer"*, 1994 edition, XII, issued by the Federal Maritime and Hydrographic Agency on a scale of 1: 375,000[2]. For electricity generated by plants commissioned no later than 31 December 2010, the fees to be paid in accordance with the first sentence shall be increased by 2.91 cents per kilowatt-hour for a period of twelve years starting from the date of commissioning. For electricity generated by plants located at least twelve nautical miles seawards and in a water depth of at least 20 metres, such period shall be extended by 0.5 months for

[2] Official information: This publication can be ordered from *Bundesamt für Seeschiffahrt und Hydrographie*, D-20359 Hamburg.

each full nautical mile beyond 12 nautical miles and by 1.7 months for each additional full metre of water depth.

(4) In derogation of Article 5(1) the grid system operators shall not be obliged to pay for electricity generated by plants that have not proved prior to commissioning that they are able to achieve at least 60 per cent of the reference yield at the intended site. The plant operator shall furnish relevant proof of this to the grid system operator by submitting a technical expertise as defined in the annex to this act and commissioned to a technical expert in agreement with the grid system operator. If the grid system operator fails to give his consent within four weeks following the plant operator's request, the Federal Environmental Agency shall name the technical expert after consulting the Fördergesellschaft Windenergie e.V. (FGW). The plant operator and the grid system operator shall bear 50 per cent of the costs each.

(5) As of 1 January 2005, the minimum fees specified in paragraph (1) above, and as of 1 January 2008, the minimum fees specified in paragraph (3) above, for new plants commissioned after these dates shall be reduced by 2 per cent annually of the relevant value for new plants commissioned in the previous year; the amounts payable shall be rounded to two decimals.

(6) For the purpose of implementing paragraphs (1) to (4) above, the Federal Ministry for the Environment, Nature Conservation and Nuclear Safety is authorised to issue an ordinance regulating the calculation and application of the reference yield.

(7) Paragraphs (1) to (6) above shall not apply to electricity generated by wind-powered plants whose construction was licensed after 1 January 2005 in an area of Germany's exclusive economic zone or coastal waters which has been declared a protected area of nature and landscape in accordance with Article 38 in conjunction with Article 33(2) of the Federal Nature Conservation Act or in accordance with Land legislation. The first sentence above shall also apply to such areas which the Federal Ministry for the Environment, Nature Conservation and Nuclear Safety has notified to the Commission of the European Communities as sites of Community importance or as European bird sanctuaries, until they have been declared protected areas.

Article 11

Fees paid for electricity produced from solar radiation

(1) The fees paid for electricity generated by plants using solar radiation shall amount to at least 45.7 cents per kilowatt-hour.

(2) If the plant is attached to or integrated on top of a building or noise protection wall, the fees shall be

1. at least 57.4 cents per kilowatt-hour up to and including a capacity of 30 kilowatts,

2. at least 54.6 cents per kilowatt-hour for a capacity 30 kilowatts and over, and

3. at least 54.0 cents per kilowatt-hour for a capacity of 100 kilowatts and over.

The minimum fees in accordance with the first sentence above shall each be increased by 5.0 cents per kilowatt-hour if the plant is not integrated into the roof or designed to be the roof of the building and if it forms a substantial part of the building. Buildings shall be understood as meaning roofed building structures that can be independently used and entered by humans and are suitable for or designed for the purpose of protecting humans, animals or objects.

(3) In cases where the installation is not attached to or integrated on top of a building structure used primarily for purposes other than the generation of electricity from solar radiation, the grid system operator shall only be obliged to pay fees if the installation was commissioned prior to 1 January 2015

1. within the scope of application of a local development plan within the meaning of Article 30 of the Federal Building Code or

2. on a site for which a procedure in accordance with Article 38 first sentence of the Federal Building Code was carried out.

(4) For electricity generated in an installation in accordance with paragraph (3) above erected within the scope of application of a local development plan drawn up or amended at least also for this purpose after 1 September 2003, the grid system operator shall only be obliged to pay fees if the installation is located on

1. plots of land which were already sealed when the decision on drawing up or amending the local development plan was adopted,

2. land converted from economic or military use or

3. on green areas designated for the construction of this installation in the local development plan and used as cropland at the point in time when the decision on drawing up or amending the local development plan was adopted.

(5) As of 1 January 2005, the minimum fees pursuant to paragraph (1) and paragraph (2) first sentence above paid for new plants commissioned after that date shall be reduced by five cent annually of the relevant value for new plants commissioned in the previous year; the amounts payable shall

be rounded to two decimals. As of 1 January 2006, the relevant percentage pursuant to the first sentence above for plants specified in paragraph (1) above shall be increased to 6.5 per cent.

(6) For the purposes of calculating the amount of the fees in accordance with paragraph (2) above for the installation which was commissioned most recently, and in derogation of Article 3(2) second sentence, several photovoltaic installations attached to or on top of the same building and commissioned within six consecutive calendar months shall be deemed to be one installation even if they are not directly attached to building structures and installations that are commonly used and are technically necessary for operation.

Source: Excerpted from the Web site of Germany's Federal Ministry for the Environment, Nature Conservation, and Nuclear Safety (http://www.erneuerbare-energien.de/inhalt/6465/36356).

2005 Report by the Federal Republic of Germany on achievement of the indicative target for electricity consumption from renewable energy sources by 2010, Berlin, October 2005

Germany's energy industry as a whole has agreed to invest 30 billion euros in constructing new power plants and other forms of energy infrastructure by 2012, and 40 billion euros in the expansion of renewable energy use. For a report on the progress Germany had made in obtaining a greater amount of electricity consumption from renewable sources, see this report published in 2005 by the German government.

Germany is aiming to increase the proportion of electricity generated from renewable sources to at least 12.5% by the year 2010, and to at least 20% by the year 2020. After 2020, these targets will be substantially upgraded. By the year 2010, it is hoped that renewable energies will account for at least 4.2% of total energy consumption, and around half of energy consumption by the middle of the century. The Federal Government is keen that renewable energies should become competitive in the internal energy market in the medium to long term. Renewable energy sources will only be able to play a permanent supporting role in the energy market when they are able to assert themselves on the market without the aid of financial support. Consideration of the various external costs (in particular, long-term environmental and climate damage) of both conventional and renewable energy sources while ensuring economic viability remains an important goal at both national and international level.

The Renewable Energy Sources Act (EEG) is a pivotal element of the Federal Government's raft of environmental and energy policy measures. The Act was adopted by the *Deutscher Bundestag* (Lower House of Parliament) on 29 March 2000, and entered into force on 1 April 2000, replacing the Act on the Sale of Electricity to the Grid (StrEG) in force since 1991, which likewise contained a supply and fee payment system in favour of regenerative electricity. Hence, when developing the EEG, we were able to draw on 10 years of experience with the StrEG. The EEG was adapted in line with the conditions of the liberalised electricity market and a number of significant improvements were added. In particular, the new system of fees, differentiated according to the various renewable energy segments, served to balance out the remaining competitive disadvantages of renewable energies versus conventional electricity generation, and cleared the way for the use of biomass, solar power and geothermal energy in electricity production.

In 2004, the EEG was comprehensively updated for the first time, bringing it into alignment with the Federal Government's sustainability strategy, implementing Directive 2001/77/EC, and offering longer-term prospects to the affected players. Specifically, the amendments

- improved the incentives for improved energy efficiency and technical innovations to renewable energy plants, e.g. by offering a bonus for cogeneration plants
- created better supply conditions for the energetic use of biomass and geothermal energy, and included the modernisation of large hydropower plants in the compensation scheme
- adjusted the fees for windpower, and in part significantly increased the rates for solar power, to compensate for the expiry of the 100,000 roofs programme
- increased the annual degression in the fees for new installations to strengthen the incentives for technical innovations and cost-cutting,
- incorporated a number of provisions designed to increase transparency and consumer protection, and
- made enforcement of the law easier.

With particular regard for the polluter pays principle, the provisions of the law serve to implement the requirement to protect the natural bases of life in view of our responsibility toward future generations, as stipulated in Article 20a of Germany's Basic Law, as well as implementing the environmental protection targets set out in Articles 2, 6, 10 and 175 of the Treaty Establishing the European Community.

In recent years, the EEG has effected a significant increase in the use of renewable energy sources to generate electricity, by offering guaranteed fee rates secured over a period of many years. This is particularly true in the case of windpower, which has developed to become the leading source of electricity generation ahead of hydropower.

The emission savings that are made by generating electricity from renewable energies are particularly beneficial from an environmental and climate policy viewpoint: As a result of the EEG and the annexes initiated by its preceding law and later incorporated into the EEG, some 33 million tonnes of carbon dioxide were saved in 2004 alone.[1] Once the set targets have been met, this figure will rise to more than 50 million tonnes of carbon dioxide in the electricity sector alone by the year 2010. The growth in renewable energies is a major factor in reducing emissions of greenhouse gases in Germany by 21 percent during the 1st commitment period of the Kyoto Protocol, within the context of burden sharing within the EU.

In 2004, some 52.2 million tonnes of CO_2 were avoided in total thanks to the generation of electricity from renewable energies (installations with and without entitlement to fee payment under the EEG).[2] Emissions of other pollutants were also avoided as a result of electricity generation from renewable energies. In 2004, these included some 28,700 tonnes of SO_2, 32,900 tonnes of NO_x, 16,700 tonnes of CO and 700 tonnes of non-methane volatile organic compounds (NMVOC).

Additionally, CO_2 savings of around 14.6 million tonnes were achieved in the heating sector in 2004 thanks to the use of renewable energies, while the use of biofuels led to savings of around 3 million tonnes in the same year. Overall, then, the use of renewable energies in Germany led to CO_2 savings of around 70 million tonnes.[3]

---

[1]The significant increase in this figure compared with that given in the previous report of 2002 has arisen as a result of the growth in the use of renewable energies for the generation of electricity under the EEG, and current figures on the CO_2 emissions avoided in the electricity sector as a result of renewable energies.

[2]The significant rise in this figure compared with that given in the previous report of 2002 has arisen as a result of the growth in the use of renewable energies for the generation of electricity under the EEG, and current figures on the CO_2 emissions avoided in the electricity sector as a result of renewable energies.

[3]CO_2 emission factor with the current energy mix in Germany: Reduction in emissions via the use of renewable energies by 0.935 kg CO_2/kWh for electricity, 0.229 kg CO_2/kWh for heat and 0.351 kg CO_2/kWh for biofuels.

Source: Excerpted from the Web site of Germany's Federal Ministry for the Environment, Nature Conservation, and Nuclear Safety (http://www.bmu.de/english/renewable_energy/doc/5305.php).

DOCUMENTS OF IRAN

Excerpts from Iranian president Mahmoud Ahmadinejad's statement at the news conference in Tehran, Iran, January 14, 2006

Iranian president Mahmoud Ahmadinejad has consistently defended Iran's decision to conduct nuclear research and has contended that Iran has the right to research peaceful uses of nuclear technology. For an example consider his statement at the news conference in Tehran on January 14, 2006.

A nation which has culture, logic and civilization does not need nuclear weapons. The countries which seek nuclear weapons are those which want to solve all problems by the use of force. Our nation does not need such weapons.

They ask us why we have started [nuclear] research. Our reply is that there is no limitation to research. There are no limits imposed on research in the Nuclear Non-Proliferation Treaty nor in the Additional Protocol. Nor have we made any such commitment.

They say openly that they are opposed to this research. By what right do you make such a statement? Is this not a fundamentalist medieval perspective? We believe that, unfortunately, despite their technological and scientific development, in certain parts of the world, several Western countries still have an ideological and intellectual perspective which belong to the medieval age.

Today, the language of hegemony, of force, of nuclear, chemical and biological weapons is no longer applicable. These things are no longer effective in international dealings. Today, nations have awakened, and they will determine their own future.

We have adhered to international laws and continue to do so. Over 1,400 days of inspection have been carried out at our facilities—that is unprecedented.

We have put forward a proposal calling for the UN to set up a disarmament committee. But some countries hold a stick over our head during the negotiations, threatening to refer our case to the UN Security Council unless we bend to their will. Why do they tarnish the integrity of international institutions? Why do they force the IAEA [International Atomic Energy Agency] to get involve with politics?

Our path is very clear. We shall continue with our activities within the regulations of the IAEA and the NPT. According to international law, Iran has the right to the peaceful use of nuclear energy. We are acting in the framework of the IAEA's regulations. Today, we are carrying out research, I should also say that our people will not accept that others should impose on us whatever they want. Our nation has a definite right to peaceful nuclear technology and will achieve it.

We are the only nation that has invited all others to come and join us. If you say that you do not trust us, you can come and become our partner. We are ready to become your partner in your [nuclear] technology. We can each supervise the other's activities. We can watch that you do not deviate towards nuclear weapons, and you can also become our partners and monitor our activities directly.

How do you justify this logic of having a full arsenal of nuclear weapons, but when it comes to nations such as ours, you do not even allow research? This logic cannot rule the world today. Even if the Security Council gets involved in this subject, it will not help solve the equation. We do not want to move in this direction. But those who insist on undermining our rights should know that this will not happen.

Source: Excerpted from the Web site of BBC News (http://news.bbc.co.uk/1/hi/world/middle_east/4613644.stm).

"Zanganeh stresses increased recovery, collecting associate gas" as reported by *Iran Petroleum Magazine,* December 2005

As reported in this article, Iran's oil minister, Bijan Namdar Zanganeh, has called for increased oil production and improved oil recovery in Iran. Plans included doubling national oil production to more than 7 million barrels per day by around 2015.

TEHRAN—Minister of Petroleum Bijan Namdar Zanganeh yesterday stressed the need to increase oil recovery, collection of associated gas as well as scientific and engineering research in oil industry. Addressing the inaugural ceremony of six oil and gas projects in south oil regions, he noted that the government has paid due attention to the above goals during recent years.

The minister stated that associated gas collection project at Bangestan oil field, know as Amak project, will soon become operational and will lead to daily collection of 300 million cu. ft. (8.5 million cu. m.) associated gas as well as production of 30,000 barrels gas liquids.

Zanganeh also stated that the government gives priority to establishing desalting plants and noted that studies on 14 reservoirs in south oil regions including Bibi Hakimeh, Ahvaz Asmari, Maroun Asmari, Ahvaz Bangestan, Ab Teimur, Mansouri, Kerenj, Parsi, Rag Sefid, Shadegan, Pazenan, Maroun Bangestan and Mansouri Asmari have been finished.

"Those reservoirs account for more than 85 percent of production in south oil regions," he said.

Stressing the need to increase oil recovery, the stated that keeping production level has been an important step taken in south oil regions during past years because out of 300,000 barrels annual reduction in production, about 250,000 barrels is accounted for by south oil regions.

He noted that increased gas injection into oil wells was another important step taken in south oil regions, adding that gas injection into oil wells has increased by 50 percent during recent years.

The minister stated that gas injection averaged 57 million cu. m. in 1996 and will reach about 90 million cu. m. by early 2005 with current year's figure being estimated at 100 million cu. m.

Referring to 2-billion-dollar contract for building Bidboland 2 refinery he said the plant will sweeten about 2 billion cu. ft. (56.65 million cu. m.) natural gas, which will be used as petrochemical feedstock after separating ethane and heavy hydrocarbons.

He also explained [the importance of] oil engineering operations in south oil regions over the past years and noted that many steps have been taken over preceding years to uplift the status of oil engineering through delegating maximum powers to National Iranian South Oil Company and affiliated companies.

Zanganeh further noted that Iranian economy and budget heavily rely on oil, adding, "That dependence is not good, but it must not be used as a pretext to weaken oil sector or insult its workers. If we want to reduce dependence on oil, we must develop other economic sectors by using oil revenues. . . . If we invested 20 billion dollars in [the] oil industry over the upcoming 10 years, we would be able to reduce our dependence on oil in foreign trade by developing [the] petrochemical industry," he added.

Source: "Zanganeh stresses increased recovery, collecting associate gas." Iran Petroleum Magazine 24, December 2005. Available online. http://www.iranpetroleummag.com. Accessed on January 29, 2006.

DOCUMENTS OF SAUDI ARABIA

"A Roadmap to the Energy Future: Saudi Arabia's Perspective," speech by Ali I. Al-Naimi, Minister of Petroleum and Mineral Resources in Saudi Arabia, CERA's Fourth Annual Global Oil Summit, Houston, Texas, February 7, 2006

After the U.S. invasion of Iraq in March 2003, which placed Iraqi oil production on hold, Ali bin Ibrahim al-Naimi, Saudi Arabia's minister of petroleum and mineral resources since 1995, asserted Saudi Arabia's commitment to stepping up its own oil production to fill the void in Middle Eastern oil exports. In this speech in February 2006 he declared that Saudi Arabia would increase its production while keeping oil prices competitive.

Ladies and gentlemen,

There is much talk these days about roadmaps. In the energy industry, the great uncertainty and risk we face as consumers and producers are generating talk of the need for both energy supply and demand roadmaps. Understandably, there has been great interest in Saudi Arabia's own roadmap, specifically our plans for adding capacity in both the upstream and the downstream. In light of this, I would like to share with you Saudi Arabia's views about the energy future and the roadmap that outlines the journey on which we have embarked.

As we all know from personal experience, roadmaps are invaluable tools for getting us from where we currently are to where we want to go. As the saying goes, "Whoever travels without a guide needs 200 years for a 2 day journey."

Unfortunately for all of us, our energy future is not as well-defined or as clearly laid out as the maps that you can buy at your local gas station. The roadmap to our energy future is dynamic and evolving. We know that change is inevitable, but we are unsure of its pace. Challenges will arise and require adjustments along the way. Unforeseen events could create roadblocks requiring temporary detours.

I want to begin by telling you what you won't see in our roadmap: a forecast of oil prices.

Predicting oil prices in a world of imperfect information is merely speculation. Experience has repeatedly shown us that our ability to predict accurately the movement of oil prices is extremely limited and inconsistent. The fact is, none of us knows what the price of oil will be next month, next year or 10 years hence.

Oil prices are the product of the complex interaction of numerous forces and the random events that operate daily in international oil markets. Our comprehension of these forces and how they affect prices is limited by lack of timely information and by incomplete understanding of the processes at work. The task is made all the more difficult by our inability to predict the unpredictable: those seemingly random political, economic and natural events that can turn the status quo on its head.

Instead of trying to guess where prices are headed, let us look at those forces that will shape the future of our industry. Furthering our understanding of these forces and their implications increases our knowledge of the road ahead.

Let me first talk about where we currently are as an industry. Then I will offer my thoughts on where we are going, and the challenges along the way.

Ours is an industry which has been blessed with an abundant and readily accessible resource that can be produced in large quantities at relatively low costs. It is therefore not surprising that we have operated for much of our history in an environment of abundance, where too much supply was chasing too little demand.

As a result, the oil industry, more often than not, has faced the problem of managing excess capacity in the supply system. Today, we face a different environment, one where there are a myriad of constraints on supply. Those constraints are the product of the cyclical nature of investment patterns in the oil industry and the changing cost structure of oil supplies.

In the early days of the industry, chaos was, more often than not, the rule of the day. New oil discoveries were plentiful, as wildcatters, drawn by the prospects for instant wealth, rushed to cash in on the new "black gold." In this get-rich-quick environment, there was little thought given to the limits of the resource base and to the sustainability of prevailing practices. Wells were drilled with impunity, without consideration for sound reservoir management. The guiding force was short-term gain. Many fields were damaged and their long-term potential compromised.

The watershed event was the discovery and development of the huge East Texas oil field in the early 1930s. Unrestricted production from this giant caused world oil prices to plummet and created chaos in the oil industry. A coalition of oilmen, scientists and public officials recognized that it was imperative to halt this self-destructive behavior.

After a protracted political struggle, the Texas Railroad Commission was given the authority to pro-rate oil and gas production in the State of Texas, thereby restoring order to the industry.

By limiting the production from each well in Texas, the Commission succeeded in supporting prices and conserving the State's valuable resource.

Fast forward to the post-war period and we see that the experiences of the early days of the U.S. industry were being played out on the world stage. The giant oil finds of the 1950s and 1960s were no longer in the U.S., but primarily in the Middle East and Africa. But the results were similar to the U.S. experience. Oil was plentiful and prices were depressed. There was no incentive for conservation and efficiency, a fact that would have a profound impact on the development of the post-World War II economic infrastructure.

In this environment, oil use and the industry grew by leaps and bounds. The problem was that this rapid expansion of the industry, like those early days in Texas, was built upon an unsustainable price that did not reflect the true long-term value of a depleting resource.

But unlike in Texas, there was no global Railroad Commission to step in to help conserve this valuable resource. That role fell to the members of OPEC, who assumed stewardship of their own petroleum resources in the 1960s and 1970s, and in doing so, gained control of a large portion of the world's reserves.

By the end of the 1960s it was becoming apparent that the low price environment of the 1950s and 1960s was no longer sustainable. Demand was rising rapidly and spare crude oil production capacity was quickly diminishing. Pressure began to grow for prices to rise. From OPEC's perspective, the longer prices remained "artificially" low, the greater the waste of the resource base.

The market was clear: things could not continue on as they were. Higher prices were needed to encourage conservation and efficiency and to spur the development of new supply. Without such an adjustment, the oil market was headed for a so-called "train wreck."

The adjustment process in the early 1980s was both rapid and painful for the oil industry. Demand fell as inefficiency and waste were wrung out of the system. At the same time, higher prices encouraged the industry to bring to market previously undeveloped crude oil reserves.

Unfortunately, the adjustment process brought on by higher prices was not precise and the pendulum swung back toward oversupply. With demand contracting and supply rising, the oil industry's ability and desire to provide the market with refined products far exceeded what consumers actually wanted. The only way balance could be restored was through growth in demand and/or capacity reduction. There was no quick solution. In fact, it took almost 20 years. But spare capacity has now been removed from the system. So much so, that we can no longer accommodate a significant increase in demand without major new investments.

Now I would like to turn to the future and share my thoughts on where we as an Industry are headed.

Some would say that the outlook for the oil industry is quite favorable: prices are high, demand is growing, capacity across the supply chain is tight, and there appears to be no serious challenge to oil's role as the dominant transportation fuel.

Current conditions are positive, if one's perspective is merely short-term. A healthy long-term market is one where demand and supply are in balance and there is sufficient spare capacity to ensure the system can handle unexpected surges in demand or disruptions to supply smoothly, and without sharp price spikes. In my opinion, the current tight capacity conditions are not conducive to this outcome. In the current environment, market volatility is exacerbated. The lack of global spare capacity magnifies the price impact of relatively minor supply disruptions or demand surges.

A healthy oil market in balance is one where prices benefit both consumers and producers. It is imperative that prices be high enough to provide sufficient return to producers, but not so high that they harm economic growth. When oil prices are too high or too low they become unsustainable. Oil prices should always provide an incentive to conserve and to use this valuable resource efficiently.

The question facing the oil industry today is: can it achieve a balanced market that both ensures a bright future for the industry, while at the same time fulfilling the aspirations of the world's people? I'm convinced the answer is "Yes," because the two go hand-in-hand.

We need each other. I believe the future of our industry is inextricably tied to mankind's aspirations for a better way of life. Without this, our industry stagnates. In turn, I believe that the future economic well-being of the world's people requires an oil industry that is vibrant with the capability of delivering the most cost-effective energy source known to mankind.

We in the oil industry can and will provide the safe and clean energy products required to achieve these aspirations. But it will not happen without skill, effort and some good old-fashioned entrepreneurial risk-taking. There is no denying that there are significant risks on both sides of the equation, demand and supply.

I will now take a closer look at the major risks and hurdles that we are likely to face as we move forward. I will first discuss demand and then supply.

On the demand side of the equation, the primary reason for optimism is the underlying long-term trend growth in the world economy. Aided and enhanced by globalization, experts expect three simple factors: population growth, productivity gains and the desire for a better life to continue to propel the world economic expansion in the coming years.

A growing world economy is good for business. In fact, experts project rising economic activity to significantly boost oil demand in the coming years. But a crucial question is "By how much?" On this point the experts disagree, in part, because of their differing assumptions about the strength of the economic expansion and the rate of efficiency gains in energy use and the role of alternative fuels.

At this time, oil appears to be well-positioned to capture its fair share of any increase in world energy demand. Given current and anticipated technologies, it appears highly likely that oil will remain the fuel of choice in the transportation sector for many decades to come.

While merely knowing that oil is going to remain important in the energy mix is crucial, and I might add gratifying, it is not enough for an industry with massive capital requirements.

Projections of future demand are not merely number exercises, they are critical to our business. We in the oil industry face the prospect of undertaking long lead-time mega-projects requiring massive commitments of capital with long payback periods. There is little room for error when one is dealing with projects on such a scale.

A healthy oil industry will require that the growth in supply roughly matches the growth in demand, not only in the aggregate, but also in terms of specific products. Our knowledge of demand and its potential growth is probably the weakest link in our efforts to understand the future. The root of the problem is the inadequacy of existing data collection and analysis in much of the world. Good data is crucial for developing a baseline for projecting the future. In addition, our ability to know and predict the drivers of oil demand growth in many countries is weak at best.

We believe better demand data and analytical capabilities are crucial to a smooth transition to the future. As I said previously, the industry is being asked to commit huge amounts of capital to meet the future energy needs of the world's people. Unfortunately, we often are forced to do so with inaccurate or missing data on demand. I therefore wholeheartedly support efforts by consuming countries to develop energy demand roadmaps. The development of these roadmaps would help us all prepare for the future.

To this point, my focus has been on the quantity of demand, but there is another aspect of demand, what I will call the quality of demand. It is just as crucial a driver of investment decisions as quantity. By every account, the demand for clean energy products will continue to rise in the coming years. We in the oil industry are strong supporters of a clean environment and we are investing large sums in research to make sure that our products meet the highest environmental standards.

However, we need to provide these products to consumers at reasonable costs so that they are both economically and environmentally friendly. To do this, the industry requires certainty and rationality with regard to environmental mandates and product specifications. Constantly changing environmental standards or allowing a proliferation of product specifications will undermine industry efforts to provide products that are the most economically and environmentally beneficial to consumers.

More broadly speaking, I believe that we should not impoverish people in the name of a cleaner environment. There is an economic cost associated with improving the environment that must be borne by consumers. We must always strive to achieve our goals in a manner that is good for both the environment and the economic well-being of the world's population. Lowering living standards or limiting peoples' ability to rise out of poverty in order to improve the environment trades one potential health hazard for another.

This is exactly what we would be doing by mandating, as some propose, that consumers forego using oil for some less efficient and more costly alternative that under normal circumstances could not meet the test of the market place. A more sensible approach is to improve the products that we use and the way we use them. That's why we in Saudi Arabia, for example, are supporting efforts to research and develop carbon sequestration techniques and processes. We believe this is a superior approach because it can be effective in reducing CO_2 concentrations in the atmosphere at substantially lower risk than a radical new approach like moving away from the use of hydrocarbons.

We should also recognize that consuming government policies aimed at reducing oil demand create another element of uncertainty for producers. This added risk is detrimental to timely investment decisions.

Finally, before I turn to the supply, I would like to address one other demand uncertainty which lurks in the background. This uncertainty is technological innovation. It is potentially the most disruptive force we face. The risk is all the greater due to its unpredictability.

The one thing we can be certain of is that humans will continue to innovate and create. It is in our nature. The potential impact on energy demand, on the fuels we use and how we use them is beyond compare. I am certain that one day we will innovate ourselves right out of our current hydrocarbon-based energy model. Today we use hydrocarbons, tomorrow who knows? The difficulty for the oil industry is that we cannot predict when and how fast the transition will occur.

Currently, we do not see anything on the immediate horizon, but the potential for a major technological upheaval is always there.

Even if the transition is still a long way off, we can expect technological innovations to continue to improve the efficiency of our energy use, making a valuable contribution to meeting our future energy needs. Governments can make a positive contribution to our energy future by promoting energy efficiency, not through the tax code, but by investing in research and development of new technologies.

Now I want to turn our attention to the outlook for energy and oil supplies. There are the doubters out there who say that oil production is past its peak and that the industry will not be able to keep pace with the increases in demand.

As I told the delegates at the recent World Petroleum Congress in South Africa, what we face is a deliverability problem, not a problem of availability. Deliverability is the capacity of the oil industry to develop, produce, refine, transport and deliver to consumers the energy they require in their daily lives. This is primarily an investment issue, whereas availability is a resource issue.

With regard to availability of crude reserves, there is plenty of conventional oil left to be found and produced. My assessment is based on a careful analysis of USGS resource estimates and a conservative assumption about the likely positive impact of technological innovation and prices on finding and recovery rates. The future looks even brighter when one considers the vast quantities of unconventional oil reserves that exist.

The power of technology is sometimes not given proper consideration. Let me offer a simple example which has great significance. Two months ago, I was in the Divided Zone between Saudi Arabia and Kuwait, witnessing a demonstration of an experimental method to extract more oil from heavy petroleum fields. This process is called "steam injection." It is expected to increase the recovery rate of heavy oil fields from about 6 percent to more than 40 percent This represents the kind of innovation that holds great promise for meeting the world's energy needs.

Now I will turn my focus to Saudi Arabia and our plans for the future. Our overriding objective in the future is to continue to be a source of stability for world energy markets. To this end, we in the Kingdom are addressing the problem of deliverability head on. We are doing this by committing an unprecedented level of resources to projects aimed at increasing upstream and downstream capacity.

In the upstream, Saudi Arabia is undertaking a very ambitious program to increase output capacity. Our plan for the next four years is to increase gradually our production capacity from the current level of 11 million barrels a day to 12.5 million. As a first step, we are currently bringing online an additional 300 thousand barrels per day of light crude from the Haradh field.

In the downstream, Saudi Arabia is also doing its part to ease the refining capacity bottleneck by investing locally and internationally. We are planning to build two new grassroots joint venture export refineries—one in Jubail on the Kingdom's east coast and the other in Yanbu' on the west coast. Each of these facilities will have a 400 thousand barrel-per-day capacity. We will also be expanding our Ras Tanura Refinery, and are studying the possibility of transforming it into an integrated refining and petrochemical complex.

We have already embarked on a similar expansion at our west-coast Rabigh Refinery, in partnership with Sumitomo Chemical in a $9 billion program to transform the existing facility into one of the world's largest refining and petrochemical complexes. Upon completion in 2008, this refinery will be able to produce fuels that meet European and U.S. market specifications.

In addition, we are upgrading our existing Yanbu' Refinery to increase its complexity and boost capacity by 100,000 b/d. As is the case with other refineries, we are also looking at petrochemical integration plans for that site.

Our international downstream investments will also be substantial. For example, here in the U.S., Motiva is looking at expanding one of its Gulf Coast refineries by up to 300,000 b/d.

Our plans are even more ambitious in Asia. In China, we are partnering with ExxonMobil and Sinopec to expand an existing refinery in China's Fujian Province, and to add petrochemical facilities to the complex. We are also working with Sinopec to study the possibility of partnering in a grassroots refinery in Qingdao, Shandong province.

Our plans also include the possibility of building a new grassroots refinery in South Korea that would give us the ability to supply additional refined products to the Korean market and to fast-growing markets like China.

Taken together, these projects mean that over the next five years, we will be boosting our total refining capacity by almost 50 percent, to some six million b/d.

As you can see, Saudi Arabia's roadmap is an ambitious one. I am convinced that the investments we have and will undertake are crucial to future oil market stability.

* *** *

Source: Excerpted from the Web site of Saudi Aramco, the national oil company of Saudi Arabia. Available online. URL: http://www.saudiaramco.com/bvsm/JSP/content/articleDetail.jsp?BV_SessionID=@@@@1388990870.1166647372@@@@&BV_EngineID=cccdaddjjjekelmcefeceefdfnkdfhl.0&datetime=12%2F20%2F06+23%3A43%3A48&SA.channelID=11748<0x 0026>SA.programID=19469&SA.contentOID=1073765890. Accessed on August 2, 2006.

"Saudi Nuclear Intentions and the IAEA Small Quantities Protocol" by Alex Pojedinec, Center for Defense Information (CDI), June 30, 2005

Saudi Arabia signed the International Atomic Energy Agency (IAEA) Small Quantities Protocol in 2005, which allows countries with a low risk of nuclear weapons proliferation to opt out of extensive inspections for nuclear weapons if they make a disclosure about their nuclear activities. By signing this protocol, Saudi Arabia has raised suspicion in the global community that it is in the process of developing nuclear weapons and is trying to avoid detection of its nuclear activities.

Saudi Arabia's recent signing of the International Atomic Energy Agency (IAEA) Small Quantities Protocol on June 16, 2005, raises larger issues of possible nuclear ambitions on the Arabian Peninsula. With the cooling off of relations between Riyadh and Washington, and the resultant retreat of U.S. security assurances to the region, it would seem a logical conclusion for the

Saudi government to begin to look for other means of self-preservation, especially when faced with two regional nuclear powers: Israel with a rumored 200-plus weapons, and Iran with a nascent program purportedly in development.

The controversy centers on Saudi Arabia's request to the IAEA to sign the Small Quantities protocol in May, after an internal IAEA document came to light calling for a change in the status of the Protocol over concerns it may pose a proliferation risk. The Protocol allows states considered to be of low risk to opt out of more intensive inspection regimes in return for a declaration of their nuclear activities. In addition, the Protocol allows for the possession of up to 10 tons of natural uranium or 20 tons of depleted uranium, and 2.2 pounds of plutonium without reporting. While it does not appear that Saudi Arabia aspires to develop a domestic weapons grade uranium or plutonium-processing ability, 10 tons of natural uranium is still enough by most standards to produce between one and four nuclear devices (depending on their design). In theory, the Protocol is supposed to allow the IAEA to focus its efforts on other nations viewed as being of higher proliferation risk. Despite strong concerns of IAEA board members, the European Union, the United States and Australia that the Protocol may provide a loophole for would-be proliferators, the IAEA approved Saudi Arabia's request to join the 75 other nations that are already Protocol signatories. Consequently, Saudi Arabia is exempt from normal UN inspections, effectively ceasing nuclear monitoring there, in exchange for a declaration tantamount to a self-policing agreement.

Although Saudi Arabia is not traditionally considered a regional proliferation threat, the dynamic changes presently taking place in the Middle East necessitate taking a closer look at the motives behind the Saudi request. The timing is particularly suspect, since the validity of the Small Quantities Protocol itself is currently in question by the IAEA. After years of advocating a nuclear-free Middle East, a Saudi request to join the Protocol at this point seems opportunistic considering the circumstances surrounding both the Protocol's status and Saudi Arabia's tumultuous nuclear past.

There is a slew of evidence that Saudi Arabia sought to acquire nuclear capabilities as early as 1975 when a nuclear research center at Al-Suleiyel was created. Further evidence points to a transfer of up to $5 billion to Iraq from 1985 until just prior to the first Gulf War in a deal to further the Iraqi nuclear program in exchange for weapons, should the program prove successful. There was apparently also an offer on the table to pay for reconstruction of the Osirak reactor destroyed by Israel, whose covert nuclear capabilities make it a mutual concern of Iraq and Saudi Arabia. Lastly, several high-level exchanges between Saudi and Pakistani officials and a general warming of relations between these two countries points to Saudi Arabia not only having the intent, motivation, and impetus to procure nuclear weapons, but now also the means.

Interestingly enough, any nuclear threat Saudi Arabia may face from Iran may actually have been proliferated by those whose nuclear program was also funded by the Saudis and whose help the Saudis are now seeking: Pakistan. After the mid-1994 defection to the United States of a former Saudi ambassador to the United Nations, Muhammad Khilewi, thousands of documents were uncovered, some of which hinted at an agreement by which Saudi Arabia partially funded Pakistan's bomb project in exchange for retaliation with these nuclear weapons in the event of nuclear aggression against the Saudis. In 1999, the reciprocity of this nuclear alliance became even more apparent as Saudi Second Deputy Prime Minister Prince Sultan bin Abdul Aziz visited Pakistan's Kahuta uranium enrichment plant with Pakistan's Prime Minister Nawaz Sharif and both were personally briefed by Dr. A.Q. Khan. Then, in 2002, a son of Crown Prince Abdullah attended the firing of the Ghauri, Pakistan's new nuclear-capable medium-range missile. Further attesting to the cordial nature of the alliance, Nawaz Sahrif the prime minister of Pakistan deposed by Pervez Musharraf's 1999 coup, was given amnesty in Saudi Arabia through a deal worked out between Islamabad and Riyadh.

Other evidence for having nuclear intentions stems from Saudi Arabia's 1988 purchase of between 50 and 60 Chinese CSS-2 missiles. While these missiles are now largely considered obsolete, it is the purchase of a nuclear capable missile with a 3,500 km range and 2,500 kg capacity that is damaging to Saudi claims of innocence. Apparently of concern is the gross inaccuracy of the Chinese missile, rendering it completely ineffective for use with traditional warhead payloads. This points to a possible conclusion that one intended use could be with nuclear warheads, whose destructive radius negates the inherent inaccuracy of the missile. In addition, there has been recent speculation of prospective purchases of more modern Chinese missile systems (such as the CSS-5 and CSS-6) by Saudi Arabia.

The recent decline in U.S.-Saudi relations may be indirectly responsible for increased pressure felt by the Saudis to find alternative security arrangements in the region. With the movement of Central Command to Qatar in 2002, and resultant decline in U.S. military presence in Saudi Arabia, the Saudis are feeling hard pressed when faced with the geopolitical realities of the Middle East, especially when the recent behavior of Iran is taken into account. The Small Quantities Protocol exempts Saudi Arabia from IAEA inspections at a time of perceived threat to regional stability, and therefore the Protocol may provide ample opportunity for Saudi Arabia to actively consider a nuclear option as a viable alternative or supplement to current security assurances from the West. The Saudis face the possibility of a nuclear armed adversary, Iran, whose Shiite government opposes the Sunni monarchy in place, as well as the Sunni oppression of Shiite groups within Saudi Arabia. The threat of an Iranian nuclear program coming on-line, an already viable Israeli nuclear program, and a decline in relations with the traditional protector of the region may provide Saudi decision-makers with enough incentive to consider procuring non-traditional means to assure their security.

Source: Alex Pojedinec. "Saudi Nuclear Intentions and the IAEA Small Quantities Protocol," Center of Defense Information, June 30, 2005. Available online. URL: http://www.cdi.org/program/document.cfm?DocumentID=3050&from_page=../index.cfm. Accessed on August 2, 2006.

DOCUMENTS OF NIGERIA

Federal Republic of Nigeria Official Gazette, 1st March, 2004: A Bill for An Act to Provide . . . (i.e., the Electric Power Sector Reform Act)

The Electric Power Sector Reform (EPSR) Act, signed into law in March 2004, aims to reform the nation's unreliable and disorganized electricity generation system. The law enables private power companies to participate in electricity generation, transmission, and distribution and separates NEPA into eleven electricity distribution firms, six electricity-generating companies, and an electricity transmission company. The law calls for all of these operations to be privatized. Nigeria's passage of the EPSR Act set in motion the privatization of NEPA and a long-awaited reform of the nation's unreliable and disorganized electricity generation system.

Federal Republic of Nigeria, Official Gazette, 1st March, 2004:

. . .

A Bill

for An Act to Provide for the Formation of Companies to Take Over the Functions, Assets, Liabilities, and Staff of the National Electric Power Authority; Develop Competitive Electricity Markets; Establish the Nigerian Electricity Regulatory Commission; Provide for the Licensing and Regulation of the Electricity; Enforce Such Matters as Performance Standards, Consumer Rights and Obligations; Provide for the Determination of Tariffs; and Related Matters

(Electric Sector Policy Reform)

Enacted by the National Assembly of the Federal Republic of Nigeria—
Part I—Formation of Initial and Successor Companies and the Transfer of Assets and Liabilities of the National Electric Power Authority

1. The National Council on Privatization shall, not later than six months after the coming into force of this Act, take such steps as are necessary under the Companies and Allied Matters Act to form a company, limited by shares, which shall be the initial holding company for the assets and liabilities of the Authority.

2. The shares of the initial holding company on its incorporation shall be held by the Ministry of Finance Incorporated and the Bureau of Public Enterprises in the name of and on behalf of the Federal Government of Nigeria.

3.— (1) The National Council on Privatisation, shall by written notice fix the date (in this Act known as the "initial transfer date"), on which the assets and liabilities of the Authority as specified in Subsection (2) shall be transferred to the initial holding company.

(2) The assets and liabilities vested in the Authority as at the initial transfer date and appearing on the audited balance sheets of the Authority, shall vest in the initial holding company.

(3) The vesting of assets and liabilities under Subsection (2) shall be provisional until final orders are issued by the National Council on Privatisation under this section.

(4) Within eight months from the initial transfer date, the National Council on Privatisation shall, on the finalisation of the opening balance sheet of the initial holding company, issue final order either confirming or specifying the transfer of assets and liabilities under Subsection (2) and this Part applies with necessary modifications to the final order.

(5) The transfer of a liability or obligation under this section releases the authority from the liability or obligation.

(6) All bonds, hypothecations, securities, deeds, contracts, instruments, documents and working arrangements that subsisted immediately before the initial transfer date and to which the Authority was a party shall, on and after that date, be as fully effective and enforceable against or in favour of the initial holding company as if, instead of the Authority, the initial holding company had been named therein.

(7) Any cause of action or proceeding which existed or was pending by or against the Authority immediately before the initial transfer date shall be enforced or continued, as the case may be, on and after that date by or against the initial holding company in the same way that it might have been enforced or continued by or against the Authority.

(8) An action or other proceeding shall not be commenced against the initial holding company in respect of any employee, asset, liability, right or obligation that has been transferred to the initial holding company if, had there been no transfer, the time for commencing the action or other proceeding would have expired.

(9) A transfer under Subsection (2) shall not be deemed to—

(*i*) constitute a breach, termination, repudiation or frustration of any contract, including a contract of employment or insurance;

(*ii*) constitute a breach of any Act, regulation or municipal by law, or

(*iii*) constitute any event of default or force majeure;

(*iv*) give rise to a breach termination, repudiation or frustration of any licence, permit or other right;

(*v*) give rise to any right to terminate or repudiate a contract, licence, permit or other right; and

(*vi*) give rise to any estoppel.

(10) Subsection (9) does not apply to the contracts as maybe prescribed by the regulations.

(11) Subject to Subsection (10) nothing in this Act and nothing done as a result of a transfer under Subsection (2) creates any new cause of action in favour of—

(*a*) a holder of a debt instrument that was issued by the Authority before the initial transfer date;

(*b*) a party to a contract with the Authority that was entered into before the initial transfer date.

(12) Any guarantee or surety ship which was given or made by the Federal Government of Nigeria or any other person in respect of any debt or obligation of the Authority, and which was effective immediately before the initial transfer of the principal debt or obligation, shall remain fully effective against the guarantor or surety on and after the initial transfer date in relation to the repayment of the debt or the performance of the obligation, as the case may be, by the initial holding company to which the principal debt or obligation was transferred.

* *** *

9. The National Council on Privatisation shall, not later than eight months after the formation of the initial holding company under Section I of this Act, take such steps as are necessary under the Companies and Allied Matters Act to form such number of additional companies, limited by shares, as the National Council on Privatisation may deem appropriate, which shall be the successor companies for assuming the assets and liabilities of the initial holding company including, but not limited to, companies with functions

relating to the generation, transmission, trading, distribution and bulk supply and resale of electricity.

10. All the respective shares of each of the successor companies from the dates of their incorporation shall be held jointly in the name of the Ministry of Finance Incorporated and the Bureau of Public Enterprises for and on behalf of the Federal Government of Nigeria.

11.— (1) The National Council on Privatisation shall, not later than one year from the initial transfer date, by an order, require the initial holding company to transfer employees, assets, liabilities, rights and obligations of the initial holding company to a successor company, as specified in the order.

* *** *

18. The National Council on Privatisation may, at any time within one year after making a transfer order, make a further order amending the transfer order in any way that the National Council on Privatisation considers necessary or advisable, including such order or orders as may be necessary to rectify the transfer of any of the employees, assets, liabilities, rights and obligations pursuant to a verification or an audit of the employees, assets and liabilities of the initial holding company as at the date of the relevant transfer order, and this Part applies with necessary modifications to the amendment.

Source: Excerpted from the Web site of Nigeria's Bureau of Public Enterprises (http://www.bpeng.org/10/add doctool.asp?Doc10=103&MenuID-43#).

"Man who brings gas to your homes" by Agha Ibiam, *This Day* (Nigeria), January 12, 2006

For a commentary on natural gas policy in Nigeria, including Nigeria's efforts to reform natural gas infrastructure and end natural gas flaring, consider this article by Agaha Ibiam, published on January 12, 2006 in This Day Online.

With Nigeria intensifying efforts to curtail flaring of natural gas, some organisations have also introduced innovative and sure methods of storing the product in a highly technological device for home use.

President Olusegun Obasanjo recently remarked that government has identified six core investment areas that needed special attention. These areas include oil, gas, solid minerals, agriculture, manufacturing and tourism. With that, he urged investors to take advantage of new and viable investment opportunities that exist in the country.

In natural gas, Nigeria is endowed with the 10th largest proven gas reserves. And according to the World Bank, gas flared in Nigeria is equivalent to total annual power generation in sub-Saharan Africa. The country also faces a number of difficulties in harnessing its abundant gas reserves, which are estimated at about 120 trillion cubic feet, but could total 300 trillion cubic feet. It also lacks gas utilisation infrastructure. When most of its oil facilities were built in the 1960s and 1970s, at a time when gas was not a popular energy source in the world, little thought was given to gas collection facilities.

The extent of health problem attributable to gas flaring is unclear but doctors have found an unusually high incidence of asthma, bronchitis, and skin and breathing problems in communities in oil-producing areas. Moreover, flaring is a global source of greenhouse gas emissions, contributing to global warming. The World Bank estimates that gas flaring in the Niger Delta releases some 35m tonnes of carbon dioxide annually into the air.

However, when Shell Nigeria Gas, (SNG) was launched in Nigeria, many believed that the event would mark a new chapter in the development of a domestic market to tap Nigeria's enormous gas reserves. The company's objective then was to be the major catalyst in the efforts to promote the use of natural gas as a preferred fuel for industrial use by the year 2010.

SNG also hoped that by taking a lead, they could help to stimulate investment in gas development projects. It was also expected that sometime within the next century, gas will become Nigeria's main resource. The company plans to construct gas delivery lines to the factory gate, from where prospective users will provide required infrastructure within their premises.

The government and its joint-venture oil partners acknowledge the need to clean up oil-production activities, including substantially reducing the proportion of associated gas that is flared. Most of Nigeria's oil comes from reservoirs containing gas, so associated gas is produced with the oil. Though current levels of gas flaring remain unhealthy and high, Nigeria has made some progress in reducing flaring when less than two per cent of the country's gas production was utilised.

Government now aims to eliminate unnecessary gas flaring in Nigeria by 2010. Using a carrot-and-stick strategy with the oil companies, government in 1998 annual budget expanded fiscal incentives for investment in gas production, and at the same time raised the penalty paid by oil companies.

All the major oil companies operating in Nigeria have resolved to reduce gas flaring. Their programmes include increasing the volume of associated gas re-injected into oil reservoirs, reducing oil production from wells with high ratios of gas to oil, and fitting more efficient flare tips to flares.

Associated gas, which is produced at low pressure, compressed and treated in facilities specifically built for the purpose, is one of the most difficult and expensive gas sources to harness. In Nigeria, it requires an expensive network of compression facilities and pipelines to link scattered fields that do not produce sufficient quantities of gas to be commercially viable on their own.

Probably the biggest constraint to gas development in Nigeria is the lack of ready markets for the commodity. Domestic gas demand is a meagre 300 mn cubic feet per day, in a country where few households have modem cooking appliances and most still use traditional, cheaper forms of energy. Exports of Nigerian gas involve high transport costs because of the distance to the major international gas markets of Europe and Asia.

Despite the difficulties, there are several projects, started in recent years to boost the commercial use of Nigeria's natural gas. Most are being undertaken by multinational oil firms, operating as minority equity partners in joint ventures with the state-owned Nigerian National Petroleum Corporation (NNPC).

In the league of industrialists that want to ensure that natural gas is adequately utilised in Nigeria is an economist, Mr. Ige Belo, managing director of Firstgas Distribution Services Limited. The company has envisioned to become Nigeria's best preferred present and future source of energy for domestic and mechanical applications.

Belo believes that oil has not benefited the vast majority of Nigerians. Every year it is estimated that the country loses more than N3 billion from importing gas (LPG) with the price of one metric tonne rising to as much as N50,000 not to talk of the cost of importing petrol. These product, observers believe could be produced in abundance if government had given serious attention and thought to it in the past.

The company, though relatively young, has positioned itself to accomplish its vision through attractive value added products and support services by using competencies developed and acquired for domestic, automotive and other applications through provision of superior quality products and services.

Belo said Nigeria has reached the stage of using gas in apartments, kitchens and even to drive automobiles as obtained in different parts of the world. "Most of us have traveled a lot to know that gas is essential used for heating rooms, cooking and doing so many things, not only in Europe, but also in other parts of the world, Africa inclusive."

"Nigeria has reached that stage too. Just open up to your burner and gas will flow. We keep saying that gas is for the future. But it's evident that from the escalating prices of petrol that we have to face the alternative available to us by converting most of the things we do now with petrol to gas", Belo said.

Change may be slow in terms of cylinder use, but Belo said Firstgas is pioneering the use of new lightweight propane cylinder in the country. The cylinder which is translucent in nature would give the user the benefit of being able to determine gas level at all time, unlike the metal cylinder.

The cylinders according to him are a product of the strong design and development philosophy of Ragasco Company, a Norway-based company. The company he said strives for quality product solutions optimised on a balanced mix of right material combinations, rational production processes and efficient logistic solutions.

The display of the cylinders at Firstgas office, Lagos he said is made from high quality materials consisting of three layers with different functions as liner, composite layer and outer casing. The liner, he explained is the seamless blow molded liner that provides the inner gas barrier. It is made up of a tough polymeric material and is chemical resistant.

The composite layer, he described as load carrying segment consists, of a mix of fibre-glass and resin. The material, he added is applied in a filament-winding process with great capacity to withstand pressure. An advanced curing process which ensures the see-through features of the cylinder is also added, while the last outer casing is injector-molded and provides protection for the pressure vessel as well as for the valve. It has ergonomic handles for easy lifting and stacking function. The casing, he explained also creates the distinct design of the cylinder.

Following due process, Belo explained that for the cylinder to be allowed into the country, it needed to be certified by the Standard Organisation of Nigeria (SON). The certification which he said was important was designed to carry out evaluation and test-typing for the cylinders.

"By virtue of the certification conducted by SON, it is hoped that the transparent cylinder gas will sell like hot cake as it will soon hit the Nigerian market. And we will approach the market by first of all brokering deals with some major oil companies," he said.

The cylinder, according to Belo has also gone through an extensive test regime. Some of the tests the director explained are fire, drop, cycling, high temperature creep, shotgun, flawed cylinder with cuts in the wall and permeability. Quality control, he said is an integrated part of the manufacturing processes, while the production is equipped with statistical tools that collect the process data.

He commended government policy on gas production and flaring which he hoped will improve gas usage and reduction of import duty on cylinders. Belo hoped that as the coast is clear for gas industry to thrive in Nigeria in the near future, Ragasco will set up a cylinder manufacturing company to service most of the oil industries in the country.

Source: Agha Ibiam. "Man who brings gas to your homes." This Day Online, *This Day* (Nigeria), January 12, 2006. Available online by subscription. http://www.thisdayonline.com/nview.php?id=37952. Accessed on January 29, 2006.

DOCUMENTS OF VENEZUELA

Decree with Force of Organic Law of Hydrocarbons (November 2, 2001)

The Hydrocarbons Law (Decree with Force of Organic Law of Hydrocarbons) was signed into Venezuelan law by Hugo Chávez, president of the Bolivarian Republic of Venezuela, on November 2, 2001. This law increased the royalties that private companies must pay to the government for producing oil in Venezuela and required foreign oil investments to be in the form of joint ventures with Venezuela's state-owned oil company, Petroleos de Venezuela, SA (PdVSA). The law is controversial because it guarantees PdVSA a majority share of any new projects, while there are many doubts about PdVSA's ability to sufficiently fund investment in expanding crude oil production. The climate for foreign investment in Venezuela's oil sector has grown even more unfriendly under Chávez; in November 2004 he announced a new royalty rate that was the highest rate allowable under Venezuela's prior hydrocarbons laws.

Bolivarian Republic of Venezuela,
DECREE WITH FORCE OF ORGANIC LAW OF HYDROCARBONS,
signed into law November 2, 2001

CHAPTER I
FUNDAMENTAL PROVISIONS

Section I
Scope of the Decree-Law

Article 1. Everything concerning the exploration, exploitation, refining, industrialization, transportation, storage, commercialization and conservation of hydrocarbons, as well as of their by-products and the works required by these activities, shall be governed by this Decree-Law.

Article 2. The activities associated with gaseous hydrocarbons shall be governed by the Organic Law of Gaseous Hydrocarbons.

Section II
Reservoir Ownership

Article 3. Hydrocarbon reservoirs existing in the national territory, whatever their nature may be, including those under the territorial seabed, in the continental platform, in the exclusive economic zone and within national boundaries belong to the Republic, are public domain assets and, therefore, are inalienable and may not prescribe.

CHAPTER II
HYDROCARBON ASSOCIATED ACTIVITIES

Section I
General Provisions

Article 4. The activities to which this Decree-Law refers, as well as the works required for their performance, are hereby declared of public utility and social interest.

Article 5. The activities governed by this Decree-Law are primarily dedicated at promoting the integral, organic and sustained development of the country, taking into account the reasonable use of the resource and the environmental preservation. For this purpose, the strengthening of the national production sector and the transformation wlthin the country of raw materials from hydrocarbon, as well as the incorporation of advanced technologies, shall be promoted.

The income received by the Nation from hydrocarbons shall be used to finance health and education and to create funds for macroeconomic stabilization and to make productive investments in order to achieve an appropriate link between oil and the national economy, all in favor of the welfare of the population.

Article 6. The decisions adopted by the Republic by reason of the international agreements or treaties on hydrocarbon matters entered into by the Republic shall apply to the persons performing the activities which this Decree-Law covers.

Article 7. The activities indicated in this Decree-Law are subject to both the provisions herein and the provisions contained in other laws, decrees or resolutions issued, or to be issued, as applicable.

Section II
Competition

Article 8. It is the competence of the Ministry of Energy and Mines to formulate, regulate and enforce policies; and plan and perform and supervise hydrocarbons activities, which comprise the development, conservation, usage and control of such resources, as well as market studies, the analysis of prices of hydrocarbons and their products. The Ministry of Energy and Mines is, in such regard, the national body with competence in all matters associated with the administration of hydrocarbons and, accordingly, is empowered to inspect the inherent works and activities, and to audit the operations that give rise to the taxes, duties or charges set forth in this Decree-Law and review the respective accounting books.

The Ministry of Energy and Mines shall carry out the planning function to which this article refers to in accordance with the National Development Plan. For purposes of compliance with this function, the National Executive may provide the necessary resources pursuant to the pertinent legal provisions.

The officials and individuals shall provide the national employees in charge of the aforementioned duties, the broadest means to facilitate their compliance.

Section III
Primary Activities

Article 9. The activities associated with exploration in search of hydrocarbon reservoirs to which this Decree Law refers to their production in their natural state, gathering, and initial transportation and storage of such hydrocarbons shall be called primary activities for purposes of this Decree-Law.

In accordance with the provisions in Article 302 of the Constitution of the Bolivarian Republic of Venezuela, the aforementioned mentioned primary activities, as well as those associated with the works required for such primary activities are reserved to the State in the terms provided in this Decree-Law.

Section IV
Refining and Commercialization Activities

Article 10. The activities associated with the distillation, purification and transformation of the natural hydrocarbons set forth in this Decree-Law, performed to add value to these substances and the commercialization of the products obtained, constitute refining and commercialization activities and may be performed by the State and private persons, either jointly or separately, in compliance with the provisions of Chapter VIII of this Decree-Law.

The existing facilities and works, as well as their expansions and modifications, owned by the State or companies exclusively owned by the State, dedicated to national hydrocarbon refining activities in the country and to the basic transportation of products and gas, are reserved to the State under the terms provided in this Decree-Law.

Article 11. The refineries to be built shall adhere to a national plan for their installation and operation and shall be linked to specific projects approved by the National Executive through the Ministry of Energy and Mines. Such refineries shall aim mainly at the upgrading of the processes to be used and at obtaining clean fuels.

Article 12. The companies wishing to perform natural hydrocarbon refining activities shall obtain a license from the Ministry of Energy and Mines, who may grant such license based on the prior the definition of the pertinent project and in accordance with the provisions in this Decree-Law and its Regulations.

The assignment, transfer or encumbrance of the licenses require the prior approval of the Ministry of Energy and Mines and, without such approval, shall have no effect. In the event of compelling transfers due to foreclosure, the State may replace the creditor on prior payment of the applicable sum.

Article 13. The following minimum requirements shall be met in order to obtain the license referred to in the previous article:

1. Identification of the companies and their representatives.

2. Description of the project indicating the applicable technology and the destination of the products, as well as the economic resources to be employed in the project.

3. Duration of the company or project, which shall not exceed twenty-five (25) years, renewable for a term to be agreed on, which shall not exceed fifteen (15) years, provided that project requirements have been complied with.

4. Indication of the special advantages offered to the Republic.

Article 14. Those engaged in natural hydrocarbon refining activities within the country shall register with the registry kept for such purpose by the Ministry of Energy and Mines. Likewise, assignments, transfers, encumbrances or foreclosures of licenses shall be recorded in such registry.

Article 15. The licenses granted for activities associated with the refining of natural hydrocarbons shall expressly include the provisions comprised in letters a and b, item 3, Article 34 of this Decree-Law and if such provisions are not expressly stated, they shall be considered as having been inserted in the wording of the license.

Article 16. The assignment, encumbrance and foreclosure of the rights granted by the licenses for natural hydrocarbon refining activities, shall require prior authorization from the Ministry of Energy and Mines.

Article 17. The licenses granted pursuant to this Decree-Law may be revoked by the Ministry of Energy and Mines, if the causes for revocation set forth in the license occur or if the license is assigned, encumbered or foreclosed without authorization from the Ministry.

Section V

National Capital Participation and Use of National Goods and Services

Article 18. The National Executive shall adopt measures aimed at the creation of national capital to stimulate the incorporation and consolidation of companies engaged in the operation, service rendering, manufacturing and supply of goods of national origin for the activities governed by this Decree-Law. In this regard, the State, the entities and the companies to which this Decree-Law applies shall incorporate in their contracting processes the participation of national capital companies in such terms as to ensure the optimum and effective use of goods, services, human resources and capital of Venezuelan origin.

Section VI
Obligations Arising from Hydrocarbon Activities

Article 19. The persons engaged in the activities to which this Decree-Law applies shall do so in a continuous and efficient manner, in compliance with the applicable regulations and the best available scientific and technical practices on security and health, environmental protection and rational use of hydrocarbons, conservation of hydrocarbon energy and maximum final recovery of hydrocarbon reservoirs.

Article 20. The persons engaged in the activities to which this Decree-Law applies shall provide the National Executive all the information which it may require in relation to the performance of such activities. For this purpose, the persons carrying out primary activities together with industrial and commercial activities shall keep and submit separate accounting for such activities. The National Executive shall maintain the confidentiality of the information supplied when requested by the interested party, and such request is according to law.

Article 21. The persons engaged in the storage, transportation and distribution activities covered in this Decree-Law are obliged to allow the use of their facilities to other storage, transportation and distribution entities, when such facilities have available capacity and it is required by the social or public interest. Such facilities shall be used pursuant to the conditions agreed upon by the parties. In the absence of an agreement, the Ministry of Energy and Mines shall establish the conditions for the provision of such service.

CHAPTER III
PERFORMANCE OF PRIMARY ACTIVITIES

Section I
Manner and Terms for the Performance of Primary Activities

Article 22. The primary activities mentioned in Article 9 shall be performed by the State, either directly by the National Executive or through exclusively owned companies. Likewise, these activities may be performed by companies in which the State has decision making control by holding over 50% of the capital stock, which for purposes hereof shall be denominated joint venture companies. Companies engaged in the performance of primary activities shall be operating companies.

Article 23. The National Executive, through the Ministry of Energy and Mines, shall set the boundaries of the geographical areas where the operating companies shall perform the primary activities. These areas shall be divided into lots with a maximum surface of one hundred square kilometers ($100km^2$).

Article 24. The National Executive, by Decree, may assign to operating companies the right to perform primary activities. Likewise, the National Executive may assign to operating companies the ownership or other rights on real or personal property of the private domain of the Republic, required for the efficient performance of these activities. The National Executive may revoke such rights when the operators fail to comply with their obligations in such a way that it will prevent them from achieving the purposes for which such rights were assigned.

Article 25. Operating companies may carryout the necessary actions to perform the assigned activities and enter into the pertinent contracts, all in accordance with the provisions of this Decree-Law or any other applicable provisions.

Article 26. Operating companies may establish or contribute to maintain experimentation, research and technological development institutes and universities, that provide technical support for their operation, as well as create and maintain training centers for the personnel involved in the activities contemplated In this Decree-Law, in due harmony with the operation and development of other centers and institutes which may exist in the country for similar purposes.

Section II

State-Owned Companies

Article 27. The National Executive, by decree in Council of Ministers, may create companies exclusively owned by the State to perform the activities set forth in this Decree-Law giving them the legal structure it deems convenient, including that of a corporation with a sole shareholder.

Article 28. Without prejudice to the reserve set forth in this Decree-Law, the companies referred to in the previous article may create other companies for conducting their activities, subject to the prior approval of their respective Shareholders Meeting. Likewise, such approval shall be obtained to modify the purpose of the created companies, as well as to merge, associate, dissolve or liquidate them, or to make any amendments to their by-laws.

Such authorization shall also be necessary for companies to be created by the affiliates.

Article 29. State-owned oil companies shall be governed by this Decree-Law and its Regulations, their own by-laws, the provisions issued by the National Executive through the Ministry of Energy and Mines, and by applicable general law provisions.

Article 30. The National Executive, through the Ministry of Energy and Mines, shall inspect and supervise the State-owned oil companies and their affiliates, both with national and international reach, and shall issue the guidelines and policies that must be complied with regarding the matters to which this Decree-Law applies.

Article 31. The incorporation and the capital stock increases of State-owned companies or its affiliates, due to asset revaluation or dividends, implying the issuance of shares to be subscribed by the State or such companies, as well as the merger of State-owned companies or their affiliates and the transfer of assets among them, shall not be subject to payment of the registration fees due for such operations.

Article 32. With the exception of members of the Board of Directors of the companies, the workers of State-owned oil companies shall enjoy labor stability and may only be dismissed for the reasons expressly set forth in the labor legislation. Likewise, the State shall guarantee the current labor union contract regime and the enjoyment of social, economic, welfare, union and professional improvement benefits, and any other benefit set forth in the Collective Bargaining Agreement and labor laws, as well as the bonuses or premiums and other remuneration and compensation given as performance incentives and that have been granted due to usage and practice and due to the application of personnel administration rules, which workers have traditionally enjoyed according to the policy followed by companies on such matter.

Moreover, the State shall guarantee the enjoyment of retirement plans and their respective annuities to workers who retired before the enactment of this Decree-Law. These retirement plans, as well as any other worker benefit plans implemented by companies, including workers' saving funds, shall be maintained in force without prejudice to Collective Bargaining Agreement.

The provisions contained in the law creating the National Institute for Cooperative Education shall continue to be applicable to the companies created according to the Law Reserving to the State the Industry and Commerce of Hydrocarbons.

Trusts created for the benefit of workers shall be governed by the agreed Collective Bargaining Agreement.

Section III
Joint Venture Companies

Article 33. The incorporation of joint venture companies and the conditions governing the performance of primary activities shall require the prior approval of the National Assembly, for which purpose the National Executive, through the Ministry of Energy and Mines, shall inform the National Assembly of all the pertinent circumstances involving said incorporation and conditions, including the special advantages provided in favor of the Republic. The National Assembly may amend the proposed conditions or establish those deemed convenient.

* *** *

CHAPTER IV
SUPPLEMENTARY RIGHTS

Section I
Temporary Occupation, Expropriation and Servitudes

Article 38. The persons authorized to perform the exploration, production, gathering, initial transportation and storage, processing and refining activities of natural hydrocarbons shall be entitled to request temporary occupation or expropriation of assets, as the case may be, as well as the granting of servitudes in favor of the activity.

Article 39. Regarding expropriation, the provisions in the special law regulating the matter shall apply.

Section II
Procedures

Article 40. When servitudes are going to be granted on private property lands, the authorized persons shall enter into the necessary agreements with the owners. If no agreement is reached, the interested persons may resort to a First Instance Civil Court, with jurisdiction in the location, for such Court to authorize the commencement of the works. The applicant shall accurately state the areas and assets to be affected and the work to be performed, and shall fill out in the application all the proper requirements. The application for the granting of a servitude shall include: 1. The name of the owner, as well as the names of every person, if known, having any rights over the asset that shall be subject to the servitude. 2. The assets to be affected by the servitude, as well as the areas needed and the works to be performed. Data regarding the title and liens that may exist on the assets shall also be provided. 3. Servitude duration and other terms. 4. Other data deemed necessary by the concessionaire to illustrate the judge.

Upon receipt of the application, the Court shall, on the same day, summon the affected party to appear on the third working day after the summon, at the proceedings for appointment of the experts who will assess, the possible damage. If summons cannot be served, the Court shall order the publication of a notice in the newspaper of widest national and regional circulation, summoning the affected party to appear on the third working day after publication, and on such third day the experts responsible for assessing the possible damages and the pertinent indemnification shall be designated. At the occasion fixed for appearance of the affected party, the applicant shall appoint one expert and the affected party shall appoint a second expert. If the affected party does not appear or refuses to appoint an expert, the Court shall do it on its behalf. The Court shall appoint the third expert.

The appointed experts may attend the appointing proceedings for purposes of accepting their appointment and to be sworn in, otherwise the Court shall appoint their substitutes. The experts shall submit their report within five (5) working days after their appointment.

Upon submittal of the report, the applicant, within the following five (5) working days, shall deposit with the Court the estimated indemnification amount and the Court, within the following five (5) working days, shall authorize the commencement of the works. If the affected party accepts the indemnification, the Court shall grant the servitude under the terms requested. In the event of a disagreement, the process shall continue pursuant to the standard trial procedure, and consequently, the application shall become the complaint and the term to answer the complaint shall begin as of the declaration of the disagreement. The applicant may, within this term, make the changes and improvements to the application it may deem convenient.

Article 41. For the granting of servitudes on public lands, the authorized persons shall enter into the necessary agreements with the National Executive and pay the agreed upon consideration, unless the National Executive decides to exempt the applicant from payment.

If the lands under servitude have improvements made by private persons, the indemnification to which such persons are entitled shall be paid by the beneficiary of the servitude and such indemnification shall be established according to the procedure set forth in the previous article.

* *** *

CHAPTER VI
ROYALTY AND TAX REGIME

Section I
Royalties

Article 44. The State is entitled to a thirty per cent (30%) participation, in the form of a royalty, of the volumes of hydrocarbons extracted from any reservoir.

If proven to the satisfaction of the National Executive that a mature reservoir or an extra-heavy oil reservoir from the Orinoco Belt cannot be economically produced with the thirty per cent (30%) royalty set forth in this Decree-Law, such royalty may be reduced up to twenty per cent (20%) to make the production economically feasible, and the National Executive is likewise empowered to increase such royalty, fully or partially, to bring it again to thirty per cent (30%) when proven that the production of the reservoir is economically feasible with such increase.

If proven to the satisfaction of the National Executive that projects for bitumen blends from the Orinoco Oil Belt are not economically feasible with the thirty per cent (30%) royalty set forth in this Decree-Law, such royalty may be reduced up to sixteen and two thirds per cent (16 2/3%) for the production of such projects to be economically feasible, and the National Executive is likewise empowered to increase such royalty, fully or partially, to bring it again to thirty per cent (30%) when proven that project profitability may be maintained with such increase.

Article 45. The royalty may be requested by the National Executive in kind or in cash, fully or partially, unless otherwise requested, it shall be understood that the National Executive has chosen to receive the royalty in full and in cash.

Article 46. When the National Executive decides to receive the royalty in kind, it may use, for transportation and storage purposes, the services of the producing company, which shall render the services up to the location indicated by the National Executive, who shall pay the price agreed upon for such services.

Article 47. When the National Executive decides to receive the royalty in cash, the producer shall pay the price for the corresponding volumes of hydrocarbons, measured at the production field and at market value, or at the agreed value; or if no such values exist, at an assigned fiscal value set by the tax authority. For this purpose, the Ministry of Energy and Mines shall issue a payment form, which shall be paid to the National Treasury within five (5) working days after being received.

Section II
Taxes

Article 48. Without prejudice to the provisions in other national laws regarding tax matters, the persons performing the activities to which this Decree-Law applies shall pay the following taxes:

1. Surface Tax: For the granted area which is not being produced, a sum equal to one hundred Tax Units (100 TU) per km^2 or a fraction of a km^2 per

year. This tax shall increase annually by two per cent (2%) during the first five (5) years and by five per cent (5) during subsequent years.

2. Fuel Consumption Tax: Ten per cent (10%) of each cubic meter (m^3) value of hydrocarbon by-products produced and consumed as fuel in its own operations, calculated based on the sales price to the final consumer. If this product is not sold in the domestic market, the Ministry of Energy and Mines shall fix its price.

3. General Consumption Tax: Per each liter of hydrocarbon by-products sold in the domestic market, between thirty per cent and fifty per cent (30% and 50%) of the price paid by the final consumer; the rate between both limits shall be annually fixed through the Budget Law. This tax, to be paid by the final consumer, shall be withheld by the supplier and paid monthly to the National Treasury.

The National Executive may, fully or partially for any specified time, exonerate the payment of the General Consumption Tax with respect to specific activities of public or general interest in order to stimulate such activities. This tax may also be restored to its original rate when exoneration causes cease.

* *** *

CHAPTER VII
INDUSTRIAL ACTIVITIES

Section I
Manner and Terms of the Activities

Article 49. The Industrialization of refined hydrocarbons comprises the separation, distillation, purification, conversion, blending and transformation of such hydrocarbons aimed at adding value to these substances by obtaining special oil products or other hydrocarbon by-products.

Article 50. Industrial activities with refined hydrocarbons may be performed directly by the State, by companies owned exclusively by the State, by joint venture companies with any proportion of private and State equity participation, and by private companies.

Article 51. The National Executive shall adopt the measures necessary for industrializing refined hydrocarbons in the country and such measures shall:

1. Stimulate the greatest and deepest transformation of the refined hydrocarbons.

2. Promote the investments in substance-generating projects that support the development of the national industrial sector.

3. Ensure that refineries and hydrocarbon processing plants under State control prioritize over exports the timely supply, for subsequent processing, of the basic substances in quantities and qualities and with the prices and commercial terms that will allow the development of competitive companies in the international markets.

4. Develop industrial parks in areas located near to the refineries and in areas which allow for easy access for the supply of hydrocarbons and hydrocarbon by-products.

5. Stimulate the creation and participation of financial entities in hydrocarbon industrialization processes in the country.

6. Make those companies that perform hydrocarbon industrialization activities in the country to promote the downstream industrialization of their products.

7. Any other stated in the regulations.

Article 52. The National Executive shall give priority to refined hydrocarbon industrialization projects that stimulate the creation of national capital and link such capital to creating greater added value to the processed hydrocarbons, and to those projects which products are competitive in the external market.

Article 53. Private companies engaged in refined hydrocarbon industrialization activities in the country are required to obtain a permit granted by the Ministry of Energy and Mines, which permit shall be granted after compliance with the following requirements:

1. Identification of the companies and their representatives
2. Indication of the source of supply for raw material.
3. Project definition, stating the destination of the products.

Article 54. Those persons engaged in refined hydrocarbon industrialization activities in the country shall register with the registry kept for this purpose by the Ministry of Energy and Mines.

* *** *

CHAPTER VIII
COMMERCIALIZATION ACTIVITIES

Section I
Persons Who May Perform Them

Article 56. The commercialization activities to which this Decree-Law applies comprise domestic and foreign trade, of both natural hydrocarbons and their by-products.

Article 57. The commercialization activities of natural hydrocarbons, as well as the commercialization of those hydrocarbon by-products specified by the National Executive through Decree, may only be performed by the companies referred to in Article 27 of this Decree-Law.

Article 58. The commercialization activities of the by-products which are excluded from the application of the previous article may be performed directly by the State or by companies owned exclusively by the State, or by joint venture companies with any proportion of private and State equity participation, and by private companies.

Section II

Domestic Trade

Article 59. Those hydrocarbon by-products specified by the National Executive by Resolution issued through the Ministry or Energy and Mines, shall be subject to the domestic trade regulations set forth in this Decree-Law.

Article 60. The supply, storage, transportation, distribution and retail activities of the hydrocarbon by-products specified by the National Executive pursuant to the previous article, used for domestic collective consumption, constitute a public service. The National Executive, through the Ministry of Energy and Mines, shall fix the prices of the hydrocarbon by-products and adopt measures necessary to guarantee supply, service efficiency and to prevent service interruption. In fixing such prices, the National Executive shall take into account the provisions of this Decree-Law and those set forth in the Regulations. Such prices may be fixed in bands or using any other system appropriate for the purposes of this Decree-Law, taking into account the investments and their profitability.

Article 61. Individuals or corporate bodies wishing to perform supply, storage, transportation, distribution and retailing activities of hydrocarbon by-products shall previously obtain a permit from the Ministry of Energy and Mines. Permit granting shall be subject to the provisions set forth in this Decree-Law, its Regulations and the respective Resolutions. The individuals or corporate entitles engaged in the above-mentioned activities may undertake more than one activity, provided that there is legal and accounting separation between such individuals and/or entities.

Source: Excerpted from the Web site of the Bolivarian Republic of Venezuela's Ministry of Energy and Mines (http://www.leydehidrocarburos.gov.ve/ProyectoLeIngles.doc).

"Expert warns that Venezuela's threats could be the greatest danger to oil prices," The Daily Reckoning, http://www.form.dailyreckoning.com/Media/ release 081805.htm. August 18, 2005

Venezuelan president Hugo Chávez is outspoken about his disdain for United States foreign policy, often threatening to cut off oil supplies to the United States. This article that appeared in the online Daily Reckoning, *a daily e-letter written by* New York Times *authors Bill Bonner and Addison Wiggin, reports on the potential impact on oil prices should Venezuela make good on its threats.*

BALTIMORE, Aug. 18 /PRNewswire/ — Venezuelan President Hugo Chavez is becoming increasingly outspoken with his distrust of the United States, accusing Washington of planning attacks against his country. Likewise, the U.S. government has expressed public concern about Venezuela's plans to buy 100,000 rifles and ammunition from Russia.

Venezuela's energy minister, Rafael Ramirez, has said in recent reports that if the U.S. shows any signs of aggression toward his country, Caracas is "ready and willing" to cut off its oil supply to the United States. Ramirez went on to say that "the US market is not indispensable to us" and that Caracas, the only Latin American member of the Organization of the Petroleum Exporting Countries, had other clients to court, including China.

According to energy expert and frequent contributor to The Daily Reckoning, Kevin Kerr, if the two daily boats of Venezuelan oil head somewhere other than the United States, it would lead to big troubles for oil prices. Kerr warns, "Venezuela and the US eroding relationship couldn't come at a worse time. This threat could easily drive oil prices to $75 and $80."

Venezuela exports about 1.5 million barrels a day to the US and is the world's fifth largest oil producer. Reports for December 2004 showed that Venezuela supplied more oil to the US than any other country, making it the number one oil supplier to the US. "Obviously," Kerr added, "losing Venezuela would be a devastating blow to U.S. oil supplies."

Kevin Kerr's expertise in futures and commodities has made him a regular contributor to news outlets like CNN fn, CNBC and MarketWatch, where he's been quoted in over 500 articles. With 15 years of experience, and his own trading service, Resource Trader Alert, Kevin Kerr is a true veteran of the commodities markets dealing with everything from cotton to currencies to oil and natural gas.

Source: Bill Bonner and Addison Wiggin, "Expert warns that Venezuela's threats could be the greatest danger to oil prices," *The Daily Reckoning*, August 18, 2005. Available online. URL: http://www.dailyreckoning.com/Media/release081805.html. Accessed on August 2, 2006.

PART III

Research Tools

6

How to Research Global Energy Issues

There is such a wealth of information about energy and natural resources that it may be difficult to know where to begin when researching issues related to these subjects. In addition, the dimensions of energy-related issues are ever-changing: Each day's news includes reports of new advancements in energy technologies, energy-related environmental findings, energy-related political strategies, energy policy decisions and laws, resource supply problems, and economic developments. Occasionally such developments entirely change the outlook on an issue. For example, the Arab oil embargo of the 1970s had a profound impact on U.S. views of energy supply and oil imports. For this reason keeping abreast of current events is a vital first step for any researcher of energy-related issues.

The tips in the following section may serve as a good launching pad from which to begin your research into issues surrounding energy and natural resources.

GETTING STARTED

Step 1: Be attentive to current events. This is especially important if your goal is to research energy and resource issues that are of global importance or that relate specifically to a country that is not your native country. Why? Outside reference materials such as encyclopedias or almanacs, news reports will probably be your primary exposure to comprehensive accounts of events happening in other parts of the world; more in-depth information may often only be available in a foreign language or in sources that are not easily accessible.

When you hear or read current news reports of energy- and natural resource–related happenings, always jot down what you learn, especially noting names, dates, places, organizations, meetings, and figures and statistics, and perhaps even writing down quotes that seem important to you. You might wish to keep records of where you obtained all of this information, and be sure to verify the accuracy of what you hear or read by investigating other sources for the same facts. However, for this early stage of your research, your main goal is to keep your eyes and ears open for current energy-related events so that you can build your base knowledge of the issues at hand. As you progress in your research later, you will conduct a targeted search into specific information sources, many of which are covered in this chapter.

(But before you reach this stage, you will need to know how to weed out any biases or factual inaccuracies in the sources you discover. The next section, "Focusing on the Facts," provides suggestions for doing so.)

Step 2: Read part I of this book. Skim chapter 1 for background on basic energy concepts and the various energy-related issues that confront the world. In this chapter you will find the historical context for the energy-related issues discussed in this book. You may wish to highlight material related to particular issues that interest you and that you wish to make the focus of your research.

If you are interested in specifically United States–related energy issues, go on to read chapter 2 thoroughly. For your convenience, the information in chapter 2 is organized according to the following five energy-related issues that concern the United States:

1. U.S. Energy Crises and Shortfalls
2. U.S. Energy Affordability and Economics
3. U.S. Foreign Oil Dependence, Political Instability, and Foreign Relations
4. U.S. Infrastructure-Related Limitations on Energy Production
5. U.S. Energy-Related Environmental Issues and the Potential of Renewable and Alternative Sources of Energy

You can read the entire chapter closely if you wish to gain an overall understanding of the issues faced by the United States and the ways the nation has responded to these issues. Reading the chapter closely will also be useful if you need to identify one particular U.S. energy issue on which to focus your research. You can read only the section(s) that pertain to your particular issue(s) of interest if you have already narrowed the focus of your research.

Next, read chapter 3 to gain a perspective on other countries that have dealt with the same five energy-related issues that the United States has faced and to learn how these countries approached these issues. Chapter 3 is arranged by country, with each country's section focusing on the specific energy-related issue(s) that have especially affected that country, as follows:

1. Energy Issues in China: Crises, Shortfalls, Environmental Issues, and the Potential of Renewable and Alternative Sources of Energy

2. Energy Issues in Germany: Shortfalls, Foreign Oil Dependence, and the Potential of Renewable and Alternative Sources of Energy
3. Energy Issues in Iran and Saudi Arabia: Affordability, Economics, and Nuclear Energy
4. Energy Issues in Nigeria: Infrastructure-Related Limitations on Energy Production
5. Energy Issues in Venezuela: Political Instability and Foreign Relations

Again, if you have already narrowed the focus of your research to a particular country, you may find it useful to read only that country's section in chapter 3 closely. Otherwise, carefully read the entire chapter to get an overall picture of how energy issues faced in the United States have also affected other parts of the world.

As you read both chapters 2 and 3 it may be helpful to look up concepts in the Glossary at the back of the book; find the profiles of any key people who are mentioned in chapter 8, "Key Players A to Z"; and locate events in the Chronology at the back of the book. Referring to these sections and reviewing the appropriate entries may help you to commit important concepts, names, and events to mind.

Step 3: Explore part II of this book. An important part of becoming familiar with energy issues in the United States and other countries is gaining an understanding of the key documents that have shaped or been shaped by these energy issues. Examples of such documents include the U.S. Energy Policy Act of 2005, China's Renewable Energy Law, and the Kyoto Protocol. Throughout chapters 2 and 3 you will find references to such documents. The documents will help to give you a legal and historical overview of the issue.

Identify possible sources of information by using part III of this book. Chapter 10 is an annotated bibliography that is organized by region and issue. Find and skim some of the resources that are listed for the particular energy issue or region that concerns you. Reading the forewords, introductions, and contents pages of some of the books you find, as well as looking up concepts in their indexes, may also be helpful for determining whether these books are relevant to the topic that interests you and how focused and objective they are. It may also allow you to develop a broader understanding of your topic, which will be useful later as you undertake a more targeted search into books, articles and papers, Web documents, and other media.

Also scan the Web sites of the organizations and agencies listed in chapter 9. These Web sites can be particularly valuable in giving you a sense of the various sides to an issue since many of these organizations and agencies advocate a particular viewpoint on the energy issue at hand. The sites are often rich with articles, papers, conference proceedings, timelines, news feeds, newsletters, downloadable reports, and even e-books, all of which may assist you in understanding an issue and ways it has affected society. The sites also include contacts for obtaining more information if you are so inclined.

The preceding preliminary steps are especially useful if you need to narrow the focus of your research or even decide upon a topic to study. When you have determined the goal of your research, you can then move on to more detailed research efforts. But before you do, a word of caution: Because energy and natural resources have become so important to many people as a measure of the welfare, quality, and prosperity of their lives, the issues surrounding energy and natural resources can arouse vastly different, even divisive, opinions and viewpoints. Global warming is an example of a divisive energy issue: Some environmental activist groups blame global warming directly on human energy production and use, while some scientists call for a balanced view of global warming that also takes into account natural ecological processes that release carbon dioxide and other greenhouse gases into the atmosphere. So, when researching global warming and many other energy issues, you may have difficulty finding consensus, not to mention sources of information that are truly unbiased and objective in reporting the facts. Especially on the Internet, you may encounter an incredible amount of opinion-based misinformation. Objectivity and accuracy standards tend to fall by the wayside among those who put up Web sites to advocate a particular viewpoint on an issue. Thus, the following tips may help you to navigate your way through all the opinions to uncover the truth.

FOCUSING ON THE FACTS

Whether you are viewing a Web site, skimming a book, watching a television news show or documentary, scanning a newspaper, listening to a conference presentation, or reading a newsletter or research report, the following are some ways to verify facts and identify any biases that may have shaped the source of information you are exploring.

Consider the source. An essential part of researching is actually researching your sources: that is, not simply relying on these sources for information on your topic. Learn more about the organization, group, or person from which the information in question was generated. For example, if you are exploring a Web site, find out who is responsible for the site. Once you find out who is responsible for generating this information, conduct some research, sort of a "background check," on this person or group. Does this person or group have any biases? What is the reputation of this person or group? Is the authority of the person or group well known? Do other sources often cite this person or group as a source of facts and information, or mainly as a source of opinions?

Examine the facts. Next look at the "hard facts" reported in your source: dates, statistics, names, locations, specific descriptions of occurrences, sequences of events, and so on. Where did all of these facts come from? Your resource should state its sources clearly; the absence of such citations is a red flag for possible inaccuracies. Once you learn which sources were consulted for the facts, check on the background and reputation of each source. Then compare sources—find out whether the first source has correctly reported the facts and information in the original source. Also see whether the facts can be backed up by other sources, at least three.

Separate fact from opinion. Finally, read through the resource carefully, studying it for statements that seem to reflect a certain bias or opinion. It may be helpful to underline such statements so that you can be sure to separate them from any of the facts. Check to see whether any opinions are stated as facts. For example, *Human energy use is the direct cause of global warming* is an opinion that is stated as if it were a fact. Opinions ought to be reported in a way that makes it clear that they are the opinions of another person or group—for example: *Many scientists attribute some level of global warming to natural processes* or *Many environmentalists believe that human beings have done significant damage to the environment through their energy use.* Statements like this are clear indications that your source is attempting to provide a balanced, objective view. Another good indication of a source's objectivity is its inclusion of multiple points of view on a topic. If, on the other hand, your source states opinions flatly without providing any other points of view, be cautious about trusting the information it provides.

Now, with this guidance in mind, you are ready to begin the intensive phase of your research.

CONDUCTING DETAILED RESEARCH

Part III of this book is not only a valuable resource to help you get started in your research but is also a useful tool when you are ready to conduct more detailed research. For example, you should scan through chapter 10 again to identify books, articles, papers, Web documents, and other media that pertain directly to your particular topic. Once you have identified these sources, review them for whatever valuable information they can add to your research. Then extend your search beyond this book to find additional books, articles, papers, Web documents, and other resources. There are three main avenues through which to go about this: the Internet, a library, or a bookstore. The section that follows is a guide to conducting research on the Internet, in a library, or in a bookstore. Provided for each of these domains are tips for making your research go smoothly, a description of the research tools available (for example, directories, catalogs, and other information-search media), and a listing of the kinds of energy resources you might find.

I. Conducting Research Online

There is an enormous amount of information on the Internet, and it is easy to become overwhelmed by what you find. Sometimes it is difficult to determine what, out of all that you find, is the most suitable to your purposes. A good starting point would be to visit the Web sites listed with the organizations and agencies in chapter 9. These represent just a sampling of the many sites on energy issues, however. Moreover, groups whose intent is not to report energy-related facts but to take a stand on certain energy-related issues also operate many of these sites. So, they may be more informative in terms of offering you a sampling of opposing viewpoints on an issue than in terms of presenting hard facts.

Furthermore, keep in mind that the Internet is a self-publishing medium; anyone with access to a computer and an Internet service provider can set up a Web site. With all of this in mind, here are some tips for finding Web sites that will meet your research needs.

TIPS FOR RESEARCHING ONLINE

The most important step in your detailed research phase is probably evaluating Web sites for their value and appropriateness to your research. Here are some ways of making an accurate assessment of a Web site's quality and its usefulness to you:

- Identify the site's intended audience. Clues to the intended audience include the site's language, type of content, tone, and style. Do you fall into this intended audience? If not, the Web site may not be very useful to you.
- Evaluate the purpose and the nature of the informational resources presented on the site. You might find everything from scholarly articles to financial reports to news briefs to promotional press releases. Do the materials offered really suit your purposes? Would you be able to cite them as sources when you present your research? If not, exploring the site may not be very worthwhile.
- Identify the author, sponsor, or producer of the Web site. This point was touched upon earlier in "Focusing on the Facts." It bears repeating here because the identify of your online source is a crucial part of whether the site is of value to you; the author, sponsor, or producer of the site may be extremist or disreputable, and you will want to avoid using the site's information in those cases.

The identity of the source is not always obvious. One of the main clues, however, is the site's own Web address (technically known as its URL, or uniform resource locator). Sites with a URL ending in *.gov* are operated by government organizations and agencies. Educational institutions, such as universities, operate sites with URLs ending in *.edu*. And sites with a URL ending in *.com* are mainly operated by commercial or commercially sponsored entities, or by individuals. Additionally, kindergarten through 12th-grade educational institutions sometimes operate sites that include *k12* in their URLs.

Another way of learning more about the operator behind a site is to look for a statement of mission or purpose on the site or to read the "About us" section, if available.

Once you have identified the person or group behind a site and have read what the site itself has to say about this person or group, conduct your own research into the credentials of the person or group, using some of the search tools described in "Research Tools Available Online."

- Bear in mind that Internet content can never be taken at face value. Unlike printed publications such as books, scholarly materials, and newspaper articles, many Web sites are rarely fact-checked or reviewed for accuracy and balance. Keeping in mind the points from "Focusing on the Facts" will help you to avoid incorporating unreliable Web content into your research.
- Determine how current a Web site is. You should be able to find a date on the home page indicating the last time the site was updated. Then ask yourself whether the information is current and timely enough for your research purposes.
- Be careful about relying too heavily on any links to other sites that are included. Sometimes these are present because of an advertising arrangement between the main site and the linked site. Always try to find additional sites on your own rather than simply depending on links. Additionally, if a site contains "broken" links (links that do not connect you to another page or site as they should) or has other navigational defects and operational glitches, take this as an indication of the site's overall quality.
- Beware of "blogs" (short for *Web logs*). In reality, these are nothing more than online diaries that cannot be considered trustworthy resources for serious research, no matter who sponsors them or how many people contribute to them.

RESEARCH TOOLS AVAILABLE ONLINE

There are three major ways of conducting research online:

1. Using online subject directories or Web indexes
2. Using online search engines
3. Surfing Web content in a self-directed manner

The following is a brief guide to each method, its advantages, and its limitations.

Online Subject Directories or Web Indexes

Online directories, also called Web indexes, are services that offer a structured hierarchical listing of topics and subtopics. They enable you to click on these topics and subtopics to be directed to a Web portal that offers a series of links to other sites and/or to online guides and other information about your selected topic. Topics and subtopics are usually listed in a logical, commonsense manner, with related topics and subtopics grouped together. Many subject directories are also searchable—rather than simply browsing through all the topics, you can enter a specific term in a search box and receive a listing of all the related directory links and guides.

A dedicated staff of researchers is usually responsible for compiling the links in the directory or for writing the informational guides that are available. These researchers also update the directory regularly. And, usually, they only choose links to be included after they have evaluated the corresponding sites for their usefulness and quality.

The following are some of the largest and most widely accessed online subject directories:

- **About.com (http://www.about.com),** which offers not only links related to specific topics but something to the effect of a home page for each topic. This page contains an overview of the topic or a brief guide to the topic, which is usually an article written by a person who is identified as an expert in the field.
- **Infomine (http://infomine.ucr.edu),** which is an "academic" Web index service, meaning that it caters to the needs of researchers. Its compilation of more than 100,000 links is fully annotated and mainly relates to university-level subjects and disciplines.
- **The Internet public library (http://www.ipl.org),** a searchable subject index that was originally created by a class at the University of Michigan's School of Information. Maintained by schools, it is a valuable tool for the student researcher.

- **Librarian's Internet Index (http://www.lii.org),** another academic subject directory. It offers a carefully researched collection of links on various topics, organized into a well-structured, self-explanatory hierarchical directory. The site, which receives funding from the U.S. Institute of Museum and Library Services, is a trustworthy resource, especially useful for academic and scholarly research.
- **Suite101.com (http://www.suite101.com),** which is a portal of discussions, articles, and guides to hundreds of topics written by experts, whose background and credentials are provided online. This subject directory is less academic in nature, and it includes links to blogs, so the quality of its offerings may be uneven.
- **The WWW Virtual Library (http://vlib.org),** a subject index on topics mostly academic in nature. Topics can be browsed by subject category or according to an alphabetical list.
- **Yahoo! (http://www.yahoo.com),** which is a "commercial" subject directory, meaning that it is geared to the general public and sometimes includes links to sites that compete with one another for commercial traffic on the Web. The Yahoo! subject directory offers a comprehensive main index of topics on its home page. You can click on one of these topics to be directed to more specific topics and to portals of links on each topic. Yahoo! also offers a feature that allows you to search its subject index directly.
- **Netscape (http//www.netscape.com), LookSmart's "Vertical Search" (http://search.looksmart.com),** and **Google Directory (http://www.google.com/dirhp)** are other commercial sites with directory services that are similar to Yahoo! but more limited in scope.

Be discerning when relying on information from commercial subject directories. For serious research, an academic or professional subject directory, such as one of the directories previously described, may be better suited to your purposes.

The advantage of using a subject directory or Web index service is that its researchers have done part of your work for you. They have evaluated Web sites beforehand and arranged them into a logical, user-friendly topical index. The disadvantage, however, is that you are relying on their judgment as to which sites are valuable to you, and that can be a particular problem if some of the sites are listed for commercial reasons, as in Yahoo! or LookSmart.

Search Engines

A search engine is an online service that utilizes a special computer program to crawl, or browse, the Internet, traveling from link to link and reading and collecting Web pages. The service compiles and indexes these pages into a large database, which can then be searched by you, the Internet user, by entering a keyword or keywords and clicking "Search." In return you receive an inventory of various Web pages, which are usually listed in the order of their relevancy to the keywords you entered.

There are hundreds of search engines to choose from; the largest and most widely accessed ones include the following:

- **AltaVista (http://www.altavista.com)**
- **Excite (http://www.excite.com)**
- **Google (http://www.google.com)**
- **Hotbot (http://www.hotbot.com)**
- **LookSmart (http://search.looksmart.com)**
- **Lycos (http://www.lycos.com)**
- **Mamma (http://www.mamma.com)**

One advantage of using a search engine is that it allows you to search the entire World Wide Web for resources almost in the same way you would search a library's catalog, while allowing you a greater number of keywords by which to conduct your search. Disadvantages include the fact that search engines spit out an overwhelming flurry of Web addresses in response to your keyword search, and it is up to you to figure out how suitable each site is to your research. When you find yourself having difficulty with this, return to the advice in "Tips for Researching Online."

Another difficulty with search engines is that search engines catalog Web pages differently, so relying on only one search engine will limit your results. However, services called metacrawlers or metasearch engines allow you to submit a keyword search to several search engines at once. Examples of metacrawlers include Dogpile (http://www.dogpile.com), which includes major search engines such as Google in its database and allows you to compare the results of all the sites that are included in your search, and Surfwax (http://www.surfwax.com), which includes governmental, news, institutional, and educational search engines in its searching capabilities and so may be particularly helpful to researchers of energy issues. Overall, however, metacrawlers are not recommended for serious research because they often incorporate sponsored sites and duplicate Web pages in their results.

In addition, here are some general tips for using search engines:

- Enter quotations around the key phrases for which you are searching if you want the search engine to find the exact phrasing. For example, entering *"oil crisis"* will return Web pages that contain or point to that exact phrase, whereas entering *oil crisis* will return some Web pages that contain either *oil* or *crisis,* along with pages containing both terms. (Sites containing both words will usually be listed first for your convenience, however.)

- When you are searching for a general topic rather than a specific term or phrase, enter related, descriptive nouns instead of a single keyword. For example, entering the term *electric power* will return better results than simply entering *power,* which will end up generating some Web pages about authority and control.
- If the search engine that you are using supports Boolean operators (the words *and, or,* and *not*), use them between your keywords or phrases either to broaden or to narrow your searches. Use *and* between your keywords or phrases when you would want to receive Web pages that contain both of the words or phrases; use *or* when you would like results that contain either; and use *not* between your keyword and another word that you wish to exclude from your results (for example, if your keyword is associated with another meaning outside the one you are actually researching).
- If "cached" versions of Web pages are available within the results that a search engine has generated, always click the word *cached* (usually found beneath a result's URL and brief description). Why? A search engine's "cached" page is the last version that contained references to the keywords for which you are searching; it will not reflect the page as it exists in "real time," with any recent updates applied by its Web master. The cached version is useful because it assures you that the page you are looking at relates directly to your search. On the other hand, if you simply click the main link provided in your results, more often than not you will be directed to the latest version of the page, which may or may not contain or refer to any of the terms you were looking for.

Surfing Web Content in a Self-directed Manner

Surfing the Web in a self-directed manner means entering various URLs you have obtained in advance and reviewing the sites on a case by case basis. This may take longer than conducting automated searches, but it can also be very productive if you know you have obtained URLs for appropriate, high-quality sites. That is why it is important to read chapter 9; it provides URLs for organizations and agencies that will fit the scope of your research. In addition, the next section identifies some online sites and services that may also be useful to your energy-related research.

ONLINE SOURCES OF ENERGY INFORMATION

Good online sources of information related to energy and natural resources include the Web sites listed for the Web documents in the annotated bibliography in chapter 10, as well as the Web sites of the organizations and agencies included in chapter 9. The following is a guide to some of these sites, plus a few additional energy-related sites, presented according to region.

U.S. Energy Issue Sites

- **American Council on Renewable Energy (http://www.acore.org),** which offers information on a broad range of strategies, policies, and programs related to the research and development of renewable energies in the United States; it covers all forms of renewable energy.
- **American Nuclear Society (http://www.ans.org),** which includes a public information page that offers news, press releases, and educational materials about nuclear energy in the United States.
- **American Petroleum Institute (http://www.api.org),** which offers a portal of information for energy consumers, including descriptions of technological information, analyses, news, policies, environmental issues, and basic information on oil and natural gas.
- **American Solar Energy Society (http://www.ases.org),** which offers many in-depth reports on solar energy options as well as policy initiatives for establishing solar energy in the U.S. energy market.
- **American Wind Energy Association (http://www.awea.org),** which provides information on wind power technology, plus wind power news and publications.
- **Association for the Study of Peak Oil and Gas (http://www.aspo-usa.org),** which presents research and statistics related to the challenge of shrinking oil supplies.
- **Consumer Energy Center (http://www.ConsumerEnergyCenter.org),** which provides tips on improving energy efficiency and conservation in homes and businesses, plus facts on alternatively fueled vehicles and renewable energy.
- **Edugreen (http://www.edugreen.teri.res.in),** which teaches elementary and middle school students the fundamentals about energy, energy sources, climate change, air pollution, and other energy-related topics.
- **Electric Power Supply Association (http://www.epsa.org),** which offers documents on electricity generation for the general public.
- **Energy Information Administration of the U.S. Department of Energy (http://www.eia.doe.gov),** which offers an enormous amount of infor-

mation on energy resources, supply, and use in the United States (as well as energy profiles for many other countries). Data and statistics on energy consumption, production, and reserves are organized by resource type, region, sector, and price, with historical data and projections to 2020.

- **Energy Trends (http://energytrends.pnl.org),** a site belonging to the U.S. Department of Energy's Pacific Northwest National Laboratory, which analyzes energy policies and data in industrialized countries.
- **Hydrogen and Fuel Cell Newsletter (http://www.hfcletter.com),** which provides news about developments in fuel cell and hydrogen technology.
- **National Hydropower Association (http://www.hydro.org),** which provides information, news, position statements, and publications related to hydropower.
- **National Renewable Energy Laboratory of the U.S. Department of Energy (http://www.nrel.gov),** which provides educational resources and research findings on all forms of renewable energy.
- **Office of Energy Efficiency and Renewable Energy of the U.S. Department of Energy (http://www.eere.energy.gov),** which offers links to Web sites and online documents on energy efficiency and renewable energy and downloadable reports on federal programs related to energy efficiency, conservation, and alternatives.
- **Office of Fossil Energy of the U.S. Department of Energy (http://www.fe.doe.gov),** which includes facts and information on petroleum, coal, and natural gas, plus hydrogen and other fossil fuel alternatives. It offers information on government programs for oil and gas supply and delivery, natural gas regulation, electricity regulation, and petroleum reserves.
- **Solar Energy Industries Association (http://www.seia.org),** which describes solar energy technology, solar energy projects, and news stories related to solar energy in the United States.
- **U.S. Department of Energy (http://www.energy.gov),** which provides news about the U.S. DOE and its activities, as well as information on energy supply, energy sources, energy efficiency, and energy-related environmental issues, science, and technology, with many materials developed specifically for researchers, students, teachers, and children.
- **U.S. Environmental Protection Agency (http://www.epa.gov),** which disseminates information on global climate change, oil spills, acid rain, and other energy-related environmental issues, as well as facts on energy conservation and environmental regulations.

ENERGY SUPPLY AND RENEWABLE RESOURCES

International Energy Issue Sites

- **Alliance to Save Energy (http://www.ase.org),** which offers educational materials and fact sheets about conserving energy from the perspective of a nonprofit advocacy group.
- **Climate Action Network (http://www.climatenetwork.org),** which discusses nuclear energy issues, emissions trading, and environmental compliance, as well as issues related to global warming and the Kyoto Protocol, from the perspective of a coalition of nongovernmental organizations concerned with energy and climate.
- **International Energy Agency (http://www.iea.org),** which offers publications on energy-related topics that are of global concern, including nuclear energy and sustainable energy use, as well as facts and statistics related to world energy supply, consumption, and production.
- **International Energy Agency Coal Research Centre (http://www.iea-coal.org.uk),** which offers information on global clean coal technologies.
- **International Solar Energy Society (http://www.ises.org),** which offers information on worldwide initiatives and projects related to the field of solar energy.
- **World Energy Council (http://www.worldenergy.org),** which provides energy news briefs, worldwide energy statistics, and energy market, production, and technology data, from the perspective of a worldwide coalition of energy producers, policymakers, and environmental groups.
- **Worldwatch Institute (http://www.worldwatch.org),** which offers details and background information on climate change, energy sources, and environment-related energy facts.

Chinese Energy Issue Sites

- **China Watch (http://www.worldwatch.org/taxonomy/term/53),** a service of the Worldwatch Institute that reports on energy, agriculture, population, water, health, and the environment in China, emphasizing business and government policy decisions. It also offers a news feed on important Chinese events and issues.
- **Global Environmental Institute (http://www.geichina.org/en/index.htm),** a nonprofit, nongovernmental Beijing-based organization that advocates energy conservation and ecologically sound, sustainable development in China. It offers extensive information on climate change and energy-related environmental issues in China and includes a library of relevant documents.

- **State Environmental Protection Administration of China (http://www.zhb.gov.cn/english/index.php3),** which offers an English version of all of its news, reports, and regulations covering the protection of China's natural resources and environment.
- **U.S. Energy Information Administration Country Analysis Brief: China (http://www.eia.doe.gov/emeu/cabs/china.html),** an extensive report on the energy resources, demand, supply, production, and consumption in China, with an analysis of the relation of all of these factors to the economic, political, and environmental well-being of the country itself. The end of the report lists links to other sites with valuable information on China and its energy resources.

German Energy Issue Sites

- **German Energy Agency (http://www.deutsche-energie-agentur.de/page/index.php?id=717&L=4),** which offers a wealth of information (in British English) on developments in Germany's energy industry, such as news briefs, reports, descriptions of laws and policies, and historical background.
- **German Federal Ministry for the Environment, Nature Conservation, and Nuclear Safety (http://www.bmu.de/english/aktuell/4152.php),** the Web site of a German federal regulatory agency, which provides, in English, news briefs, a newsletter, downloads, and in-depth analyses of Germany's climate and energy issues, waste, soil, water, and transportation issues, as well as information on international environmental policy.
- **U.S. Energy Information Administration Country Analysis Brief: Germany (http://www.eia.doe.gov/emeu/cabs/Germany/Background.html),** an extensive report on the energy resources, demand, supply, production, and consumption in Germany, with an analysis of the relation of all of these factors to the economic, political, and environmental well-being of the country itself. The end of the report lists links to other sites with valuable information on Germany and its energy resources.

Iranian Energy Issue Sites

- **Iran Petroleum Magazine (http://www.iranpetroleummag.com),** a monthly online periodical with the latest news regarding Iran's oil industry (heavily geared toward the interests of that industry). Information is provided in English.
- **Ministry of Petroleum of Iran (http://www.nioc.org),** the Web site of the regulatory agency for Iran's petroleum industry, which offers facts and information about oil production and supply in Iran. Information is provided in English.

- **U.S. Energy Information Administration Country Analysis Brief: Iran (http://www.eia.doe.gov/emeu/cabs/Iran/Background.html),** an extensive report on the energy resources, demand, supply, production, and consumption in Iran, with an analysis of the relation of all of these factors to the economic, political, and environmental well-being of the country itself. The end of the report lists links to other sites with valuable information on Iran and its energy resources.

Saudi Arabian Energy Issue Sites

- **Saudi Aramco (http://www.saudiaramco.com/bvsm/JSP/home.jsp),** the official site of Saudi Arabia's national oil company, which offers a media center and portals on Saudi Arabian energy exports, oil and gas, environment, and new energy technologies. Information is available in English.
- **Saudi Arabian Ministry of Petroleum and Mineral Resources (http://www.mopm.gov.sa),** the site of the regulatory commission that oversees Saudi Arabia's fossil fuels industry. It offers English reports on Saudi Arabian oil policy, environmental issues, and laws.
- **U.S. Energy Information Administration Country Analysis Brief: Saudi Arabia (http://www.eia.doe.gov/emeu/cabs/Iran/Background.html),** an extensive report on the energy resources, demand, supply, production, and consumption in Saudi Arabia, with an analysis of the relation of all of these factors to the economic, political, and environmental well-being of the country itself. The end of the report lists links to other sites with valuable information on Saudi Arabia and its energy resources.

Nigerian Energy Issue Sites

- **Online Nigeria: Nigeria Energy Policy (http://www.onlinenigeria.com/geology/?blurb=514),** a storehouse of information on all aspects of energy generation in Nigeria, organized as an online Web index or subject directory. Includes a "Reading Rooms" section with valuable background information on the energy-related problems in Nigeria.
- **Power Holding Company of Nigeria Plc (http://www.nepanigeria.org),** Web site of the organization given the authority, under Nigeria's 2005 Electric Power Sector Reform Act, to provide Nigeria's electricity to Nigeria. The site offers facts and statistics about electricity generation in Nigeria as well as information on current projects aimed at reforming Nigeria's electric power sector.
- **U.S. Energy Information Administration Country Analysis Brief: Nigeria (http://www.eia.doe.gov/emeu/cabs/Nigeria/Background.html),** an extensive report on the energy resources, demand, supply,

production, and consumption in Nigeria, with an analysis of the relation of all of these factors to the economic, political, and environmental well-being of the country itself. The end of the report lists links to other sites with valuable information on Nigeria and its energy resources.

Venezuelan Energy Issue Sites

- **Petroleumworld (http://www.petroleumworld.com),** a Web site devoted to news and to analysis of events and issues related to Latin American energy, oil, and gas. It offers timely news briefs and weekly reviews of the energy situation in Venezuela. It also offers maps depicting the gas and oil fields in Venezuela, plus guides to the energy industry in Venezuela, created by the Venezuelan British Chamber of Commerce.
- **Venezuela Power (http://www.venezuelapower.com),** which is a portal of news and information related to the entire energy industry in Venezuela; headlines are consistently updated.
- **U.S. Energy Information Administration Country Analysis Brief: Venezuela (http://www.eia.doe.gov/emeu/cabs/Venezuela/Background.html),** an extensive report on the energy resources, demand, supply, production, and consumption in Venezuela, with an analysis of the relation of all of these factors to the economic, political, and environmental well-being of the country itself. The end of the report lists links to other sites with valuable information on Venezuela and its energy resources.

OTHER SOURCES OF ENERGY INFORMATION ONLINE

In addition to the aforementioned energy information sites, there are other tools available on the Web to assist you with your energy-related research:

- Article search engines
- Newswires and media sites
- Legal research sites

Article search engines are a useful tool for finding information related to energy. These sites enable you to search for articles within a large collection of newspapers, magazines, journals, and other periodicals. These services are especially helpful if you know only the title of a particular article, its author, or a specific topic on which you would like to find an article. The following are a few examples of article search engines. Some provide access to the full text of articles, offering a combination of both free and paid access to these full texts, depending on the policy of each publication (different publications have different policies regarding access to their archives).

- **Energy Citations Database (http://www.osti.gov/energycitations/index.jsp),** a site maintained by the U.S. Department of Energy's Office of Scientific and Technical Information that provides citations, abstracts, and/or full text for energy and energy-related scientific and technical reports and articles published by the Department of Energy (DOE) and its predecessor agencies, the Energy Research & Development Administration (ERDA) and the Atomic Energy Commission (AEC). Records date as far back as 1948.
- **Google Scholar (http://scholar.google.com),** which includes many academic or scholarly publications in its searches.
- **LookSmart Find Articles (http://www.findarticles.com),** which searches mainly publications of interest to the general public. It also allows users to browse publications by subject.
- **Scirus (http://www.scirus.com/srsapp),** which searches scientific publications and can be useful for finding articles about ecolcgy or energy technologies.

If you already know the name of the publication in which a particular article has appeared, more often than not you can also search for the article directly on this publication's Web site. Depending on the date the article was published or the subscription policy of the publication, you may be able to access the full article (as opposed to just an abstract) without having to pay for access or hold a subscription. If you cannot access the full text of the article without paying a fee, visit the library—most libraries have paid access to a number of databases of full-text articles and reports, as well as subscriptions to numerous periodicals. See "Conducting Research in a Library."

Newswires and media sites are another important resource for researching energy issues. They can provide you with both U.S. and international news. The following media sites consistently report on energy topics of significance to both the United States and other countries:

- **Reuters, a newswire (http://www.reuters.com)**
- **Associated Press, a newswire (http://www.aponline.com)**
- **Cable News Network (CNN) (http://www.cnn.com)**
- *New York Times* **(http://www.nytimes.com),** with full news stories available only by subscription
- *Time* **magazine (http://www.time.com)**
- *Wall Street Journal* **(http://online.wsj.com)**
- *Washington Post* **(http://www.washingtonpost.com)**

Legal resource sites are another essential component of researching energy issues because of the numerous regulations, laws, treaties, agreements, and policies that govern the use of energy worldwide. Online collections of legal resources can provide you with access to the full text of many bills, regulations, and laws, and some even provide background, analysis, and commentary. It may be a bit easier to obtain the full texts of U.S. regulations, laws, treaties, agreements, and policies from such sites than it will be to obtain those of other countries. However, the informational Web sites listed for various countries, the FindLaw Web site listed in the following, and the organizations listed in chapter 9 may all be of help in this area, as well.

The following is a list of Web sites through which you can obtain full text, descriptions, or background for many of the U.S. energy-related regulations, laws, treaties, agreements, and policies that have been established. U.S. regulations and laws are part of the massive collection of U.S. laws known as U.S. Code, and most of the following sites give you the ability to search the U.S. Code by keyword, bill number, legislative title, or section number (for example: *Title 22, Section 1501*). Some sites also allow you to search for the name by which a law is commonly known (for example: *Energy Policy Act of 2005*).

- **Cornell Law School's U.S. Code database (http://www.law.cornell.edu/uscode)**
- **Ecolex, a database of U.S. and international environmental laws, court decisions, and treaties (http://www.ecolex.org/index.php)**
- **FindLaw's directory of federal, state, and international legal cases and codes (http://www.findlaw.com/casecode)**
- **The Legal Information Institute (http://supct.law.cornell.edu/supct)**
- **U.S. Government Printing Office (http://www.gpoaccess.gov/index.html)**
- **U.S. Geological Survey Guide to Environmental Laws and Regulations (http://water.usgs.gov/eap/env_guide)**
- **U.S. Library of Congress THOMAS Web site (http://thomas.loc.gov)**

Additional Legal Resources

- **Federal depository libraries (http://www.gpoaccess.gov/libraries.html)** carry federal laws, publications, and other information published by the government. The documents are available for free public use. There are federal depository libraries throughout the United States; the URL provided contains a search tool for finding one near you.

- **A "faqs.org" Web site that explains legal research (http://www.faqs.org/faqs/law/research)** may help you gain a greater understanding of confusing legal terms, legal procedures, and the hierarchy of legal systems and courts.

II. Conducting Research in a Library

Conducting your research in a library involves the use of bibliographic resources—card catalogs, indexes, bibliographies, and other guides to the library's collection of books, periodicals, or other printed materials. Bibliographic resources are like the library's version of search engines, subject directories, or Web portals. These resources may be available either in electronic/online form on the library's public-use computers or in print form in the library's file cabinet and stacks of reference materials. Here are some ways to streamline your library research experience.

TIPS FOR RESEARCHING IN A LIBRARY

- Whether you are searching for resources in the library's physical catalog or its online catalog, always tailor your search to the level of your familiarity with the topic or resources that are the focus of your research. For example, searching the catalog by author name is most appropriate when an author has written a number of works that you know would be of value to your research. Searching the catalog by title is best when you know the exact title of a work. (Remember to omit articles—*a, an, the*—when conducting such a search.)

  Searching by keyword is best when you have only a general understanding of the types of works you are interested in finding. This kind of search will generate all the works with titles that contain your keyword.

  Finally, conducting a search by subject will help you find those works that the library has assigned to a particular subject area. Use this kind of search when you are familiar with standard Library of Congress subject categories (which you can view by visiting http://catalog.loc.gov, the Web site of the catalog of the Library of Congress).

- Once you have located a particular resource, look at the other subject headings under which it falls. These are additional subject areas into which to expand your search for resources.
- It might go without saying, but always ask for help when you have trouble locating the resources you need. Many librarians have degrees in library science and are skilled researchers.

RESEARCH TOOLS AVAILABLE IN A LIBRARY

Most libraries possess a combination of print and electronic/online resources that you can use to find articles, books, and other materials. Here is a sampling of their offerings:

- Printed periodical indexes, through which you can search for articles by subject, title, or author.
- Online databases, such as InfoTrac or EBSCO Host, which are indexes to articles from hundreds of periodicals. These databases are similar to search engines in that they allow you to search for specific keywords, but they are also more advanced in that you can search by title, subject, and author, as well as browse the available publications by topic. When you conduct a search using these services, you will either receive citations for articles (listing information such as author name, title, page numbers, publication name, volume number, and issue date) and article abstracts (summaries) or full texts of the articles that you can download online.
- A physical library catalog, with the entire library's printed holdings indexed by titles, subjects, and authors.
- Online library catalogs or periodical databases. Some libraries even allow you offsite access to these resources (that is, via your home computer's Internet connection) using a username and a password linked to your library membership.
- Subscriptions to numerous publications, including not only periodicals such as magazines and journals, but also documents issued by local, federal, and state governments, courts, and agencies; conference proceedings; and even scholarly works such as theses and dissertations. A library's archives of such materials can date back several decades and more. Many libraries also have online access to electronic versions of all their publication holdings, making your research more automated and, thus, easier.

LIBRARY SOURCES OF ENERGY INFORMATION

Some libraries posses a collection of reports and documents issued by government agencies and departments, including the U.S. Department of Energy and the U.S. Environmental Protection Agency. These collections can be especially useful to your research. Energy industry periodicals, such as *Oil and Gas Journal* or *Solar Today,* or environmental organizations' periodicals, such as the monthly newsletter *Sierra,* may also be found in some libraries. Furthermore, when you locate a book on energy and natural resources in a library's bookshelves, you can browse surrounding titles and often find some additional useful resources.

II. Conducting Research in a Bookstore

Large bookstores, such as Borders and Barnes & Noble, offer an abundance of books and periodicals on energy issues. Conducting your research in a bookstore's physical storefront (as opposed to its online site) has some advantages. For example, provided that the bookstore does not sell used books, most of the books in stock will represent the latest writings on the subject.

Bookstore holdings are also organized by self-explanatory subject categories, and bookstore shelves are usually labeled with large signs indicating these subject categories. The only disadvantage of conducting energy research in a bookstore is that energy books will probably be found on shelves throughout the store—for example, on shelves labeled "Science and Technology," "Environment," "Politics," "Natural Resources," and, of course, "Energy."

Conducting research via an *online* bookstore (such as http://www.amazon.com, http://www.barnesandnoble.com, or http://www.borders.com) can be even more beneficial. These sites enable you to find books using a search engine or by browsing titles according to subject. These sites may also feature access to merchants who sell out-of-print or hard-to-obtain editions of books. In addition, they tend to include reader reviews, publisher and publication date information, and content summaries for most of their books, which make it easier for you to determine whether a book fits your research needs.

FINAL WORD ABOUT RESEARCH

With all the many ways of finding information, it is important to keep careful records of the resources you have used, whether you found them online or in print. The minimum information you should have for each resource is the title, author, publisher, publication date, and page numbers. If the resource is an article, you should also know the periodical title, volume number, and page numbers; if a Web document, the Web site URL and the date you accessed it. Having this information will be invaluable if you need to return to the source for clarification when you no longer have it on hand and for citing your resources properly, if there is a specific style you are required to follow when listing your references in a bibliography or endnotes.

7

Facts and Figures

INTRODUCTION

1.1 Global Oil Demand by Region (in millions of barrels per day)

	DEMAND		ANNUAL CHANGE		ANNUAL CHANGE (%)		
	2004	2003	2004	2005	2003	2004	2005
North America	21.14	0.47	0.57	0.23	1.9	2.3	0.9
Europe	16.47	0.20	0.26	0.10	1.2	1.6	0.6
China	6.37	0.55	0.85	0.36	11.0	15.4	5.7
Other Asia	8.54	0.22	0.44	0.21	2.8	5.4	2.5
Former Soviet Union	3.69	0.12	0.11	0.14	3.5	3.1	3.9
Middle East	5.88	0.20	0.32	0.26	3.7	5.7	4.5
Africa	2.81	0.04	0.07	0.09	1.7	2.4	3.3
Latin America	4.89	-0.09	0.16	0.10	-1.9	3.5	2.1
World	82.45	1.85	2.66	1.44	2.4	3.3	1.7

Source: Oil Market Report, December 2004, International Energy Agency.

1.2 Proven Oil Reserves at the End of 2003

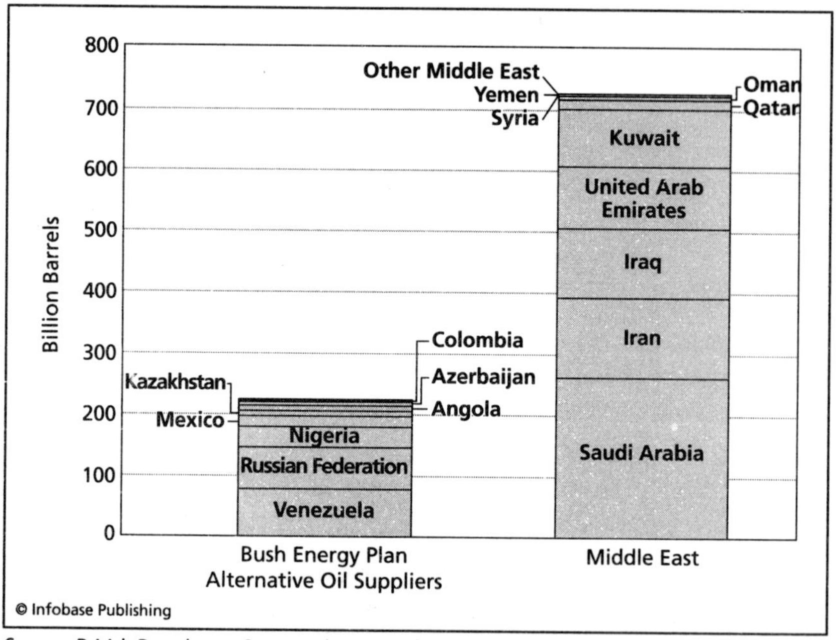

Source: British Petroleum, *Statistical Review of World Energy 2004.*

1.3 Proven Oil Reserves through 2025

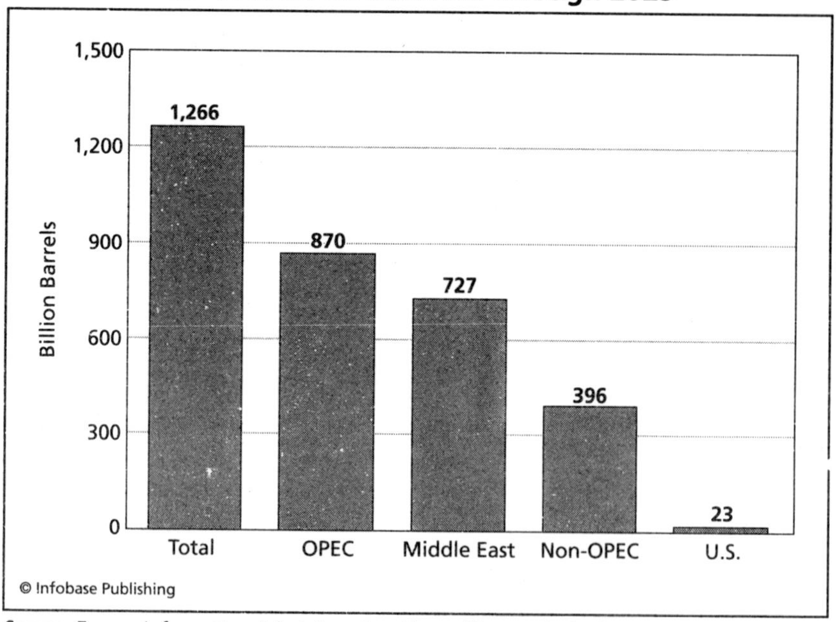

Source: Energy Information Administration, *Annual Energy Outlook 2004.*

1.4 Trends in Global Oil Production and Future Demand

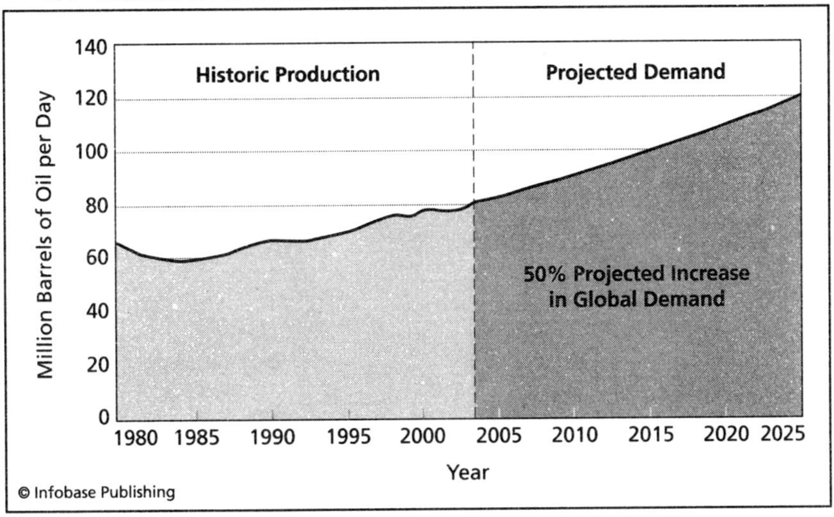

Source: Energy Information Administration, 2004.

FOCUS ON THE UNITED STATES

2.1 Growth in U.S. Energy Consumption, as Compared with Production

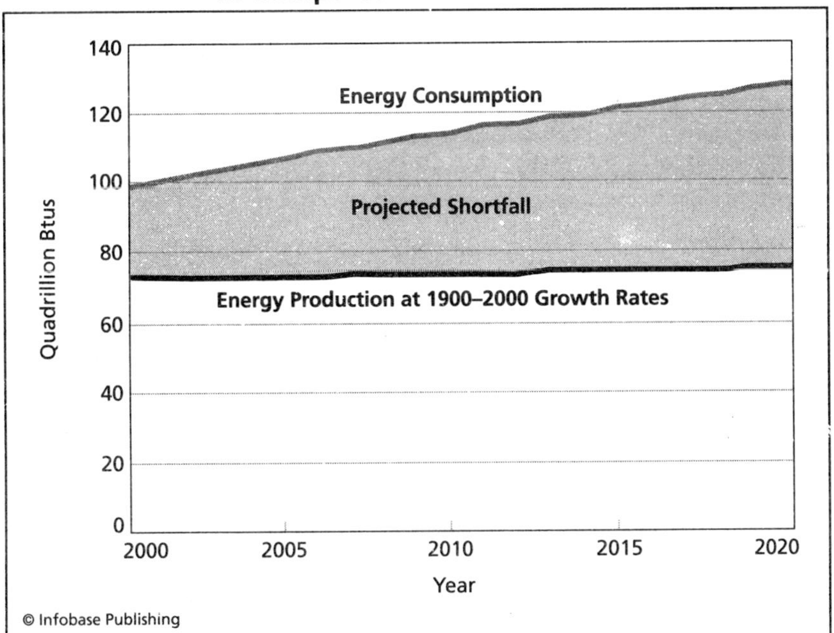

Source: Sandia National Laboratories and U.S. Department of Energy, Energy Information Administration.

2.2 U.S. Energy Overview, 1949–2004

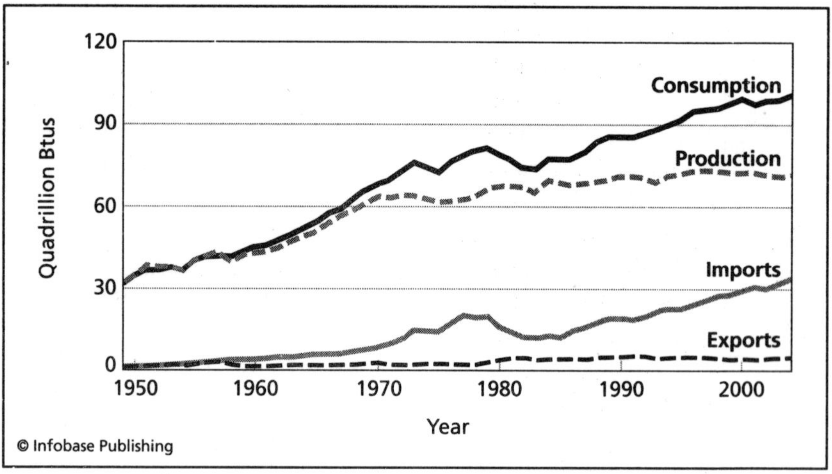

Source: Energy Information Administration, *Annual Energy Review 2004.*

2.3 Shut-in Production Comparison of Hurricanes Ivan, Katrina, Rita, and Wilma

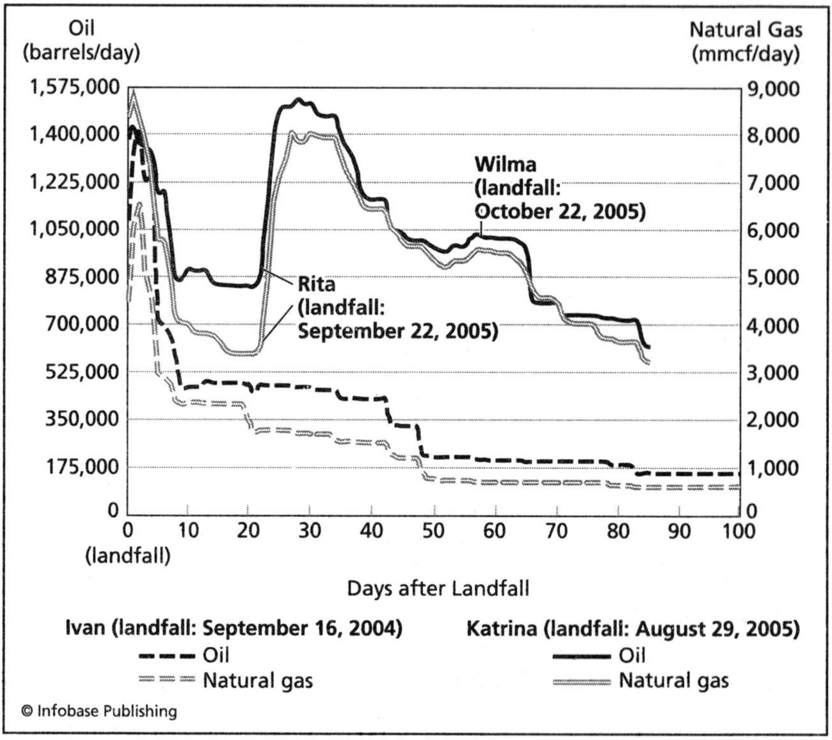

Source: Energy Information Administration.

2.4 What We Pay for in a Gallon of Regular Grade

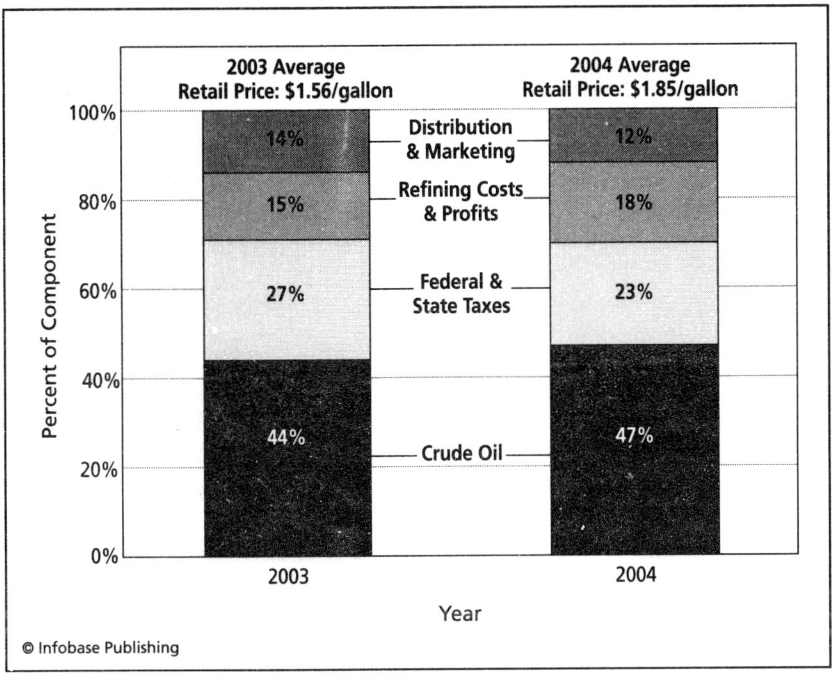

Source: Energy Information Administration, *Annual Energy Review 2004.*

2.5 Oil and the U.S. Economy: Oil's Share of GDP

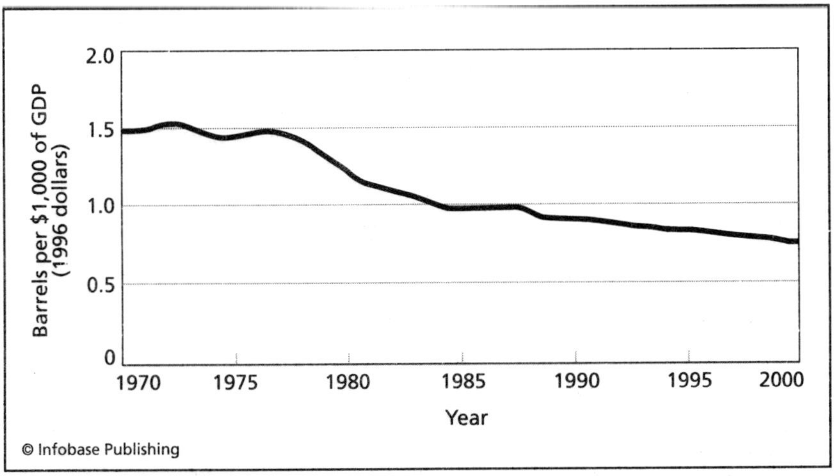

Source: Energy Information Administration, *Annual Energy Review 2004.*

2.6 U.S. Dependence on Imported Oil, 1970–2025

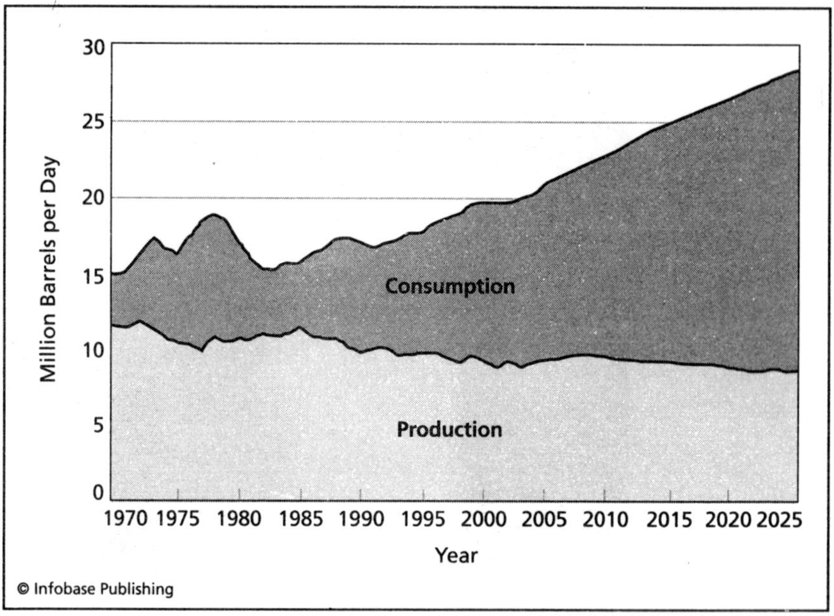

Source: Energy Information Administration.

2.7 U.S. Crude Oil Imports by Source

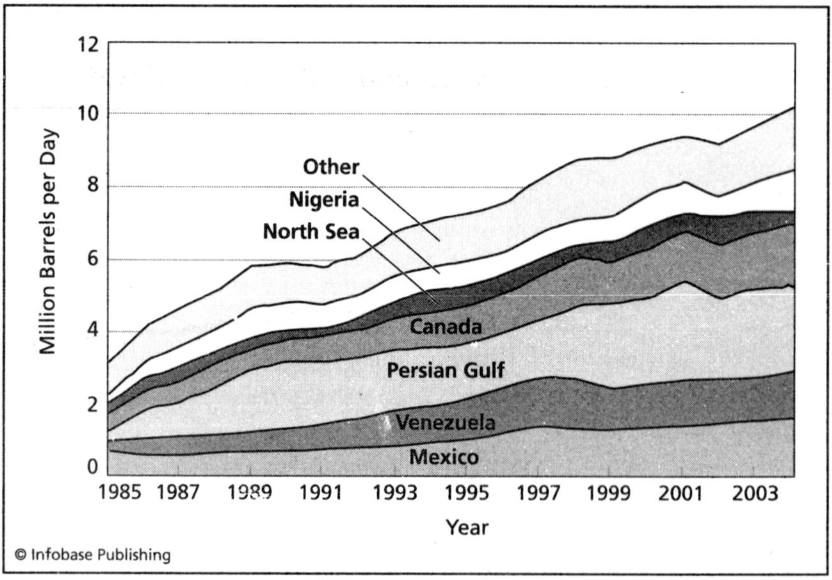

Source: International Energy Administration.

2.8 Carbon Dioxide Emissions from U.S. Energy Consumption, 2003, Total by Fuel

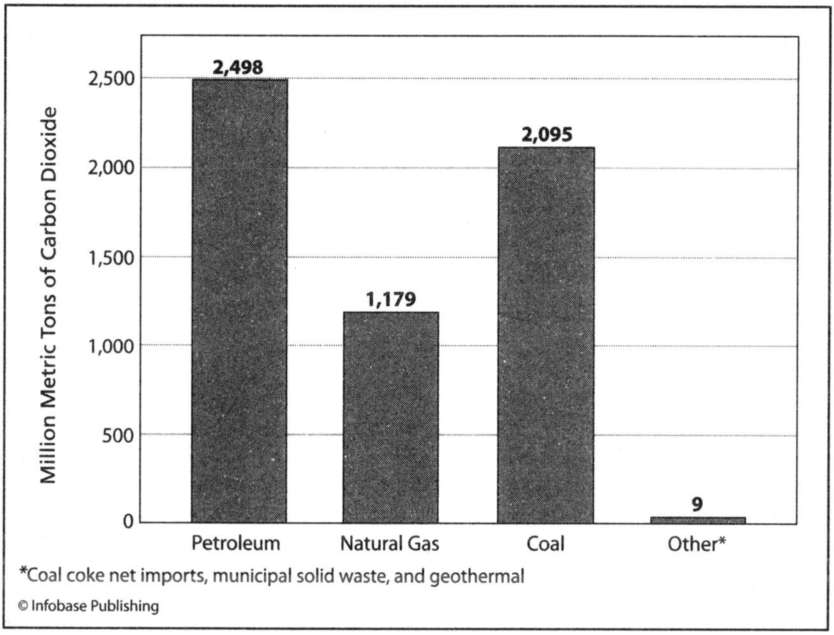

Source: Energy Information Administration.

2.9 U.S. Emissions of Greenhouse Gases Based on Global Warming Potential, 1980–2003

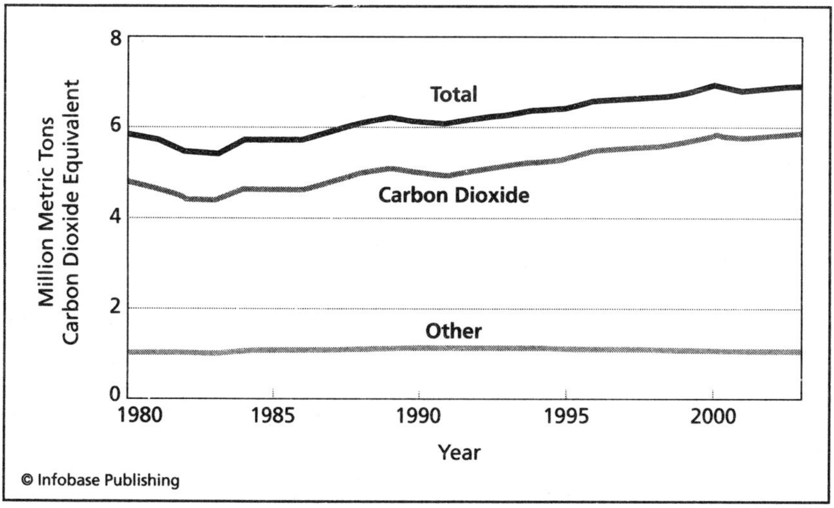

Source: Energy Information Administration.

2.10 U.S Energy Production by Fossil Fuels, Nuclear Electric Power, and Renewable Energy, 1949–2004

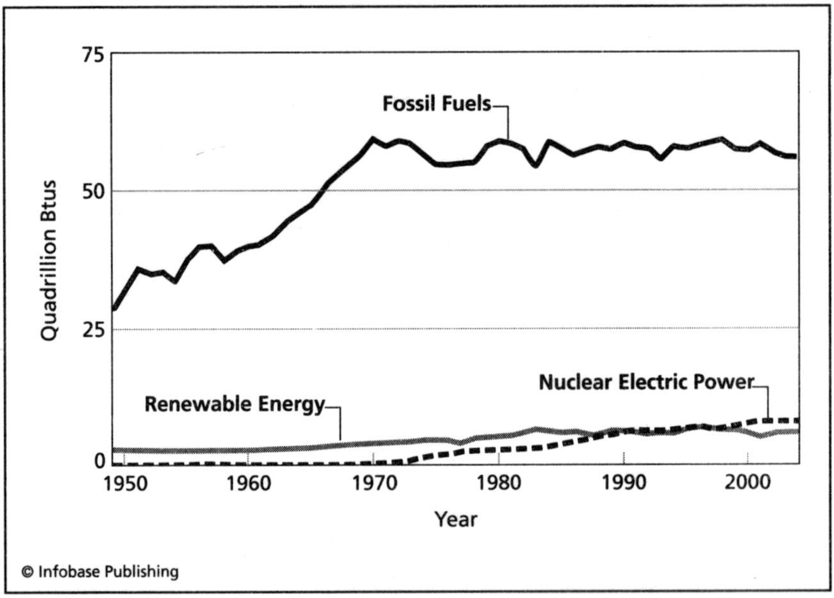

Source: Energy Information Administration, *Annual Energy Review 2004.*

GLOBAL PERSPECTIVES

3.1 China's Oil Production and Consumption, 1980–2005

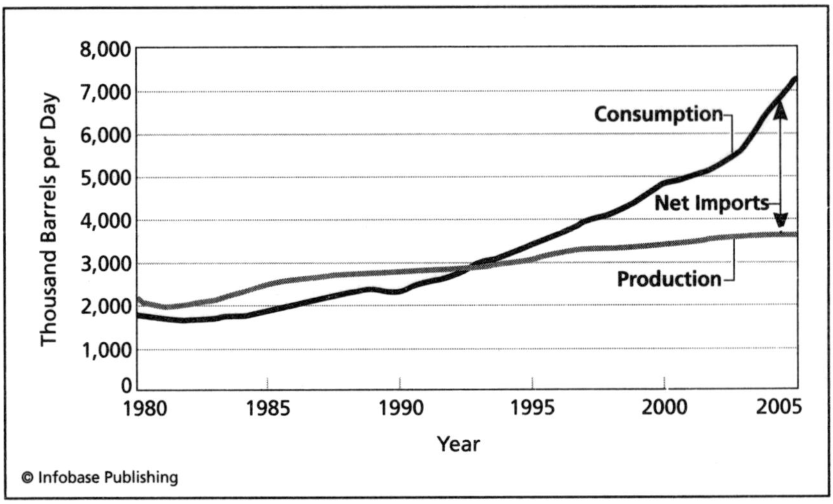

Source: Energy Information Administration.

3.2 Germany's Development of Energy Supply from Renewable Energies, 1990–2004

	HYDROPOWER	WIND POWER	BIOMASS ELECTRICITY	PHOTOVOLTAICS	GEOTHERMAL ELECTRICITY	SUM TOTAL OF ELECTRICITY GENERATION	SHARE OF GROSS ELECTRICITY CONSUMPTION	BIOMASS HEAT	SOLAR THERMAL ENERGY	GEOTHERMAL HEAT	SUM TOTAL OF HEAT GENERATION	BIODIESEL	BIO-ETHANOL	SUM TOTAL, FUELS	SUM TOTAL OF FINAL ENERGY SUPPLY	SHARE OF FINAL ENERGY CONSUMPTION
	[GWH]						%	[GWH]							%	%
1990	17,000	40	1,422	1	0	18,463	3.4	—	130	—	—	—	0	0	—	—
1991	15,900	140	1,450	2	0	17,492	3.2	—	166	—	—	2	0	2	—	—
1992	18,600	230	1,545	3	0	29,378	3.8	—	218	—	—	52	0	52	—	—
1993	19,000	670	1,570	6	0	21,246	4.0	—	279	—	—	103	0	103	—	—
1994	20,200	940	1,870	8	0	23,018	4.3	—	351	—	—	258	0	258	—	—
1995	21,600	1,800	2,020	11	0	25,431	4.7	—	440	1,425	—	310	0	310	—	—
1996	18,800	2,200	2,203	16	0	23,219	4.2	—	550	1,383	—	517	0	517	—	—
1997	19,000	3,000	2,479	26	0	24,505	4.5	48,546	695	1,335	50,576	827	0	827	75,908	2.9
1998	19,000	4,489	2,800	32	0	26,321	4.7	51,613	857	1,384	53,854	1,033	0	1,033	81,208	3.1
1999	21,300	5,528	3,020	43	0	29,890	5.4	50,951	1,037	1,429	53,417	1,343	0	1,343	84,650	3.3
2000	24,936	9,500	4,129	64	0	38,629	6.7	54,314	1,279	1,433	57,026	2,583	0	2,583	98,238	3.8
2001	23,383	10,456	5,065	116	0	39,020	6.7	55,326	1,626	1,447	58,399	3,617	0	3,617	101,036	3.8
2002	23,824	15,856	5,962	188	0	45,830	7.8	54,626	1,955	1,483	58,064	5,683	0	5,683	109,577	4.3
2003	20,350	18,919	7,982	333	0	47,584	8.0	59,248	2,465	1,532	63,245	8,267	0	8,267	119,096	4.7
2004	21,000	25,000	9,367	459	0.4	55,826	9.3	59,806	2,573	1,558	63,937	10,747	424	11,171	130,934	5.1

Source: government of Germany

3.3 Evolution of Total Production of Energy in Iran, 1971–2003

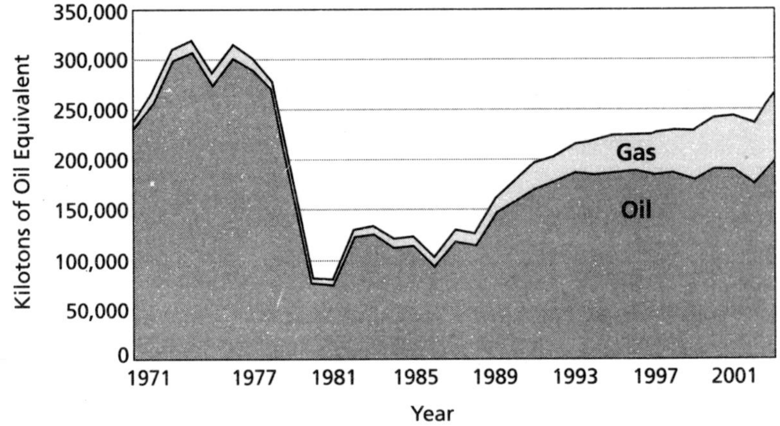

Source: International Energy Administration.

3.4 Saudi Arabia's Oil Production and Consumption, 1980–2006

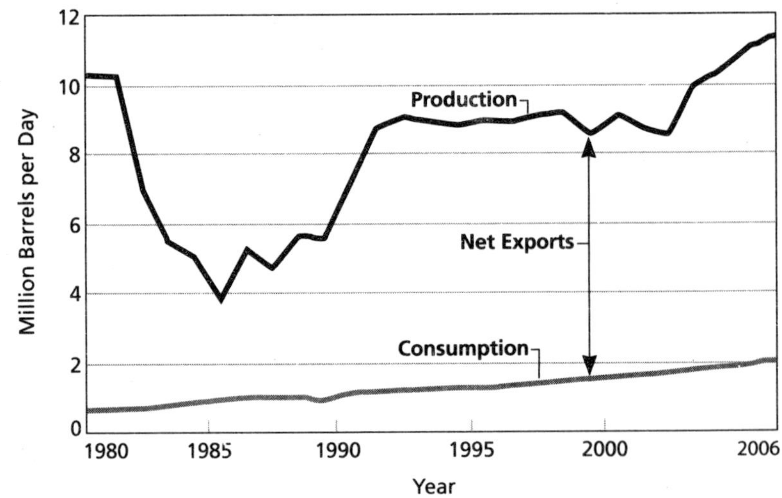

Note: Production includes crude oil, natural gas liquids, other liquids, and refinery gain.

Source: Energy Information Administration.

3.5 Venezuela's Total Oil Exports to the United States, 1960–2004*

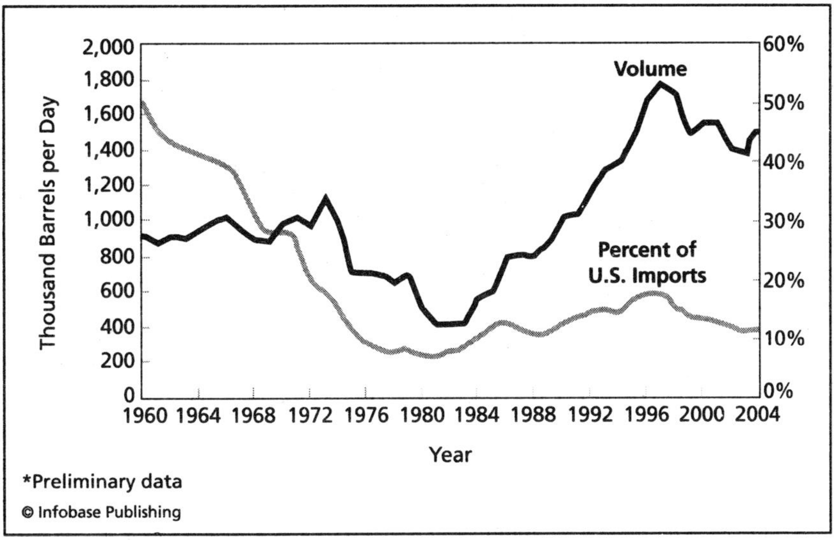

Source: Energy Information Administration.

8

Key Players A to Z

M. A. ADELMAN (1917–) professor of economics emeritus at the Massachusetts Institute of Technology (MIT) and oil economics consultant. Unlike Kenneth Deffeyes, Colin J. Campbell, and M. King Hubbert, Adelman has contended that the world is not running out of oil. He argues that calculations of peak oil production based on the world's proven oil reserves are flawed because oil reserves are continually used up and replenished.

MAHMOUD AHMADINEJAD (1956–) president of Iran since June 2005. In 2002 Iran announced plans for the construction of nuclear power plants with Russia, raising global concerns that Iran is seeking to develop nuclear weapons. Ahmadinejad has defended Iran's decision to conduct nuclear research and contends that the country is researching peaceful uses of nuclear technology. He also maintains that the research is aimed at generating nuclear power in Iran, which claims to have a severe power shortage. However, Iran was reported to the United Nations (UN) Security Council in February 2006 over charges that its nuclear program violates the Nuclear Non-Proliferation Treaty, which Iran ratified in 1970. Ahmadinejad threatened that Iran would "revise its policies" if the rights of the Iranian people were violated. The U.S. president, George W. Bush, who has called Iran part of "the axis of evil," has extended sanctions barring U.S. companies and their foreign subsidiaries from doing business with Iran or investing in the development of Iran's petroleum resources. Iran is the second-largest oil producer in the Organization of Petroleum Exporting Countries (OPEC), and its economy relies heavily on oil export revenues.

JAMES AKINS (1926–) U.S. government oil expert. He published an influential article titled "The Oil Crisis: This Time the Wolf Is Here" in 1973, the year of the first Arab oil embargo and subsequent U.S. energy crisis. During Richard Nixon's presidency (1969–74), Akins visited top officials in Venezuela to pursue an agreement guaranteeing U.S. access to Venezuela's oil.

Key Players A to Z

FRANCIS THOMAS BACON (1904–1992) British engineer. He invented the modern hydrogen fuel cell in 1959.

ANTOINE HENRI BECQUEREL (1852–1908) French professor and scientist. His 1896 discovery of radioactivity in uranium was foundational to the development of nuclear energy. He shared the 1903 Nobel Prize in physics with colleagues Marie Curie and Pierre Curie, who had conducted further research into the phenomenon.

FATIH BIROL (1958–) chief economist of the International Energy Agency (IEA), which is based in France. Birol is responsible for the IEA's *World Energy Outlook* series, which provides global energy statistics, projections, and analyses.

SAMUEL W. BODMAN (1938–) U.S. secretary of energy since February 2005.

DOUGLAS BOHI (1939–) U.S. oil economics consultant. Bohi has estimated that U.S. petroleum shortages in the 1970s reduced gross domestic product (GDP) by just .35 percent. Bohi testifies before state and federal regulatory agencies on market power and competition issues, transmission pricing and access, electric utility mergers, and energy and environmental policies.

ROBERT L. BRADLEY, JR. (1955–) president of the Institute for Energy Research, Houston, Texas, which evaluates public policies in the oil, gas, and coal markets. Bradley is the author of *Oil, Gas, and Government: The U.S. Experience.*

CAROLE BROWNER (1955–) administrator of the U.S. Environmental Protection Agency (EPA) during the administration of the U.S. president Bill Clinton (1993–2001). Under her leadership the EPA established fuel emissions standards and other requirements for engine makers to meet by 1996. Agency officials reported in 2000 that smog-forming hydrocarbon emissions had decreased by 32 percent since the regulations took effect.

GEORGE W. BUSH (1946–) president of the United States, 2001–present. The son of George H. W. Bush, U.S. president in 1989–93, he worked in the oil industry during 1975–86. Environmentalists and opponents have accused him of promoting energy and environmental policies that favor oil companies and result in increased fossil fuel use. In 2001 he introduced a massive energy bill that included a controversial plan to drill for oil on 2,000 acres within the Arctic National Wildlife Refuge (ANWR), a 19-million-acre tract of land near an oil-rich site in Alaska. In August 2005 this energy bill, with changes that included the removal of the ANWR plan, was passed and signed into law as the Energy Policy Act of 2005. The ANWR plan became a separate bill in the Senate but had yet to be approved as of this writing. The Energy Policy Act of 2005 emphasizes many alternative and renewable forms of energy, but some say it does not go far enough and also contains too many incentives for the fossil fuels industry. President Bush has also been criticized for his refusal to ratify the Kyoto Protocol, an international treaty aimed at reducing emissions from fossil fuels, which, he has said, is too lenient on developing countries whose consumption of these fuels is increasing.

RAFAEL CALDERA RODRÍGUEZ (RAFAEL CALDERA) (1916–) president of Venezuela, 1969–74 and 1994–99. At the beginning of his second term in 1994, he met with several challenges: the collapse of the country's banking sector, falling oil prices, foreign debt, and inflation. Under his presidency the government announced a plan to expand the country's gold and diamond mining operations to reduce its dependency on the oil sector.

COLIN J. CAMPBELL (1931–) British geologist, independent oil-industry consultant, and member of the American Association of Petroleum Geologists since 1959. As has Kenneth Deffeyes he has concluded that world oil production will reach its peak in the first decade of the 21st century. He has suggested that the grave economic and political consequences will cause nations around the world to experience energy scarcity and sustained increases in the price of oil.

JIMMY CARTER (1924–) U.S. president, 1977–81. Arab oil embargoes caused major oil price hikes and oil shortages in the United States in the 1970s. Under President Carter, Congress passed the Public Utilities Regulatory Policies Act, which encouraged the development of the U.S. alternative energy market by exempting small alternative energy producers from state and federal utility regulations and requiring existing local utilities to buy electricity from them. As president Carter also encouraged energy conservation and proposed a 10-year, $10-billion program to increase domestic coal production.

GUY F. CARUSO (1941–) administrator of the Energy Information Administration (EIA), a statistical agency within the U.S. Department of Energy (DOE) that provides energy-related data, forecasts, and analyses.

HUGO RAFAEL CHÁVEZ FRÍAS (HUGO CHÁVEZ) (1954–) president of Venezuela since 1999. Venezuela has the largest proven oil reserves in the Western Hemisphere, according to *Oil and Gas Journal.* It is a top oil supplier to the United States, and oil exports account for about a third of its gross domestic product (GDP). As president, Hugo Chávez has taken over Venezuela's economy, extracting higher revenues on its oil exports by scaling back oil production. In 2001 he signed a new Hydrocarbons Law. This law increased the royalties that private companies must pay to the government for producing oil in Venezuela and requires foreign oil investments to be in the form of joint ventures with Venezuela's state-owned oil company. Chávez also expanded his presidential powers through a new constitution and strengthened Venezuela's ties with Cuba and Middle Eastern oil-producing nations. Western leaders, including President George W. Bush and Secretary of State Condoleezza Rice in the United States, have called his leadership dictatorial and a threat to the international community. In late 2002 Chávez's opponents led a nationwide revolt against him. The country's oil workers went on strike, and all oil operations ceased, hurting Venezuela's economy. Later Chávez dismissed nearly half of the country's oil workforce.

DICK CHENEY (1941–) U.S. vice president, 2000–present. Cheney led a task force called the National Energy Policy Development (NEPD) Group in 2001, which made recommendations for a national energy policy. The group's report was controversial because environmentalists and watchdog groups suspected that its recommendations were influenced by oil industry interests. Cheney is a former chief executive officer (CEO) of the energy contractor Halliburton Company.

ROBERT CORRELL senior fellow of the American Meteorological Society and U.S. authority on global climate change. He led the International Arctic Science Committee, a scientific team commissioned by the United States and seven other countries to study the impact of climate change on Greenland, above the Arctic Circle. In 2005 the team produced a report, the Arctic Climate Impact Assessment. The report describes harmful effects on the food chain, ocean currents, storm systems, and indigenous life that have resulted from melting of glacial ice caused by increasing temperature, which has been blamed on fossil fuel emissions.

GRAY DAVIS (1942–) governor of California from 1999 to 2003. In early 2001, during Davis's term, California experienced rolling electrical blackouts. Davis declared a state of emergency, and the California state government put the brakes on its plans to deregulate power utilities. In an attempt to guarantee steady and reliable electric power in the future, Davis appointed the California Department of Water Resources to replace the state's utility companies as state buyer of wholesale electricity and signed long-term, expensive contracts with other energy suppliers.

KENNETH S. DEFFEYES (1931–) petroleum geologist, Princeton University professor emeritus, and former colleague of the renowned geophysicist M. King Hubbert. Using Hubbert's calculations Deffeyes has concluded that the world's oil production either is peaking now or has already peaked.

RUDOLF DIESEL (1858–1913) German mechanical engineer, who developed the diesel engine in 1895. He originally designed this engine to run on a variety of fuels, including vegetable oil. Diesel used peanut oil for fuel when he exhibited the engine at the World Exhibition in Paris in 1900. However, after his death in 1913, his engine was modified to run on the cheapest fuel available, petroleum, and oil companies labeled one of the by-products of gasoline distillation *diesel fuel.*

"COLONEL" EDWIN DRAKE (1819–1880) U.S. entrepreneur, who struck the first oil well in the United States near Titusville, Pennsylvania, in 1859.

ROBERT E. EBEL (1929–) director of the energy program at the Center for Strategic and International Studies (CSIS) in the United States. Ebel has provided commentaries on oil's status as something more than a commodity in the United States—as a determinant of well-being, national security, and international power.

THOMAS A. EDISON (1847–1931) U.S. inventor, who created the electric light in 1879. He established Edison General Electric and entered into a partnership with Werner and William Siemens in 1881 to design Europe's first public lighting network in London. In 1882 Edison established the first electricity generating plants in London and New York.

ALBERT E. EINSTEIN (1879–1955) celebrated German-born U.S. physicist. In 1906 Einstein proposed the theory of relativity, which states that energy equals mass times the square of the speed of light, or $e=mc^2$. This theory, unifying mass, energy, magnetism, electricity, and light, was the foundation for astrophysics as well as nuclear medicine and other uses of nuclear energy.

MICHAEL FARADAY (1791–1867) British scientist. In the 1830s he discovered that an electric current could flow in a wire within a magnetic field. His work demonstrated basic principles of electricity production: electromagnetism, induction, generation, and transmission.

HENRY FORD (1863–1947) U.S. automobile manufacturer, famous for producing the Ford Model T automobile. In 1899, after successfully designing his first car, Ford resigned from his job as a machinist and engineer with the Edison Company and launched the Detroit Automobile Company, later the Ford Motor Company. In 1908 he produced the first Model T automobiles. Their popularity resulted in soaring sales of gasoline, helping oil to surpass coal as the most widely consumed fuel in the United States. In 1920 the Ford Motor Company began manufacturing Model Ts in large quantities, producing nearly 17 million cars before discontinuing the model in 1928.

ADINARAYANA GOPALAKRISHNAN nuclear engineer and former chair of India's Atomic Energy Regulatory Board. Gopalakrishnan submitted a 300-page report detailing 95 safety issues related to India's nuclear energy program.

WILLIAM GROVE (1811–1896) British physicist. He invented the first hydrogen fuel cell in 1839.

HAROLD HOTELLING (1895–1973) U.S. energy author and economics authority. He wrote a foundational 1931 essay about the forecasting of natural resource supplies titled "The Economics of Exhaustible Resources." Hotelling argued that because the amount of resources is a fixed stock, as they are exploited, there is less remaining and thus they tend to increase in value.

M. KING HUBBERT (1903–1989) renowned U.S. geophysicist, who in 1956 calculated that U.S. oil production would reach its peak in the 1970s and decline thereafter. This prediction, based on his conclusion that oil production rises and falls according to a bell curve, was validated when U.S. oil production peaked in 1970. Using his bell curve model, he also calculated that world oil production would peak around the year 2000.

HU JINTAO (1942–) president of China since March 2003. Under Jintao's leadership, China has taken measures to increase the amount of its oil imports from Iran. In October 2004 China's large oil company, Sinopec Group, signed a $100 billion, 25-year oil and gas trade agreement with Iran—China's largest energy deal with the country that is the number-two oil producer in the Organization of Oil Exporting Countries (OPEC). Iran's former minister of petroleum, Bijan Namdar Zanganeh, said that Iran is China's major oil supplier and wants to be its long-term business partner.

H. L. HUNT (1889–1974) an oil entrepreneur who in the 1930s bought leases on oil-producing land in east Texas from Columbus Marion "Dad" Joiner, who was the first to strike oil on the land. Hunt sold the leases to oil companies for a huge profit, and the land eventually yielded 4 billion barrels of oil.

SADDAM HUSSEIN (1937–2006) president of Iraq, 1979–2003. As president Hussein took measures to bolster Iraq's oil industry and build up its military. He also became known for using violence to secure his power in the Arab world. Under his leadership Iraq invaded neighboring Kuwait in 1990, possibly with designs on Kuwaiti oil operations. In response a coalition of U.S. and international military forces launched Operation Desert Storm, or the Persian Gulf War, forcing Iraq out of Kuwait. In March 2003, after Hussein refused to cooperate with weapons inspections supervised by the United Nations (UN), the United States led an invasion of Iraq, contending that Iraq held weapons of mass destruction. Hussein was captured in December 2003 and placed in Iraqi legal custody. He has been executed.

ABD AL-AZIZ IBN SAUD (c.1880–1953) founder and first king of Saudi Arabia, a country whose oil deposits are among the richest in the world. In the 1930s and 1940s he built a relationship with President Franklin D. Roosevelt, thereby establishing ties between his country and the United States.

SAMUEL INSULL (1859–1938) former assistant to the U.S. inventor Thomas A. Edison and president of the Commonwealth Edison Company. In the 1900s Insull developed the largest existing electricity distribution network at that time, transmitting electricity over long distances from central power stations. His system for transmitting electricity was at first implemented in individual cities but then became the standard way to distribute electricity through entire states and multistate regions.

MARK JACCARD (1955–) British author and authority on sustainable energy. For 20 years Jaccard supported the "peak oil" theory, which maintains that oil supplies are running out. In 2005 Jaccard surprised the energy industry by concluding in his book *Sustainable Fossil Fuels* that there is enough oil, gas, and coal to sustain the world at least for another 500 years.

JIANG ZEMIN (1926–) former president of China, who in 1998 made the first visit by a Chinese head of state to Saudi Arabia. During this visit China and Saudi Arabia formally approved certain oil cooperation agreements and discussed plans for using Saudi oil in a large Chinese petrochemical complex. Earlier, in 1995, China agreed to import 3.5 million tons of crude oil from Saudi Arabia annually. Commentators have pointed to these and other arrangements with Middle Eastern oil-producing countries as a sign of China's limited ability to meet its oil needs.

COLUMBUS MARION ("DAD") JOINER (1860–1947) the man responsible for one of the last major oil discoveries in the United States. Joiner, convinced that land in east Texas held oil, leased the land so that he could drill it for oil. On October 5, 1930, beneath 140,000 acres of land, Joiner struck possibly the largest pool of oil ever found in America. An oil entrepreneur named H. L. Hunt bought Joiner's leases and sold them to oil companies, and the land eventually yielded 4 billion barrels of oil.

HENRY KELLY president of the U.S. activist organization Federation of American Scientists and former senior associate at the Office of Technology Assessment (OTA) of the U.S. Congress.

MOHAMED KHATEMI (1943–) Iranian president elected by a landslide in 1997. Khatemi's leadership was more moderate than that of his predecessors. During Khatemi's presidency the U.S. president, Bill Clinton (1993–2001), waived sanctions against Russian, French, and Malaysian firms for investing in Iran's oil industry.

MUHAMMAD KHILEWI former Saudi Arabian ambassador to the United Nations (UN), who defected to the United States in 1994 and revealed documents that hinted of an arrangement between Saudi Arabia and Pakistan through which Saudi Arabia partially funded Pakistan's development of nuclear weapons in exchange for the Saudis' ability to use such weapons should nuclear aggression be directed at them.

AYATOLLAH KHOMEINI (1900–1989) Iranian leader who assumed power in 1979 during the Iranian Revolution led by Islamic fundamentalists. Khomeini nationalized Iran's oil industry. During his reign Iran became the first modern state ruled by fundamentalist Islamic principles, and Islamist militants seized the U.S. embassy in Iran and held 52 Americans hostage for 444 days. The hostage crisis, combined with the oil crisis caused by Arab oil embargoes, seriously challenged the U.S. presidency of Jimmy Carter. The U.S. severed diplomatic ties with Iran and has considered it a rogue state since. Ayatollah Khomeini died in 1989. The current spiritual leader of Iran is Ayatollah Ali Khamenei.

HORST KOEHLER (1943–) president of Germany since July 2004. He supported Germany's new renewable energies law, which took effect in August 2004. This law improves government funding of biomass, geothermal, and offshore wind energy installations. Horst's administration also set a goal of having 12.5 percent of Germany's electricity generated by renewable energy sources by 2010, and 20 percent by 2020.

JOHANNES LACKMANN (1951–) president of Germany's federal renewable energies association, Bundesverband Erneuerbare Energien (BEE).

P. PATRICK LEAHY (1947–) has been with the U.S. Geological Survey (USGS) since 1974, acting director since June 2005.

LI PENG (1928–) former Chinese prime minister, now chair of China's National People's Congress (NPC). In 1994 as prime minister, he promoted a plan to create an alternative "people's car," the production and assembly of which China has since arranged through Volkswagen and other Western auto companies. His chief projects also included the Three Gorges Dam, the world's largest hydroelectric plant, designed to generate 18.2 gigawatts of power. In March 2002 the China Yangtze Three Gorges Electric Power Corporation was established, and in June 2003 the reservoir created by the dam began to fill, operating its initial turbines a month later.

AMORY B. LOVINS (1947–) U.S. physicist and influential environmentalist. Lovins is co–chief executive officer of the Rocky Mountain Institute, a Colorado-based independent nonprofit research center that he founded in 1982 with his then-wife, L. Hunter, a lawyer and social scientist. In his many lectures and books, Lovins has promoted energy conservation and efficiency, along with the use of renewable energy sources.

MICHAEL C. LYNCH (1954–) prominent oil economist at the Massachusetts Institute of Technology. As Peter Odell has, he has countered the claims that the world has reached or is reaching peak oil production. He has stated that these claims underestimate the oil potential of nations outside the Organization of Oil Producing Countries (OPEC) and technological changes that can result in better oil recovery.

JOSEPH MAKOJU (1948–) engineer and managing director of the National Electric Power Authority (NEPA) in Nigeria. Nigeria's passage of the Electric Power Sector Reform (EPSR) Act in 2004 set in motion the privatization of NEPA and a long-awaited reform of the nation's unreliable and disorganized electricity generation system.

JAMES CLERK MAXWELL (1831–1879) Scottish mathematician and theoretical physicist known for his work in electricity and magnetism. In the 1860s he demonstrated that electric and magnetic fields travel in waves through space. He developed a mathematical theory that expressed four basic laws of electricity and magnetism. His equations became the basis for many uses of electric power.

ANGELA D. MERKEL (1954–) chancellor of Germany since November 2005 and Germany's first female chancellor. As chancellor Merkel has forged a strong German-American friendship. In the spring of 2003 she made public her support of the U.S. invasion of Iraq. Under Merkel's predecessor Germany passed a law known as the Renewable Energy Sources Act, which contained statutes to increase Germany's production of energy from renewable sources. Merkel has expressed a commitment to continue Germany's program of promoting renewable energy and environmental protection. Under the leadership of Merkel, who was previously Germany's minister for the environment and reactor safety, the German government has announced plans to formulate an energy strategy to reduce dependency on energy imports further, prevent energy price increases, and tackle the country's energy-related environmental challenges through the year 2020. Merkel has also cautioned that her country's plan for phasing out nuclear energy will cause a gap in energy production that will create an urgent need for the development of other energy technologies.

PATRICK MOORE (1947–) cofounder of the environmental activist group Greenpeace, who later formed his own organization, Greenspirit, and now extols the power of nuclear energy, advocates the burning of certain renewable resources, and promotes some uses of carbon-dioxide–emitting fuels.

MOHAMMAD REZA PAHLAVI (1919–1980) the shah of Iran who rose to power in 1953 after the overthrow of the Iranian prime minister Mohammed Mossadegh. Whereas Mossadegh's oil policies were antagonistic to the United States and Britain, Pahlavi was an American-friendly leader who opened Iranian oil reserves to Western companies. However, Islamic fundamentalists who resented his pro-Western policies drove him out in 1979 during the Iranian Revolution. A new Islamic republic emerged, based on fundamentalist Islamic principles and led by the national spiritual leader, Ayatollah Khomeini.

MOHAMMAD MOSSADEGH (1882–1967) prime minister of Iran, 1951–53. As Iran's prime minister Mossadegh nationalized Iran's oil industry, which was originally under British control. In response Britain enforced an economic blockade against Iran and, working with the U.S. Central Intelligence Agency (CIA), helped to instigate Mossadegh's overthrow by Iranian monarchists. In 1953 the shah of Iran, Mohammed Reza Pahlavi, a pro-Western leader, replaced Mossadegh, who was sent to prison. The shah opened Iran's oil reserves to Western companies. Mossadegh died in prison in 1967.

RALPH NADER (1934–) U.S. consumer advocate and political reformer. Nader is founder of the Center for Auto Safety and the U.S. Public Interest Research Group. In 1966 the U.S. Congress passed a substantial auto safety act largely because of Nader's influence. Nader has supported increasing corporate average fuel economy (CAFE) standards. He ran for U.S. president in 1992, 1996, 2000, and 2004, campaigning as the Green Party candidate in 1996 and 2000.

ALI BIN IBRAHIM AL-NAIMI (1935–) Saudi Arabia's minister of petroleum and mineral resources since 1995. After the U.S. invasion of Iraq in March 2003, which placed Iraqi oil production on hold, al-Naimi asserted Saudi Arabia's commitment to stepping up its own oil production to fill the void in Middle Eastern oil exports. That same year he also announced the termination of negotiations with foreign oil companies on a plan that would have opened Saudi Arabia's upstream hydrocarbons sector to foreign investment for the first time since the 1970s.

THOMAS NEWCOMEN (c.1664–1729) British inventor, who created a coal-powered steam engine in 1711. His engine, an improvement over an earlier device patented in 1698, was used to pump water from English coal mines. Further improvements added by James Watt in 1765 resulted in a new engine that was capable of powering many different machines and helped to bring about the Industrial Revolution.

OLUSEGUN OBASANJO (1937–) president of Nigeria since 1999. In March 2004 Obasanjo signed into law the Electric Power Sector Reform (EPSR) Act, which aims to reform the nation's unreliable and disorganized electricity generation system. The lack of consistent power supply to support industrial growth has been debilitating to the economy of Nigeria, which is one of the world's poorest nations. Nigeria's economy is also heavily dependent on oil sector revenues. It is the largest oil producer in Africa and the 11th largest in the world. In December 2004 Obasanjo announced a plan to supply up to 15 percent of U.S. oil imports.

PETER ODELL (1930–) energy consultant and professor of international energy studies at Erasmus University, Rotterdam, England. In the 1970s he challenged the widely held view that the oil supply was running out. He argued that the oil industry would continue to expand into the mid-21st century. Odell based these ideas on increases in proven oil reserves over prior years, new technologies for recovering oil, and oil price increases that encouraged oil companies to seek out more oil than they had when oil was less expensive.

JUAN PABLO PÉREZ ALFONZO (JUAN PÉREZ) (1903–1979) former Venezuelan oil minister and cofounder of the Organization of Petroleum Exporting Countries (OPEC). Considered the "Father of OPEC," he worked to form arrangements between major oil companies and oil-exporting countries that were more favorable to the latter. He also persuaded industry leaders that the prices that oil companies declared for tax purposes were below their actual value. He established contact with Iran and the Soviet Union in the hopes of forming a partnership to limit the major oil companies' control of the international oil market. However, his country forced him into exile in 1949.

RAFAEL RAMIREZ Venezuela's minister of energy, who in recent reports said that if the United States showed any signs of aggression toward his country, Venezuela would cut off its oil supply.

RONALD REAGAN (1911–2004) U.S. president, 1981–89. The Reagan administration sharply cut funding for federal energy programs, particularly alternative energy research and development. Reagan also opposed government intervention in energy markets and refused to implement taxes on energy imports for the purposes of stimulating domestic production and encouraging conservation.

JOHN D. ROCKEFELLER (1839–1937) entrepreneur who dominated the American oil industry in the late 1800s and early 1900s. The Standard Oil Company of Ohio, which he formed in 1870 with associates, commandeered the oil refining and distributing operations of weaker businesses and developed a monopoly over U.S. oil production and distribution. In 1892, however, the U.S. government dissolved Rockefeller's Standard Oil Trust, and in 1911 a U.S. Supreme Court ruling broke up the holding company to which Standard Oil had transferred its assets. Yet Rockefeller held onto his enormous fortune, later making donations to many philanthropic causes.

FRANKLIN D. ROOSEVELT (1882–1945) president of the United States, 1932–45. His New Deal administration took a direct role in developing U.S. natural resources by establishing the Tennessee Valley Authority in 1933 and the Rural Electrification Administration in 1935. It also addressed problems caused by monopolies and enacted legislation to dissolve public utility holding companies. Roosevelt fostered U.S. ties with the oil-producing country of Saudi Arabia and personally visited its founder, Abd al-Aziz ibn Saud, in 1945.

ABDULLAH BIN ABD AL-AZIZ AL-SAUD (1924–) king of Saudi Arabia since August 2005. Saudi Arabia is the world's leading oil producer and exporter, and the United States is one of its leading oil consumers. In November 2005 the Saudi king called on leading consumer countries to cut taxes on petroleum products to help control oil prices during a time of increasing pressure for the Organization of Petroleum Exporting Countries

(OPEC) to meet global oil demand. He has also promoted privatization of his country's assets.

ANNETTE SCHAVAN (1955–) Germany's minister of education and research since November 2005. In 2006 she announced that a 2-billion euro funding increase would be implemented by 2009 to support research and development in Germany's energy sector.

GERHARD SCHROEDER (1944–) chancellor of Germany from October 1998 to November 2005. In August 2005 Schroeder and the U.S. president, George W. Bush, met in Berlin for the Working Group on Energy, Development, and Climate Change, a U.S.-German effort to cooperate on issues related to energy supply, energy efficiency, and renewable energies, according to a joint statement issued by both leaders. Germany and the United States have also indicated that they are working on a partnership to use wasted methane from coal mines and other sources as an energy source.

CLAY SELL deputy secretary of the U.S. Department of Energy (DOE) since March 2005.

WERNER VON SIEMENS (WERNER SIEMENS) (1816–1892) German electrical engineer, who introduced the first electrical railway at the Berlin Exhibition in 1878. He also partnered with his brother, William, and the U.S. inventor Thomas A. Edison in 1881 to design Europe's first public lighting network in London, England, and invented a telegraph that used a needle to point to letters instead of using Morse code. Siemens was responsible for a discovery in electrical engineering known as the dynamoelectric principle. He received a German patent for the electromechanical dynamic, or moving-coil, transducer, which the American Telephone & Telegraph Company (AT&T) adapted in the 1920s for use in a loudspeaker.

MAJOR-GENERAL GODSWILL TAMUNO military leader of the Movement for the Emancipation of the Niger Delta (MEND) in Nigeria. In early 2006 MEND blew up oil pipelines, held foreign oil workers hostage, and sabotaged major oil fields. Tamuno has stated that the group wants their land's oil wealth released from foreign interests and all foreign oil companies and their employees out of the region.

NIKOLA TESLA (1856–1943) Croatian-born U.S. electrical engineer and inventor who briefly worked for the inventor Thomas A. Edison. Tesla's own discoveries and inventions proved valuable to the development of radio transmission and electricity. These included an electrical motor capable of replacing, with greater productivity, the coal steam engine in some industries, as well as a system for transmitting electric power without wires. He also produced the first power system at Niagara Falls, New York.

JAMES WATT (1736–1819) Scottish inventor, after whom the *watt,* a unit of electrical power, is named. Watt's 1765 improvements to the coal-powered steam engine created by Thomas Newcomen resulted in a new type of engine that had many applications. It soon became the energy powerhouse for multiple British and U.S. industries, driving the Industrial Revolution that began in late 19th-century England.

GEORGE WESTINGHOUSE (1846–1914) U.S. inventor and manufacturer. In the 1890s he promoted the use of alternating current, which allowed electricity to be transmitted over long distances.

ROBERT H. WILLIAMS U.S. energy authority and senior research scientist at the Center for Energy and Environmental Studies at Princeton University. The U.S. Department of Energy (DOE) awarded Williams the Sadi Carnot Award for his contributions to the field of energy efficiency.

FRANK A. WOLAK U.S. economics professor at Stanford University. Wolak, a leading researcher on energy issues, has spoken about increasing competition in the U.S. electricity market caused by real-time pricing.

CHARLES E. WYMAN a director at the National Renewable Energy Laboratory (NREL), in Golden, Colorado, and a professor of environmental engineering at Dartmouth College. Wyman is a leading researcher in renewable and alternative fuels, such as biomass and ethanol.

XIE ZHENHUA (1949–) head of China's State Environmental Protection Administration. Zhenhua has pledged China's commitment to environmental protection and efficient, sustainable use of natural resources. The United Nations (UN) secretary-general Kofi Annan awarded Zhenhua the 2003 Sasakawa Environment Prize.

SHEIK AHMED ZAKI YAMANI (1930–) former petroleum and mineral resources minister of Saudi Arabia. In 1990 he founded the Centre for Global Energy Studies, an independent organization that analyzes energy-related developments, especially within the oil and natural gas market.

BIJAN NAMDAR ZANGANEH (1953–) Iran's most recent oil minister, who has called for increased oil production and improved oil recovery in Iran. Iran's economy is highly dependent on revenues from oil exports

9

Organizations and Agencies

The organizations and agencies that tackle issues related to natural resources and energy supply are many. They can be found at all levels of government; among the various sectors of energy suppliers, transporters, and producers; as part of consumer advocacy or public interest organizations; within worldwide governing bodies; as part of educational and research institutions; and among grassroots lobbying and environmental protection efforts. The following is a compilation of organizations and agencies that confront resource and energy issues at international, national, state, and local levels.

Alliance to Save Energy (ASE)
URL: http://www.ase.org
1200 18th Street NW
Suite 900
Washington, DC 20036
Phone: (202) 857-0666

A nonprofit coalition of business, government, environmental, and consumer groups that advocate energy efficiency programs and reduced energy use. The group conducts research and issues printed materials about energy-efficient technology. Newsletters include the *Green Schools Gazette* and the online resource *e-FFICIENCY News.*

American Association of Petroleum Geologists (AAPG)
URL: http://www.aapg.org
1444 South Boulder
Tulsa, OK 74119
Phone: (800) 364-AAPG (2274)

World's largest professional geological society. AAPG offers publications, conferences, and educational opportunities for geoscientists in order to foster scientific research, advance the science of geology, promote technology, and encourage professional conduct. It also disseminates current geological information to the general public.

American Coalition for Ethanol (ACE)
URL: http://www.ethanol.org
2500 S. Minnesota Avenue
#200
Sioux Falls, SD 57105
Phone: (605) 334-3381

National nonprofit organization drawing together ethanol producers, farmers, investors, commodity organizations, businesses supplying goods and services to the ethanol industry, rural electric cooperatives, and others supporting the increased production and use of ethanol across America.

American Council for an Energy-Efficient Economy (ACEEE)
URL: http://www.aceee.org
1001 Connecticut Avenue NW
Suite 801
Washington, DC 20036
Phone: (202) 429-8873

Nonprofit organization that promotes energy efficiency not only for environmental protection but also for a healthier U.S. economy. The group works with energy efficiency experts from universities, national laboratories, and the private sector to conduct in-depth U.S. policy assessments, advise policymakers and program managers, and organize conferences and workshops. It also publishes books, conference proceedings, and reports as well as materials to educate consumers and businesses about energy efficiency.

American Gas Association (AGA)
URL: http://www.aga.org
400 N. Capitol Street NW
Washington, DC 20001
Phone: (202) 824-7000

Professional association for U.S. natural gas utility companies that promotes growth in the natural gas industry by emphasizing the benefits of natural gas and supporting public policies favorable to increased natural gas supplies and decreased natural gas prices.

American Meteorological Society (AMS)
URL: http://www.ametsoc.org
45 Beacon Street
Boston, MA 02108-3693
Phone: (617) 227-2425

Develops and delivers information on the atmospheric, oceanic, and hydrologic sciences, including information on global climate change. AMS members include professionals, professors, students, and weather enthusiasts, and the organization publishes nine journals covering the atmospheric, oceanic, and hydrologic sciences.

American Methanol Institute (AMI)
URL: http://www.methanol.org
4100 North Fairfax Drive
Suite 740
Arlington, VA 22203
Phone: (703) 248-3636

Main U.S. organization representing the methanol industry. AMI works to expand markets for the use of methanol as a chemical commodity building block, a hydrogen carrier for fuel cell applications, and an alternative fuel. The organization has a research arm, the Methanol Foundation.

American Nuclear Society (ANS)
URL: http://www.ans.org
55 North Kensington Avenue
La Grange Park, IL 60526
Phone: (708) 352-6611

Nonprofit, international organization of engineers, scientists, administrators, and educators within the fields of nuclear science and technology. ANS concentrates on unifying the professional activities of those in the nuclear industry to encourage safe nuclear science and technology for public benefit.

American Petroleum Institute (API)
URL: http://www.api.org
1220 L Street NW
Washington, DC 20005-4070
Phone: (202) 682-8000

Trade association representing U.S. petroleum and natural gas suppliers that conducts research and participates in public policy development. Its Web site includes a consumer education portal in addition to a section for energy professionals. The site offers basic information on oil and natural gas as well as industry statistics and position papers on issues ranging from gasoline taxes to offshore oil drilling.

American Public Power Association (APPA)
URL: http://www.appanet.org
2301 M Street NW
Washington, DC 20037-1484
Phone: (202) 467-2900

Service organization representing U.S. nonprofit community-owned electric utilities.

American Solar Energy Society (ASES)
URL: http://www.ases.org
2400 Central Avenue
Suite A
Boulder, CO 80301
Phone: (303) 443-3130

U.S. arm of the American Solar Energy Society (ASES), a nonprofit organization dedicated to the development and use of renewable energy—solar and beyond. ASES organizes the annual National Solar Energy Conference and National Solar Tour; publishes newsletters, policy statements, and *Solar Today* magazine; sponsors issue roundtables; and organizes the Solar Action Network.

American Wind Energy Association (AWEA)
URL: http://www.awea.org
1101 14th Street NW
12th Floor
Washington, DC 20005
Phone: (202) 383-2500

U.S. trade association representing all aspects of the wind energy industry—power plant developers, turbine manufacturers, utility companies, consultants, financiers, insurers, researchers, and wind power advocates. To support the development of wind and other renewable energy technologies, it provides the public and the government with information on wind energy and wind energy projects. It publishes weekly and monthly newsletters.

Americans for Balanced Energy Choices (ABEC)
URL: http://www.balancedenergy.org
P.O. Box 1638
Alexandria, VA 22313
Phone: (877) 358-6699

U.S. coal-industry-funded nonprofit group that promotes dialogue on America's increasing demand for electricity and coal's role in meeting this demand, while also stressing environmental protection.

Apollo Alliance
URL: http://www.apolloalliance.org
1025 Connecticut Avenue
Suite 205
Washington, DC 20036
Phone: (202) 955-5665

Nonprofit organization formed by the Institute for America's Future and the Center of Wisconsin Strategy that supports energy efficiency initiatives and the use of renewable energy sources to create domestic jobs as well as protect the environment.

Association for the Study of Peak Oil and Gas
URL: http://www.peakoil.net
Box 25182
SE-750 25 Uppsala
Sweden
Phone: 46-70-4250604

Network of scientists affiliated with European institutions and universities who undertake studies to determine the date and possible repercussions of the peak and decline of the world's production of oil and gas.

Biodiesel International (BDI)
URL: http://www.biodiesel-intl.com
BDI Anlagenbau Ges.m.b.H
Parkring 18
A-8074 Grambach/Graz
Austria
Phone: 43-316-4009-100

Organization that serves the needs of the international biodiesel industry, offering information on building biodiesel plants; research and development services; online resource guides for methanol, ethanol, and biodiesel fuels; and other kinds of technical information related to biodiesel operations.

Biomass Energy Foundation (BEF)
URL: http://www.biomassenergyfoundation.org
554 Orchard Street
#4
Golden, CO 80401
Phone: (303) 279-3707

Nonprofit foundation promoting biomass energy and specializing in gasification.

BlueWater Network
URL: http://www.bluewaternetwork.org
311 California Street
Suite 510
San Francisco, CA 94104
Phone: (415) 544-0790

Nonprofit organization that pursues a variety of environmental protection causes, including cleaner oceans, reduced air pollution, and decreased reliance on unsafe and greenhouse-gas-emitting fuels.

The Brookings Institution
URL: http://www.brookings.edu
1775 Massachusetts Avenue NW
Washington, DC 20036
Phone: (202) 797-6000

Private, nonprofit, independent organization of researchers who study, write, and give congressional testimony on issues related to the economy; education; foreign and domestic policies, including those related to energy; and social sciences. The organization publishes a quarterly journal and several energy-related books.

Carbon Dioxide Information Analysis Center (CDIAC)
URL: http://cdiac.esd.ornl.gov
Environmental Sciences Division
Oak Ridge National Laboratory
P.O. Box 2008
Oak Ridge, TN 37831
Phone: (865) 374-3645

Greenhouse gas emissions information service operating out of the Environmental Sciences Division of Oak Ridge National Laboratory. CDIAC archives and disseminates climate and greenhouse gas emissions data. Available data include global, regional, and national annual carbon dioxide emissions from fossil fuel burning; measurements of atmospheric carbon dioxide; global and U.S. temperature and precipitation information; sea level data; trace gas emissions data; and more. CDIAC also distributes and produces reports and newsletters.

The Cato Institute
URL: http://www.cato.org
1000 Massachusetts Avenue NW
Washington, DC 20001-5403
Phone: (202) 842-0200

Public policy research foundation advocating limited government, the protection of individual liberties, and a free-market approach to issues of domestic energy supply. The organization publishes a journal and many other materials focusing on public policy issues.

Center for Auto Safety (CAS)
http://www.autosafety.org
1825 Connecticut Ave NW
Suite 330
Washington, DC 20009-5708
Phone: (202) 328-7700

Founded in 1970 by Consumers Union and Ralph Nader to represent consumer interests before the U.S. government on issues related to auto safety and quality. CAS promotes increased fuel economy standards.

Center for Energy and Economic Development (CEED)
URL: http://www.ceednet.org
333 John Carlyle Street
Suite 530
Alexandria, VA 22314
Phone: (703) 684-6292

Nonprofit group representing the interests of the coal industry. CEED propagates the view that coal is an ideal fuel for generating electricity, with promising pollution-free capabilities.

Center for Resource Solutions
URL: http://www.resource-solutions.org

P.O. Box 29512
San Francisco, CA 94129
Phone: (415) 561-2100

National nonprofit organization that undertakes projects for increasing renewable energy use and promoting sustainability. CRS supports government policies that involve tax incentives for reliance on renewable energy sources to stimulate the growth of renewable energy markets and industries.

Center for Strategic and International Studies (CSIS)
URL: http://www.csis.org
1800 K Street NW
Washington, DC 20006
Phone: (202) 887-0200

Bipartisan, nonprofit organization that researches energy issues to influence public policy. Areas of attention include rising global energy demand, the potential of liquefied natural gas (LNG), and international economic development.

Center for the Advancement of Energy Markets (CAEM)
URL: http://www.caem.org
5765-F Burke Center Parkway
PMB333
Burke, VA 22015-2233
Phone: (703) 250-1580

Nonprofit corporation that tracks and analyzes changes in domestic and global energy markets caused by technological and policy developments. CAEM also lobbies public policy, consumer, and corporate decision makers to pursue policies that will encourage energy market competition.

Centre for Alternative Technology (CAT)
URL: http://www.cat.org.uk
Machynlleth
Powys SY20 9AZ
United Kingdom
Phone: 44-0-1654 705950

Large ecological center and source of alternative energy information in Europe. CAT operates a free information service, answering energy inquiries by phone, letter, or e-mail, and publishes books on various environmental topics.

Centre for Global Energy Studies (CGES)
URL: http://www.cges.co.uk
17 Knightsbridge
London SW1X 7LY
United Kingdom
Phone: 44-0-20 7235 4334

International oil market forecasting and analysis organization started in 1990 by Saudi Arabia's minister for petroleum and mineral resources.

China Petroleum & Chemical Corporation (Sinopec Corp.)
URL: http://english.sinopec.com
No. A6 Hui xin East Street
Chaoyang District
Beijing 100029
China
Phone: 86-10-64998828

Large, publicly listed oil and gas producer and supplier in China.

Chinese National Petroleum Corporation (CNPC)
URL://http://www.cnpc.com.cn/english
6, Liupukang Jie
Xicheng District
Beijing 100724
China
Phone: 86-10-62094114

Large state-owned producer and supplier of crude oil and natural gas in China. CNPC also produces and supplies refined oil products and petrochemicals. In addition, it offers domestic marketing and international trade services, plus technical services and information.

Climate Action Network (CAN)
URL: http://www.climatenetwork.org
U.S. office:
1200 New York Ave NW
Suite 400
Washington, DC 20005
Phone: (202) 513-6240

Network of nongovernmental organizations that encourage government intervention to reduce human-induced climate change, protect the environment, and promote sustainable energy use. CAN has seven regional coordinating offices worldwide, including a U.S. office. CAN collaborates with other advocacy groups such as Greenpeace and Friends of the Earth to raise awareness and issue reports concerning the Kyoto Protocol, nuclear energy, emissions trading, and environment compliance.

Climate Program Office (CPO) of the National Oceanic and Atmospheric Administration (NOAA)
URL: http://www.climate.noaa.gov
1100 Wayne Avenue
Suite 1200
Silver Spring, MD 20874
Phone: (301) 427-2089

Office created in 2005 by the National Oceanic and Atmospheric Administration (NOAA) to coordinate the work of NOAA's Office of Global Programs, Arctic Research Office, and Climate Observations and Services Program. The new CPO leads NOAA climate change education and outreach activities and coordinates international climate projects.

Competitive Enterprise Institute (CEI)
URL: http://www.cei.org
1001 Connecticut Avenue NW
Suite 1250
Washington, DC 20036
Phone: (202) 331-1010

Nonprofit public policy organization advocating free enterprise and limited government, particularly regarding electricity supplier competition, use of natural resources, air quality, and climate change.

ConservAmerica
URL: http://www.conservationisconservative.org
3200 Carlisle Boulevard, NE
Suite 228
Albuquerque, NM
Phone: (505) 889-4576

Nonprofit, conservative institution that distributes educational materials promoting conservation and environmental protection.

EarthRights International
http://www.earthrights.org

U.S. office:
1612 K Street NW
#401
Washington, DC 20006
Phone: (202) 466-5188

U.S. division of an international organization that defends human rights in legal cases in which large corporations, oil companies, and governments have been accused of abusing people's rights.

Edison Electric Institute (EEI)
URL: http://www.eei.org
701 Pennsylvania Avenue NW
Washington, DC 20004-2696
Phone: (202) 508-5000

U.S. trade association of investor-owned electric utilities that engages in public policy development and analysis geared at ensuring its members a competitive position in the energy supply marketplace. EEI publications address such topics as electricity supply infrastructure, transmission reliability, and the environment.

Electric Power Research Institute (EPRI)
URL: http://www.epri.com
3420 Hillview Avenue
Palo Alto, CA 94304
Phone: (650) 855-2000

Independent, nonprofit organization whose scientists and engineers conduct research and development projects related to electricity generation, environmental concerns, and other worldwide challenges confronting electric utilities and consumers.

Electric Power Supply Association
URL: http://www.epsa.org
1401 New York Avenue NW
11th Floor
Washington, DC 20005-2110
Phone: (202) 628-8200

National trade association representing power suppliers not affiliated with utility companies. It aims to promote global market conditions favorable to greater competition by environmentally responsible facilities that supply electric power.

Electric Reliability Coordinating Council (ERCC)
URL: http://www.electricreliability.org
c/o Bracewell & Giuliani LLP
2000 K Street NW
Suite 500
Washington, DC 20006-1872
Phone: (202) 828-5800

Coalition of electric power generating companies that works with labor unions, consumers, and other manufacturing and service businesses. ERCC opposes New Source Review (NSR) standards enforced by the Environmental Protection Agency (EPA) regarding remodeled power plants and air pollutant emissions, contending that such rules result in unnecessary energy price hikes while providing little benefit for the environment. ERC releases fact sheets, congressional testimony, and other statements exploring these and other electric supplier issues.

Embassy of People's Republic of China in the United States of America
http://www.china-embassy.org
2300 Connecticut Avenue NW
Washington, DC 20008
Phone: (202) 328-2500

Provides information about the operations of the Chinese embassy, facts about China, and current news and issues for students and scholars about the Chinese economy and trade. The embassy also offers information on the Three Gorges Dam project, which established the world's largest hydroelectric plant in China.

Energy and Environment Group of the Bureau for Development Policy, United Nations Development Programme (UNDP)
URL: http://www.undp.org/energyandenvironment
304 East 45th Street
Ninth Floor
New York, NY 10017
Phone: (212) 906-6973

A department of the United Nations Development Programme (UNDP) that unifies the efforts of UN member countries in addressing global climate change, loss of biodiversity, ozone layer depletion, and assistance to impoverished groups in building sustainable livelihoods.

Energy Foundation (EF)
URL: http://www.ef.org
1012 Torney Avenue
#1
San Francisco, CA 94129
Phone: (415) 561-6700

Partners with charitable organizations to support financially the development of energy efficiency and renewable energy programs in areas of need. EF has a branch in China.

The Energy Foundation, China
CITIC Building
Room 2403
No. 19 Jianguomenwai Dajie
Beijing 100004
China
Phone: 86-10-6525-3765

Chinese branch of the Energy Foundation, which partners with charitable organizations to support financially the development of energy efficiency and renewable energy programs in areas of need.

Energy Future Coalition
URL: http://www.energyfuturecoalition.org
1225 Connecticut Avenue
Fourth Floor
Washington, DC 20036
Phone: (202) 463-1947

Organization that supports changes in U.S. energy policy to reduce dependence on oil, control emissions from the burning of fossil fuels, help developing nations to grow, and extend access to modern energy services to impoverished nations.

Energy Information Administration (EIA)
http://www.eia.doe.gov
1000 Independence Avenue SW
Washington, DC 20585
Phone: (202) 586-8800

Statistical agency within the U.S. Department of Energy (DOE). The EIA provides data, forecasts, and analyses regarding energy and its interaction with the economy and the environment. Comprehensive data are available in the form of graphs, tables, and charts, covering energy economics, technology, production, prices, distribution, storage, consumption, environmental impact, supply, and more, and tabulated by sector, resource, producer, and consumer types. The EIA also offers educational materials for students, country profiles, and articles that analyze energy-related issues.

Energy Intelligence Group
URL: http://www.energyintel.com
Five East 37th Street
Fifth Floor
New York, NY 10016-2807
Phone: (212) 532-1112

Publishing and information services company that produces and distributes news, data, and analysis on international energy. Publications include *Oil Daily* and *Petroleum Intelligence Weekly.*

Environmental and Energy Study Institute (EESI)
URL: http://www.eesi.org
122 C Street NW
Suite 630
Washington, DC 20001
Phone: (202) 628-1400

Nonprofit organization offering publications, briefings, workshops, and task forces that promote public policy initiatives for environmentally sustainable societies through innovative and efficient use of natural resources. Areas of concern include renewable energy, global climate change, agriculture, biofuels, smart growth, and clean bus technologies.

Environmental Defense (EDF)
URL: http://www.environmentaldefense.org
257 Park Avenue South
New York, NY 10010
Phone: (212) 505-2100

Nonprofit environmental activist organization addressing issues related to the oceans, the atmosphere, natural resources, human health and quality of life, and wildlife protection. EDF produces print and e-mail newsletters, fact sheets, and educational materials, drawing on research in science, law, and economics. EDF also features an Internet-based environmental alert system, which enables members and activists instantly to send e-mails or faxes to legislators and policymakers about pending government actions that have environmental importance.

European Renewable Energy Council
URL: http://www.erec-renewables.org
63-65, rue d'Arlon
B-1040
Brussels, Belgium
Phone: 32-2-546-1933

Umbrella organization for the leading industrial, trade, and research associations active in the European renewable energy industry, including those in the photovoltaic, wind energy, small hydropower, biomass, geothermal energy, and solar thermal sectors.

Eurosolar e.V.
URL: http://www.eurosolar.org
Kaiser-Friedrich-Strasse 11
D-53113 Bonn
Germany
Phone: 49-228-362373

Assembly of European political, scientific, technological, and industrial committees that support and promote renewable energy. Eurosolar develops political and economic plans of action and legal frameworks for the introduction of renewable energies and environmentally sustainable architecture.

Federal Energy Regulatory Commission (FERC)
URL: http://www.ferc.gov
888 First Street NE
Washington, DC 20426
Phone: (866) 208-3372

Independent government agency that regulates the interstate transmission of electricity, natural gas, and oil. FERC also reviews proposals to build liquefied natural gas (LNG) terminals and interstate natural gas pipelines, licenses hydropower projects, and carries out additional responsibilities as outlined by the Energy Policy Act of 2005.

Federation of American Scientists (FAS)
URL: http://www.fas.org
1717 K Street NW
Suite 209
Washington, DC 20036
Phone: (202) 546-3300

Established by atomic scientists, addresses a broad spectrum of national security issues especially concerning nuclear energy technology. FAS engages in research and education projects regarding nuclear arms control and global security, proliferation of weapons of mass destruction, and critical challenges in housing, energy, health, and education.

Friends of the Earth
URL: http://www.foe.org
1717 Massachusetts Avenue NW
Suite 600
Washington, DC 20036-2002
Phone: (877) 843-8687

U.S. branch of Friends of the Earth International, a global network of environmental activist groups in 70 countries. FOE campaigns for solutions to problems caused by strip mining, oil transportation, hydroelectric projects that present environmental or safety risks, and corporate pollution, among other issues.

Geothermal Education Office
URL: http://geothermal.marin.org
664 Hilary Drive
Tiburon, CA 94920
Phone: (415) 435-4574

Produces and distributes educational materials about geothermal energy to schools, energy/environmental educators, libraries, industry, and the public.

Geothermal Energy Association
URL: http://www.geo-energy.org
209 Pennsylvania Avenue SE
Washington, DC 20003
Phone: (202) 454-5261

Trade association for those who support the expanded use of geothermal energy and are developing geothermal resources worldwide for electrical power generation and direct-heat uses. The association provides assistance for the export of geothermal goods and services, compiles statistical data about the geothermal industry, and conducts education and outreach projects.

Geothermal Resources Council
URL: http://www.geothermal.org
P.O. Box 1350 2001 Second Street
Suite 5
Davis, CA 95617-1350
Phone: (530) 758-2360

Nonprofit professional geothermal education association with members worldwide. The GRC publishes a bulletin on geothermal technology and works with academic institutions, industry, and government agencies to encourage development and utilization of geothermal resources. It also offers continuing professional development for its members.

The German Embassy in Washington, DC
URL: http://www.germany-info.org/relaunch/info/missions/embassy/embassy.html
4645 Reservoir Road NW
Washington, DC 20007-1998
Phone: (202) 298-4000

Provides news, facts and figures, and information about Germany, including background papers and press releases issued by the German government and the German embassy.

Global Environment Facility (GEF)
URL: http://www.gefweb.org
1818 H Street NW
Washington, DC 20433
Phone: (202) 473-0508

Independent financial organization that provides funding for developing countries' environmental protection projects and programs, including international agreements on biodiversity, climate change, and pollutants.

Green Mountain Energy Company
http://www.greenmountain.com
P.O. Box 689008
Austin, TX 78768

Supplier of energy that has been produced by a variety of renewable and alternative means. Green Mountain also offers consumer energy education, fact sheets, and information for businesses and commercial institutions that want to reduce their carbon dioxide emissions.

Greenpeace
URL: http://www.greenpeace.org
702 H Street NW
Suite 300
Washington, DC 20001
Phone: (800) 326-0959

Well-established environmental activist group concerned with ocean preservation, wildlife protection, energy conservation, and other environmental issues, including nuclear programs, genetic engineering, and deforestation. Greenpeace publishes a magazine, reports, and books on environmental issues and has offices worldwide.

Greenspirit
URL: http://www.greenspirit.com

Organization begun by Patrick Moore, a founder and former activist of Greenpeace. Greenspirit advocates the widespread use of wood as a substitute for nonrenewable fossil fuels and as a replacement for materials such as steel, concrete, and plastic, in addition to its traditional uses in the manufacture of printing, packaging, and sanitation products. Greenspirit also supports nuclear energy as part of a sustainable energy mix.

IEA Clean Coal Centre
URL: http://www.iea-coal.org.uk
Gemini House
10-18 Putney Hill
London SW15 6AA
United Kingdom
Phone: 44-0-20 8780 2111

Center for information about clean and sustainable uses of coal for energy. The center shares its reports with governments, industry, and consumers.

The Heritage Foundation
URL: http://www.heritage.org
214 Massachusetts Avenue NE

Washington, DC 20002-4999
Phone: (202) 546-4400

Think tank that promotes conservative public policies based on free enterprise, limited government, and protection of personal liberties. It advocates increased domestic oil production and the lifting of certain regulations against oil drilling in some parts of the United States. The foundation also publishes policy analysis materials, including a magazine, online resources, and articles.

Independent Petroleum Association of America (IPAA)
URL: http://www.ipaa.org
1201 15th Street NW
Suite 300
Washington, DC 20005
Phone: (202) 857-4722

National trade association for independent crude oil and natural gas exploration and production companies in the United States. IPAA advocates its members' views before the U.S. government and provides economic and statistical information about domestic oil exploration and production.

Institute for Energy and Environmental Research (IEER)
URL: http://www.ieer.org
6935 Laurel Avenue
Suite 201
Takoma Park, MD 20912
Phone: (301) 270-5500

Organization offering scientific and technical information on environment- and energy-related issues, including assessments of the health effects and environmental impact of nuclear power facility operations, in language geared to laypeople.

Institute for Energy and Environmental Research (IEER)
URL: http://www.ieer.org
6935 Laurel Avenue
Suite 201
Takoma Park, MD 20912
Phone: (301) 270-5500

Aims to provide activists, policymakers, journalists, and the public with scientific and technical information on energy and environmental policy issues in a format and language accessible to all.

Institute of Electrical and Electronics Engineers, Inc. (IEEE)
URL: http://www.ieee.org
445 Hoes Lane
Piscataway, NJ 08854-1331
Phone: (732) 981 0060

International association for professionals in electronics, electrical energy, and electrical engineering. IEEE offers technical and professional information, resources, conferences, continuing education, journals, magazines, and other materials on subjects ranging from aerospace systems, computers, and telecommunications to biomedical engineering, electric power, and consumer electronics. IEEE also establishes professional standards for both traditional and emerging fields within telecommunications, information technology, and power generation to encourage market growth, product quality, reduced development time and cost, and sound engineering practices.

Institute of Nuclear Power Operations (INPO)
URL: http://www.eh.doe.gov/inpo

Organization affiliated with the U.S. Department of Energy (DOE) whose purpose is to ensure safety, reliability, and high operating standards among U.S. operators of commercial nuclear power plants, all of which are INPO members.

Intergovernmental Panel on Climate Change (IPCC)
URL: http://www.ipcc.ch
c/o World Meteorological Organization
7 bis Avenue de la Paix
C.P. 2300
CH-1211
Geneva 2, Switzerland
Phone: 22-730-8208

Research-oriented body established by the United Nations Environmental Programme (UNEP) and World Meteorological Organization (WMO) to assess the scientific, technical, and socioeconomic ramifications of global climate change and to make recommendations for adaptation and mitigation.

International Arctic Science Committee (IASC)
URL: http://www.iasc.se
Lilla Frescativägen, 4
P.O. Box 50003

S-10405 Stockholm
Sweden
Phone: 46-8-6739613

Nongovernmental organization whose aim is to encourage and facilitate cooperation in all aspects of arctic research in all areas of the arctic region, among all countries engaged in such research. IASC coordinates projects from different countries concerning the study of arctic ice masses and the sensitivity of glaciers to climate change.

International Association for Hydrogen Energy (IAHE)
http://www.iahe.org
5783 SW 40 Street
#303
Miami, FL 33155
ayfer@iahe.org

Organization of scientists and engineers who produce and/or support hydrogen-based energy technology. IAHE resources include a monthly scientific journal called the *International Journal of Hydrogen Energy,* a biennial international conference on hydrogen energy, and technical papers on hydrogen safety and other topics.

International Atomic Energy Agency (IAEA)
URL: http://www.iaea.org
P.O. Box 100
Wagramer Strasse 5
A-1400 Vienna
Austria
Phone: 431-2600-0

An agency founded by the United Nations to promote safe, secure, and peaceful nuclear science and technology. The IAEA inspects countries' nuclear energy research, programs, and facilities to ensure that nuclear weapons are not being developed. The agency also helps nuclear facilities with safety and security issues.

International Commission of Radiological Protection (ICRP)
URL: http://www.icrp.org
SE-171
16 Stockholm
Sweden
Phone: 46-8-729-729 8

Nongovernmental organization that provides regulatory and advisory agencies with recommendations for radiological protection. ICRP also has a scientific journal, the *Annals of the ICRP*.

International Energy Agency (IEA)
URL: http://www.iea.org
9, Rue de la Fédération
15739 Paris Cedex 15
France
Phone: 33-1-405-76500

Intergovernmental agency initially founded during the 1973–74 oil crisis to coordinate responses to oil supply emergencies. Now the organization acts as an energy policy adviser for member countries in their efforts to provide clean, reliable, and affordable energy. Staffed mainly by energy experts and statisticians from member countries, it participates in energy research, data compilation, distribution of energy-related information, and analyses of public policies on climate change, environment, and energy.

International Geothermal Association (IGA)
URL: http://iga.igg.cnr.it
c/o Samorka
Sudurlandsbraut 48
108 Reykjavik
Iceland
Phone: 354-588-4437

Nonprofit, nongovernmental association for the research, development, and utilization of geothermal resources around the world. The IGA offers scientific and technical information on geothermal energy for specialists as well as for the general public. It also contributes to renewable energy programs sponsored by the United Nations (UN) and European Union (EU).

International Hydropower Association
URL: http://www.hydropower.org
Westmead House
Suite 55
123 Westmead Road
Sutton, Surrey
SM1 4JH
United Kingdom
Phone: 20-8288-1918

Association founded by the United Nations Educational, Scientific, and Cultural Organization (UNESCO) to advance knowledge of hydroelectric power and promote good practice within the hydropower industry.

International Institute for Environment and Development (IIEED)
URL: http://www.iied.org
3 Endsleigh Street
London WC1H 0DD
United Kingdom
Phone: 44-0-20 7388 2117

International, independent policy research institute seeking environmental sustainability and more equitable distribution of the world's resources.

International Solar Energy Society (ISES)
URL: http://www.ises.org
Wiesentalstrasse 50
7911 Freiburg
Germany
Phone: 761-45906-0

Encourages the use and acceptance of solar and other renewable energy technologies and advises worldwide governments and organizations on renewable energy policies and programs.

Interstate Natural Gas Association of America (INGAA)
URL: http://www.ingaa.org
10 G Street NE
Suite 700
Washington, DC 20002
Phone: (202) 216-5900

Association of natural gas pipeline companies in the United States, Canada, and Mexico. INGAA acts as the voice for these companies before regulatory, legislative, and business bodies. It promotes natural gas as a cleaner-burning fossil fuel.

Low Impact Hydropower Institute (LIHI)
URL: http://www.lowimpacthydro.org
34 Providence Street
Portland, ME 04103
Phone: (207) 773-8190

Nonprofit organization whose goal is to minimize the potential impact of dams of hydropower plants on local fish, wildlife, and natural resources. LIHI offers a voluntary hydropower dam certification program aimed at increasing standards for hydropower providers and helping the community to identify dams that have reduced environmental impacts.

M. King Hubbert Center for Petroleum Supply Studies
URL: http://hubbert.mines.edu
Colorado School of Mines
Office of Research Development
1500 Illinois Street
Golden, CO 80401
Phone: (303) 273-3740

Nonprofit organization founded by the Colorado School of Mines and named for the geologist who produced groundbreaking research on the peak and decline of the world's production of oil and gas. The center assembles, studies, and disseminates data concerning global petroleum supplies. It distributes a quarterly newsletter about its research to individuals and organizations worldwide. Its Web site also features data on peak oil production.

National Association of Energy Service Companies (NAESCO)
URL: http://www.naesco.org
615 M Street NW
Suite 800
Washington, DC 20036
Phone: (202) 822-0950

National trade association that promotes energy conservation and efficiency through seminars, workshops, training programs, case studies, guidebooks, and industry data reports. NAESCO also offers an accreditation program for energy service providers.

National Association of State Energy Officials (NASEO)
URL: http://www.naseo.org
1414 Prince Street
Suite 200
Alexandria, VA 22314
Phone: (703) 299-8800

Only nonprofit organization representing governor-designated energy officials from each state and territory. NASEO seeks to improve the effectiveness and quality of state energy programs and policies, provide policy input and analysis, share successes among the states, and act as a repository of information on energy issues concerning states and their citizens.

National Biodiesel Board (NBB)
URL: http://www.biodiesel.org
3337a Emerald Lane
P.O. Box 104898
Jefferson City, MO 65110-4898
Phone: (800) 841-5849

National trade association representing the biodiesel fuel industry. It coordinates U.S. biodiesel research and development efforts. Members include state, national, and international feedstock and feedstock processor organizations, biodiesel suppliers, fuel marketers and distributors, and technology providers.

The National Bureau of Asian Research (NBR)
URL: http://www.nbr.org
4518 University Way NE
Suite 300
Seattle, WA 98105
Phone: (206) 632-7370

Nonprofit, nonpartisan research institution dedicated to informing and strengthening policy in the Asia-Pacific region. NBR conducts independent research on strategic, political, economic, globalization, health, and energy issues affecting U.S. relations with Asia, including China. Its reports have been published in the journal *Energy*.

National Center for Appropriate Technology (NCAT)
URL: http://www.ncat.org
3040 Continental Drive
Butte, MT 59701
Phone: (406) 494-4572

Organization that pursues sustainable ways of helping to improve the lives of low-income families. NCAT offers how-to publications and toll-free information.

National Center for Photovoltaics (NCPV)
URL: http://www.nrel.gov/ncpv/
c/o National Renewable Energy Laboratory (NREL)
1617 Cole Boulevard

Golden, CO 80401-3393
Phone: (303) 384-6469

Solar energy research and development center staffed primarily by scientists and researchers from the National Renewable Energy Laboratory (NREL) and Sandia National Laboratories (SNL), laboratories managed by the U.S. Department of Energy (DOE). NCPV issues technical papers about solar technology, holds photovoltaic conferences, and maintains an international communications system called the NCPV Hotline.

National Commission on Energy Policy
URL: http://www.energycommission.org
1616 H Street NW
Sixth Floor
Washington, DC 20006
Phone: (202) 637-0400

Organization that seeks to influence public policy reform on key energy issues through bipartisan efforts. The commission conducts research on solutions for key energy sectors such as transportation, electricity generation, and agriculture/forestry.

National Hydropower Association (NHA)
URL: http://www.hydro.org
1 Massachusetts Avenue NW
Suite 850
Washington, DC 20001
Phone: (202) 682-1700

Nonprofit association representing the interests of the U.S. hydropower industry. It advocates hydropower as an emissions-free, renewable, reliable energy source.

National Mining Association (NMA)
URL: http://www.nma.org
101 Constitution Avenue NW
Suite 500 East
Washington, DC 20001-2133
Phone: (202) 463-2600

National trade organization representing the interests of the coal industry and coal mining operations. Activities include promoting coal as a stable and secure source of energy for America through lobbying and congressional testimony, as well as publishing fact sheets, forecasts, newsletters, a magazine, and a monthly digest of current U.S. coal trade statistics showing monthly and year-to-date summaries of coal exports and imports and their value.

National Petrochemical and Refiners Association (NPRA)
URL: http://www.npra.org
1899 L Street NW
Suite 1000
Washington, DC 20036-3896
Phone: (202) 457-0480

National association for the U.S. petrochemical manufacturers and oil refining industries. NPRA represents these industries before policymakers and the public.

National Renewable Energy Laboratory (NREL)
URL: http://www.nrel.gov
1617 Cole Boulevard
Golden, CO 80401-3393
Phone: (303) 275-3000

Primary research and development laboratory of the U.S. Department of Energy (DOE) devoted to the exploration of renewable energy and energy efficiency technologies. It is home to the National Center for Photovoltaics, the National Bioenergy Center, Alternative Fuels User Facility, Solar Radiation Research Laboratory, the National Wind Technology Center, and other programs for the research and development of alternative energy technologies.

Natural Gas Supply Association (NGSA)
http://www.ngsa.org
805 15th Street NW
Suite 510
Washington, DC 20005
Phone: (202) 326-9300

Represents United States–based producers, suppliers, transporters, and marketers of natural gas in legislative and public policy initiatives. Its goal is to maintain competitive markets, ensure reliability and efficiency of natural gas transportation and delivery, and increase the domestic supply of natural gas to the U.S. market.

Natural Resources Defense Council (NRDC)
URL: http://www.nrdc.org
40 West 20th Street
New York, NY 10011
Phone: (212) 727-2700

Environmental activist organization concerned with protecting wildlife and safeguarding the environment. NRDC draws upon research in law and science to spread awareness of environmental problems. It also issues legislative alerts, biweekly bulletins covering timely environmental news, and a monthly column about sustainable living.

North American Board of Certified Energy Practitioners (NABCEP)
URL: http://www.nabcep.org
Saratoga Technology + Energy Park
10 Hermes Road
Suite 400
Malta, NY 12020
Phone: (518) 899-8186

Volunteer board of representatives of renewable energy providers. NABCEP works with renewable energy and energy efficiency corporations, professionals, and stakeholders to develop and implement credentialing and certification programs for practitioners. NABCEP offers a certification program for photovoltaic installers.

North American Electric Reliability Council (NERC)
URL: http://www.nerc.com
116-390 Village Boulevard
Princeton, NJ 08540-5731
Phone: (609) 452-8060

Nonprofit U.S. corporation uniting investor-owned utilities, federal power agencies, rural electric cooperatives, public utilities, independent power producers, power marketers, and end-use customers whose goal is the establishment of mandatory reliability standards for electricity transmission in North America.

Nuclear Energy Institute (NEI) (formerly United States Council for Energy Awareness)
URL: http://www.nei.org
1776 I Street NW
Suite 400

Washington, DC 20006-3708
Phone: (202) 739-8000

U.S. organization whose members include operators, engineers, and service companies of nuclear power plants; companies involved in nuclear medicine and nuclear industrial applications; radionuclide and radiopharmaceutical companies; universities and research laboratories; and labor unions. NEI represents the nuclear energy industry's needs in the development of legislative and regulatory issues affecting the industry and publishes information on nuclear energy for the media and the public.

Nuclear Information and Resource Services (NIRS)
URL: http://www.nirs.org
6930 Carroll Avenue
Suite 340
Takoma Park, MD 20912
Phone: (301) 270-NIRS (301-270-6477)

Information and networking coalition for citizens and environmental organizations concerned about nuclear power, radioactive waste, radiation, and sustainable energy issues.

Office of Energy Efficiency and Renewable Energy (EERE)
URL: http://www.eere.energy.gov
Mail Stop EE-1
Department of Energy
Washington, DC 20585
Phone: (202) 586-9220

Office of the U.S. Department of Energy (DOE) that develops programs for increasing energy efficiency and renewable energy in the nation. It partially funds the work of the National Renewable Energy Laboratory (NREL), creates energy-efficiency educational campaigns for the public, deploys teams to perform energy audits at energy-intensive factories and facilities, and seeks ways to reduce or even end dependence on foreign oil.

Office of Fossil Energy
URL: http://www.fossil.energy.gov
1000 Independence Avenue SW
Washington, DC 20585
Phone: (800) dial-DOE (800-3425-363)

U.S. Department of Energy (DOE) office whose focus is clean, affordable energy from conventional fossil fuels. Projects include pollution-free coal plants, carbon sequestration, more productive oil and gas fields, hydrogen and clean fuel development, and preparedness of federal emergency oil stockpiles.

Office of Scientific and Technical Information (OSTI)
URL: http://www.osti.gov
One Science.gov Way
Oak Ridge, TN 37830
Phone: (865) 576-1188

U.S. government office that provides access to U.S. Department of Energy (DOE) reports and documents, including results of research projects, conference papers and proceedings, and various forms of scientific and technical information.

Oil Depletion Analysis Centre (ODAC)
URL: http://www.odac-info.org
Second Floor
162 Skene Street
Aberdeen AB10 1PE
United Kingdom
Phone: 44-0-7812 822643

Independent nonprofit British educational institution working to increase awareness and understanding of the finite nature of the world's supply of oil.

Organization of Petroleum Exporting Countries (OPEC)
URL: http://www.opec.org
Obere Donaustrasse 93
A-1020 Vienna
Austria
Phone: 1-21112-270

Multinational organization that coordinates oil production policies and pricing among oil-producing nations. Member nations, which produce 40 percent of the world's oil, are Algeria, Indonesia, Iran, Iraq, Kuwait, Libya, Nigeria, Qatar, Saudi Arabia, the United Arab Emirates, and Venezuela.

Pacific Northwest Laboratory (PNL)
URL: http://www.pnl.gov
P.O. Box 999

Richland, WA 99352
Phone: (888) 375-PNNL (7665)

A national laboratory managed by the Office of Science of the U.S. Department of Energy (DOE) that also conducts research and development for private industry. PNL scientists and engineers work on projects to address challenges in energy, the environment, and national security for government and industry clients.

Petroleum Industry Research Foundation, Inc. (PIRINC)
URL: http://www.pirinc.org
3 Park Avenue
26th Floor
New York, NY 10016-5989
Phone: (212) 686-6470

Nonprofit, independent organization that studies energy economics, particularly in relation to oil. PIRINC researches and publishes reports on all aspects of the petroleum industry.

Petroleum Marketers Association of America (PMAA)
URL: http://www.pmaa.org
1901 North Fort Myer Drive
Suite 500
Arlington, VA 22209-1604
Phone: (703) 351-8000

Federation of state and regional trade associations representing U.S. petroleum marketers.

Pew Center on Global Climate Change
URL: http://www.pewclimate.org
101 Wilson Boulevard
Suite 550
Arlington, VA 22201
Phone: (703) 516-4146

Organization of business leaders, policymakers, scientists, and other experts who work to analyze global climate change, report on their findings, and make recommendations to reduce greenhouse gas emissions while nurturing economic growth and protecting the interests of developing nations.

Political Economy Research Center (PERC)
URL: http://www.perc.org
2048 Analysis Drive
Suite A
Bozeman, Montana 59718
Phone: (406) 587-9591

Institute whose goal is to resolve environmental problems through the promotion of free enterprise and market incentives that encourage conservation and environmental protection. PERC conducts research and policy analysis, generates articles and reports, and pursues projects on a variety of natural resource issues, including carbon sequestration, global warming, mining, and land conservation. PERC also hosts conferences for scholars, journalists, teachers, business executives, policymakers, and others.

Potomac Resources, Inc. (PRI)
1001 Connecticut Avenue NW
Suite 801
Washington, DC 20036
Phone: (202) 429-8873

Energy and natural resources policy research and lobbying organization whose clients include water and wastewater utilities, environmental organizations, and companies that provide energy- or water-saving products or services.

Public Citizen
http://www.citizen.org
1600 20th Street NW
Washington, DC 20009
Phone: (800) 289-3787

Nonprofit consumer advocacy organization founded by Ralph Nader. Among its divisions are the Critical Mass Energy Project, which opposes nuclear fuels and promotes safe, sustainable energy use, and Buyers Up, which focuses on the buying of oil for domestic heat.

Reason Public Policy Institute (RPPI)
URL: http://www.reason.org
3415 S. Sepulveda Boulevard
Suite 400
Los Angeles, CA 90034
Phone: (310) 391-2245

Nonprofit public policy research organization promoting limited U.S. government interference in people's lives, particularly in the arena of energy production and consumption. RPPI publishes a monthly magazine, *Reason,* as well as publications on air quality, climate, electricity, and transportation. It is opposed to increased fuel economy standards for automobiles, arguing that they will result in smaller, less safe cars.

Refinery Reform Campaign
URL: http://www.refineryreform.org
222 Richland Avenue
San Francisco, CA 94110
Phone: (415) 643-1870

Organization advocating decreased dependence on fossil fuels and the reform of oil refineries to reduce pollution.

Renewable Energy Access
URL: http://www.renewableenergyaccess.com
375 Jaffrey Road
Peterborough, NH 03458
Phone: (877) 650-1782

Begun by renewable energy professionals as a source of renewable energy news and information. Renewable Energy Access provides daily news reports on developments in the field of renewable energy technology, a renewable energy events calendar, and more.

Renewable Energy Policy Project (REPP)
URL: http://www.crest.org
1612 K Street NW
Suite 202
Washington, DC 20006
Phone: (202) 293-2898

Policy research organization that advocates increased use of renewable energy sources, including biomass, hydropower, geothermal, photovoltaic, solar thermal, wind, and renewable hydrogen. REPP offers policy analysis, research, renewable energy discussion groups, and information for policymakers, green energy entrepreneurs, and environmental advocates.

Renewable Fuels Association
URL: http://www.ethanolrfa.org

One Massachusetts Avenue NW
Suite 820
Washington, DC 20001
Phone: (202) 289-3835

National trade association for the U.S. ethanol industry. Members research, produce, and promote ethanol fuel. They pursue public policies and regulations that will lead to increased reliance on ethanol as an automobile fuel. The group promotes ethanol as a fuel that improves vehicle performance, reduces emissions, and results in energy security by limiting dependence on oil exports.

Republicans for Environmental Protection (REP America)
URL: http://www.repamerica.org
3200 Carlisle Boulevard
#228
Albuquerque, NM 87110
Phone: (505) 889-4544

Republican foundation that seeks to promote wider use of renewable energy sources along with conservation and efficiency measures that sustain economic growth. REP America publishes a newsletter, *The Green Elephant,* in addition to a journal, *C.E.P. Quarterly,* a joint project with ConservAmerica.

Resources for the Future
URL: http://www.rff.org
1616 P Street NW
Washington, DC 20036
Phone: (202) 328-5000

Independent institute that conducts research on natural resources, electricity, climate change, foreign oil dependence, and other energy issues in the hopes of building an environmentally sound U.S. energy policy. It also publishes reports, lectures, and transcripts of members' congressional testimony.

Rocky Mountain Institute (RMI)
URL: http://www.rmi.org
1739 Snowmass Creek Road
Snowmass, CO 81654-9199
Phone: (970) 927-3851

Nonprofit organization focusing on energy efficiency and resource conservation driven not by governmental programs but by market-oriented initiatives and awareness.

Scientific Committee for Problems of the Environment (SCOPE)
URL: http://www.icsu-scope.org
51, Bd de Montmorency
75016 Paris
France
Phone: +33 1 45 25 04 98

An interdisciplinary organization established by the International Council for Science (ICSU). It joins experts in the fields of natural and social science to study global environmental issues.

Sierra Club
URL: http://www.sierraclub.org
Sierra Club
85 Second Street
Second Floor
San Francisco, CA 94105
Phone: (415) 977-5500

Established national environmental protection organization that advocates grassroots conservation and energy efficiency efforts. It supports increased fuel economy standards; prevention of drilling in the Arctic National Wildlife Refuge (ANWR); widespread use of wind, solar, and other renewable energies; increased use of hybrid vehicles; state renewable energy standards; and programs to reduce electric and gas usage in business and commercial settings. It also publishes the bimonthly magazine *Sierra,* a newsletter, books, and fact sheets.

Solar Energy Industries Association (SEIA)
URL: http://www.seia.org
Solar Energy Industries Association
805 15th Street NW
#510
Washington, DC 20005
Phone: (202) 682-0556

National trade association of solar energy manufacturers, dealers, distributors, and contractors. SEIA hosts a national solar power conference.

Solar Energy International (SEI)
URL: http://www.solarenergy.org
P.O. Box 715

76 S. Second Street
Carbondale, CO 81623
Phone: (970) 963-8855

Educational, outreach institution that works to spread the understanding and use of renewable energy technologies worldwide. SEI's Renewable Energy Education Program (REEP) offers hands-on professional training in photovoltaic design and installation, wind power, solar water pumping, and alternative fuel technologies. SEI also helps organizations to plan, design, and implement their own clean-energy and conservation efforts, and it sponsors international projects for powering rural communities with renewable energy.

State Environmental Protection Administration (SEPA), China
URL: http://www.zhb.gov.cn/english
No-115 Xizhimennei Nanxiaojie
Beijing 100035
China
Phone: 66556004
66556005

China's national environmental protection agency, which issues and enforces standards and regulations on air quality, water quality, and pollution control. Its Web site features reports on the Three Gorges Dam project, which established the world's largest hydroelectric plant in China.

The Stella Group, Ltd.
URL: http://www.thestellagroupltd.com
1616 H Street NW
10th Floor
Washington, DC 20006-4999
Phone: (202) 347-2214

Marketing and policy firm advocating energy generation through alternative energy technologies including advanced batteries, concentrated solar energy, fuel cells, microgenerators, modular biomass, photovoltaic, and wind energy. The group generates articles and fact sheets on climate change, coal emissions, conservation, and more.

Sustainable Energy Coalition
URL: http://www.sustainableenergy.org
1612 K Street NW
Suite 202-A

Washington, DC 20006
Phone: (202) 293-2898

A coalition of U.S. business, environmental, consumer, and energy policy organizations that advocate increased federal funding for renewable and efficient energy technologies and reduced federal support of nuclear energy and energy derived from fossil fuels. The group works to contribute environmental and energy efficiency awareness to policy decisions regarding the federal budget, electric utility restructuring, pollution prevention, climate change, and tax policies.

Sustainable Energy Economy Network (SEEN)
URL: http://www.seen.org
c/o Institute for Policy Studies
1112 16th Street NW
Suite 600
Washington, DC 20036
Phone: (202) 234-9382

A network of national and global citizens' groups begun by the Institute for Policy Studies and the Transnational Institute. SEEN draws these groups together to collaborate on addressing environmental protection, justice, and human rights issues, concentrating especially on energy, climate change, environmental integrity, gender equity, and economic development.

Union of Concerned Scientists
URL: http://www.ucs.org
2 Brattle Square
Cambridge, MA 02238-9105
Phone: (617) 547-5552

A nonprofit U.S. alliance of scientists and engineers who collaborate on studies concerning renewable energy, the impact of global climate change, nuclear energy risks, and other projects that apply scientific research to environmental and social problems. The organization encourages reduction in fuel consumption and the raising of corporate average fuel economy (CAFE) standards to 40 mpg by 2012. Its research is published and shared with the news media and the public, as well as congressional and regulatory hearings.

United Nations (UN) Security Council
URL: http://www.un.org/Docs/sc
UN Headquarters

First Avenue at 46th Street
New York, NY 10017

Council of the United Nations (UN) whose primary purpose is to maintain international peace and security. The council takes actions to address and mediate UN member countries' concerns about threats to world peace and security, including, for example, other countries' pursuit of nuclear weapons programs.

United Nations Scientific Committee on the Effects of Atomic Radiation (UNSCEAR)
URL: http://www.unscear.org
Vienna International Centre
P.O. Box 500
A-1400 Vienna
Austria
Phone: 0043-1-26060-4330

Formed in 1955 by the General Assembly of the United Nations (UN) in response to widespread concerns regarding the health and environmental effects of radiation from the testing of nuclear weapons. Such testing is now banned, but UNSCEAR compiles information on natural sources of exposure to radiation, such as nuclear power production, in addition to exposure from occupational and medical sources.

United States Energy Association (USEA)
URL: http://www.usea.org
1300 Pennsylvania Avenue
Suite 550
Washington, DC 20004
Phone: (202) 312-1230

U.S. committee of the World Energy Council (WEC) and an association of public and private energy-related organizations, corporations, and government agencies. USEA represents a variety of U.S. energy sector interests and supports increasing the domestic supply of energy resources, including fossil fuels.

United States Geological Survey (USGS)
URL: http://www.usgs.gov
USGS National Center
12201 Sunrise Valley Drive
Reston, VA 20192
Phone: (888) ASK-USGS (888-275-8747)

Bureau of the U.S. Department of the Interior. USGS gathers and shares scientific data and information on energy and fossil fuels, the future sustainability of domestic oil production, and water, biological, mineral, and other natural resources.

U.S. Department of Energy (DOE)
http://www.doe.gov
1000 Independence Avenue SW
Washington, DC 20585
Phone: (800) dial-DOE (800-3425-363)

U.S. government agency that oversees the protection of national, economic, and energy security through programs that aim to diversify the supply and delivery of energy, promote safe and conscientious use of natural resources, advance scientific knowledge, safeguard the environment, and address nuclear energy security and safety issues. The DOE also researches new energy-related technologies, such as alternative fuels, in addition to pursuing advancements in science and medicine.

U.S. Environmental Protection Agency (EPA)
URL: http://www.epa.gov
Ariel Rios Building
1200 Pennsylvania Avenue NW
Washington, DC 20460
Phone: (202) 272-0167

U.S. government agency responsible for protecting the environment and public health. The EPA researches and sets national standards for various environmental programs, funds programs and research that support environmental protection, and enforces regulations related to the nation's environmental laws, including the Clean Air Act. The EPA may also play a lead role in various environmental protection agreements between the United States and other countries, such as the Carbon Sequestration Leadership Forum (CSLF), a voluntary climate initiative that includes 20 nations and the European Commission.

U.S. Global Change Research Information Office (GCRIO)
URL: http://www.gcrio.org
1717 Pennsylvania Avenue NW
Suite 250
Washington, DC 20006
Phone: (202) 223-6262

Government office that acts as a clearinghouse for documents and reports from federal agencies and committees involved in the U.S. Global Change Research Program (USGCRP). The office also provides access to data and information on climate change research, strategies and technologies for adapting to and mitigating climate change, and educational resources on climate change for governments, institutions, researchers, educators, students, and the general public.

U.S. House of Representatives Committee on Resources
URL: http://resourcescommittee.house.gov
1324 Longworth House Office Building
Washington, DC 20515-6201
Phone: (202) 225-2761

Committee of U.S. congressional lawmakers whose jurisdiction includes natural resource conservation projects, government-protected lands, mining and mineral resources, coastal zone management, petroleum conservation on public lands, conservation of the radium supply in the United States, and the Trans-Alaska Oil Pipeline.

U.S. Nuclear Regulatory Commission (NRC)
http://www.nrc.gov
One White Flint North
11555 Rockville Pike
Rockville, MD 20852-2738
Phone: (800) 368-5642

Independent agency established by the Energy Reorganization Act of 1974 to regulate civilian use of nuclear materials and ensure the safety of nuclear power production.

U.S. Public Interest Research Groups (USPIRG)
URL: http://uspirg.org
218 D Street SE
Washington, DC 20003
Phone: (202) 546-9707

Watchdog organization created by the National Association of State Public Interest Research Groups, a network of independent, state-based, citizen-funded organizations that aim to represent the interests of the public. U.S. PIRG monitors and addresses public policies concerning arctic wilderness, clean air, clean water, endangered species, energy, protection of forests, global climate change, and toxic waste, among other issues.

U.S. Senate Committee on Energy and Natural Resources
URL: http://energy.senate.gov/public
364 Dirksen Office Building
Washington, DC 20510
Phone: (202) 224-4971

Committee of U.S. senators who consider, report on, and oversee important legislation enacted by the U.S. Congress on issues including energy resources and development, environmental regulation, conservation, strategic petroleum reserves, nuclear energy, public lands and their renewable resources, surface mining, and water resources.

The Veggie Van Organization
URL: http://www.veggievan.org
8033 Sunset Boulevard
#154
Los Angeles, CA 90046
Phone: (310) 496-3292

Resource for biodiesel education programs, information, and discussion.

World Association of Nuclear Operators (WANO)
URL: http://www.wano.org.uk
WANO-Coordinating Centre
First Floor
Cavendish Court
11-15 Wigmore Street
London W1U 1PF
United Kingdom
Phone: 44-0-20 7478 9200

Organization established by nuclear plant operators after the Chernobyl accident in 1986 to ensure the safety and reliability of nuclear power plants from that point forward. WANO encourages communication, standards comparisons, and information sharing among its members.

The World Bank Group Energy Program
URL: http://www.worldbank.org
1818 H Street NW
Washington, DC 20433
Phone: (202) 473-1000

Energy services division of the World Bank, a source of financial and technical assistance to developing countries with the goal of reducing global poverty and improving living standards. The Energy Program finances energy supply reform projects and attempts to help developing countries establish reliable access to clean, modern, and affordable energy services. The Energy Program also offers information on the energy sectors of countries throughout the world.

World Business Council for Sustainable Development (WBCSD)
URL: http://www.wbcsd.ch
4, chemin de Conches
1231 Conches-Geneva
Switzerland
Phone: 41-22-839 3100

Coalition of international companies advocating sustainable development that takes into account economic growth, ecological balance, and social progress. WBCSD networks with national and regional business councils and other organizations worldwide.

World Conservation Union (formerly the International Union for the Conservation of Nature and Natural Resources)
URL: http://www.iucn.org
Rue Mauverney 28
Gland 1196
Switzerland
Phone: 41-22-999-0000

Large conservation network consisting of governments, government agencies, nongovernmental organizations, and scientists and experts throughout the world. It pursues projects related to helping nations and societies conserve and protect nature and use their resources in an equitable and ecologically sustainable way.

World Energy Council (WEC)
URL: http://www.worldenergy.org
Regency House
Fifth Floor
1–4 Warwick Street
London W1B 5LT
United Kingdom
Phone: 44-20-7734 5996

Independent nonprofit organization whose members include energy producers and consumers, policymakers, and environmental organizations around the world. Members are interested in issues related to all types of energy, including coal, oil, natural gas, nuclear energy, hydropower, and renewable energy.

World Information Service on Energy (WISE)
URL: http://www10.antenna.nl/wise
P.O. Box 59636
1040 LC Amsterdam
The Netherlands
Phone: 31-20-612-6368

International group that opposes nuclear energy and distributes a magazine solely devoted to issues regarding nuclear energy.

World Nuclear Association (WNA)
URL: http://www.world-nuclear.org
Carlton House
22a Street James's Square
London SW1Y 4JH
Phone: 44-0-20-7451-1520

International association of nuclear power producers advocating safe nuclear energy as an indispensable part of sustainable global development.

World Petroleum Council
URL: http://www.world-petroleum.org
Fourth Floor
Suite 1
1 Duchess Street
London W1W 6AN
Phone: 44-0-20-7637-4958

Global organization representing the petroleum industry. It sponsors congresses, reports, regional meetings, and workshops to address issues and challenges faced by the industry.

World Resources Institute (WRI)
URL: http://www.wri.org
10 G Street NE
Suite 800

Washington, DC 20002
Phone: (202) 729-7600

Institute that participates in a variety of research and development projects covering global needs related to both the environment and economic development. Projects include assisting developing countries with new agricultural and environmental regulations and exploring the potential of a large-scale transition to biomass-derived feedstock for the production of energy, fuels, and products.

World Wind Energy Association
URL: http://www.wwindea.org
Charles-de-Gaulle-Strasse 5
53113 Bonn
Germany
Phone: 228-369-40-80

Represents an international league of wind power associations, scientific institutes, and companies. WWEA promotes wind power through conferences, exhibits, and publications.

Worldwatch Institute
URL: http://www.worldwatch.org
1776 Massachusetts Avenue NW
Washington, DC 20036-1904
Phone: (202) 452-1999

Nonprofit environmental and policy research organization that receives funding from private foundations as well as United Nations (UN) programs. It publishes the bimonthly *World Watch* magazine and papers covering environmental policies that intersect with global economic concerns. It also promotes alternative energy.

10

Annotated Bibliography

The following annotated bibliography focuses on works about global issues in energy and natural resources. It is organized by region. Entries covering issues of interest to the United States are listed first, followed by entries with international applicability. Entries for China, Germany, Iran and Saudi Arabia, Nigeria, and Venezuela follow.

Entries for regional categories are then arranged according to energy-related issues, where applicable, as follows:

Energy Affordability and Economics

Energy-Related Environmental Issues/Renewable and Alternative Energies

Energy-Related Infrastructure Issues

Energy Supplies, Crises, and Shortfalls

Foreign Oil Dependence/Energy-Related Political Issues

Each of the categories is then subdivided into the following sections:

Books

Articles and Papers

Web Documents

Other Media

UNITED STATES

U.S. Energy Affordability and Economics

BOOKS

Beaudreau, Bernard C. *Energy and the Rise and Fall of Political Economy.* Westport, Conn.: Greenwood, 1999. Examines the impact of advancements in energy use on the U.S. political economy in the 19th and 20th centuries.

Annotated Bibliography

Hunt, Lester C., ed. *Energy in a Competitive Market: Essays in Honour of Colin Robinson.* Northampton, Mass.: Edward Elgar, 2003. Compilation of essays on energy-related economics issues. Emphasizes issues related to the market allocation of energy resources.

Leeb, Stephen, and Donna Leeb. *The Oil Factor: How Oil Controls the Economy and Your Financial Future.* New York: Warner Business Books, 2004. Proposes that the United States is headed for a substantial economic collapse due to its dependency on shrinking supplies of oil. Analyzes ways for individuals to make financial choices that, collectively, may help the U.S. economy become less dependent on oil imports.

Leeb, Stephen, and Glen Strathy. *The Coming Economic Collapse: How You Can Survive When Oil Costs $200 a Barrel.* New York: Warner Business Books, 2004. Predicts that as the fast-growing economies of developing countries such as China and India push world oil supply to its limits, the United States will experience a permanent energy crisis that far exceeds that experienced in the 1970s.

Peirce, William Spangar. *Economics of the Energy Industries,* 2d ed. Westport, Conn.: Praeger, 1996. Presents an overview of U.S. energy industries, including the coal, oil, natural gas, electricity, nuclear power, and alternative energy industries, and their significance to the economy, as well as economic analyses of the problems associated with energy production and consumption.

ARTICLES AND PAPERS

Ball, Jeffrey. "As Prices Soar, Doomsayers Provoke Debate on Oil's Future," *Wall Street Journal,* 21 September 2004. Commentary on the increasing U.S. consumption of increasingly higher-priced oil.

Bezdek, Roger H., and Robert M. Wendling. "The U.S. Energy Subsidy Scorecard: Surprises Abound." *Issues in Science and Technology* 22 (March 22, 2006): 83–85. Analysis of the tax subsidies that the U.S. government grants toward the research and development of various energy technologies. Reports findings that a relatively small portion of tax subsidies are allotted toward research and development, with the largest share granted to oil production in comparison with other forms of energy.

The Economist. "Unstoppable? How OPEC's Fear of $5 Oil Led to $50 Oil," *The Economist,* 21 August 2004, 59–60. Gives an example of the way an OPEC production policy decision can impact oil prices around the world.

Guidi, Marco G. D., Alexander Russell, and Heather Tarbert. "The Effect of OPEC Policy Decisions on Oil and Stock Prices." *OPEC Review* 30 (2006): 1–18. Presents evidence of the effects of OPEC policy decisions on the U.S. and UK stock markets and oil prices during the period 1986–2004. Shows a correlation between times of conflict and OPEC policy–related market changes.

LeClair, Mark S. "Achieving Gasoline Price Stability in the U.S.: A Modest Proposal." *The Energy Journal* 27 (April 1, 2006): 41–54. Proposes a plan for reducing the volatility of gasoline prices in the United States that involves adjusting federal tax policies.

WEB DOCUMENTS

Dooley, J. J. "Unintended Consequences: Energy R&D in a Deregulated Energy Market." Pacific Northwest National Laboratory, February 6, 1997. Available online. URL: http://www.pnl.gov/er_news/04_98/unincons.htm. Accessed on January 29, 2006. Presents an overview of the impact of energy sector deregulation on industrialized nations' research and development into emerging energy technologies and studies the implications of this trend in the United States.

Stronberg, Joel B. *Common Sense: Making the Transition to a Sustainable Energy Economy.* American Solar Energy Society (ASES), May 2005. Available online. URL: http://www.ases.org/print_catalog/ases_reports/PS_Common_Sense.pdf. Accessed on June 2, 2006. Proposes that the U.S. Congress implement programs such as a national renewable energy standard and a renewable fuel standard to help foster the economic growth of sustainable energy. Recommends that federal, state, and local government leaders begin a transition to a clean domestic energy–based economy.

U.S. Energy-Related Environmental Issues/Renewable and Alternative Energies

BOOKS

Asmus, Peter. *Reaping the Wind: How Mechanical Wizards, Visionaries, and Profiteers Helped Shape Our Energy Future.* Washington, D.C.: Island Press, 2001. Details the history of U.S. commercial wind power. Discusses the technologies, challenges, and policies behind the development of this form of renewable energy in the United States.

Berman, Daniel M., and John T. O'Connor. *Who Owns the Sun? People, Politics, and the Struggle for a Solar Economy.* White River Junction, Vt.: Chelsea Green, 1996. Explores the history of solar energy in the United States and proposes economic and environmental reasons for greater reliance on solar energy and reduced dependence on fossil fuels.

Blackwood, John R., ed. *Energy Research at the Cutting Edge.* Hauppauge, N.Y.: Nova Science, 2002. Analyzes the impact and characteristics of both conventional and alternative fuels in the energy industry today as well as their likely effects in the future. Favors conventional fuels as more practical sources of energy than alternative fuels.

Bosselman, Fred, Jim Rossi, and Jacqueline Lang Weaver. *Energy, Economics and the Environment: Cases and Materials.* New York: Foundation Press, 2000. Discusses energy laws, utility regulation, and environmental laws related to energy and their impact. Explores the issue of global warming and laws designed to address it.

Buell, Frederick. *From Apocalypse to Way of Life: Environmental Crisis in the American Century.* New York: Routledge, 2003. An analytical account of American environmental crises of all forms, including those caused by energy use.

Chirase, Daniel D., Oliver S. Owen, and John P. Reganold. *Natural Resources Conservation: Management for a Sustainable Future,* 9th ed. Upper Saddle River, N.J.:

Prentice Hall, 2004. Provides general history, background, and future projections related to U.S. energy resources, mainly from the perspective of conservationism.

Hostetter, Martha. *Energy Policy.* New York: Reference Shelf, 2002. A compilation of essays and articles on fossil fuels, energy alternatives, nuclear energy and hydrogen, and energy and the environment designed to give researchers a well-rounded understanding of U.S. energy policy.

Kemp, William. *The Renewable Energy Handbook: A Guide to Rural Energy Independence, Off-Grid and Sustainable Living.* Tamworth, Ontario, Canada: Aztext Press, 2006. Discusses the technologies that are available to consumers to enable them to live "off-the-grid," that is, without needing to obtain their energy from a power utility company. Focuses on energy conservation; alternative heating and cooling; photovoltaic, wind, and hydropower; voltage regulation, inverters, and backup systems.

Komp, Richard J., *Practical Photovoltaics: Electricity from Solar Cells.* 3d ed. Ann Arbor, Mich.: AATEC, 2002. Reference guide on solar electricity that covers not only the theory, manufacture, and current and future prospects of solar electricity but technical information such as solar system design, installation, and maintenance.

Koppel, Tom. *Powering the Future: The Ballard Fuel Cell and the Race to Change the World.* Etobicoke, Canada: John Wiley & Sons Canada, 1999. Details the history of Ballard and Company, the pioneer of fuel cell development. Portrays fuel cells as an up-and-coming technology with great promise.

Kruger, Paul. *Alternative Energy Resources: The Quest for Sustainable Energy.* Hoboken, N.J.: John Wiley & Sons, 2006. Examines the history of energy use, with special emphasis on the boom in energy use that began with the Industrial Revolution. Assesses the viability of various alternative energy sources, such as hydroelectric, solar, wind, biomass, geothermal, nuclear, and hydrogen, then proposes the use of hydrogen, produced through a combination of renewable and nuclear energy, as the fuel to sustain the future.

Peavey, Michael. *Fuel from Water,* 11th ed. Louisville, Ky.: Merit Products Inc., 1998. Explains how hydrogen is generated by electrolysis, how an internal combustion engine is converted to run on hydrogen, and how hydrogen can be used in home appliances. Advocates hydrogen fuel as the most practical alternative to fossil fuels over the long term.

Rifkin, Jeremy. *The Hydrogen Economy.* New York: Jeremy P. Tarcher/Penguin, 2003. Reviews the history of oil dependence and then proposes that hydrogen holds the key to cleaner, safer, and more sustainable energy production.

Schaeffer, John. *Real Goods Solar Living Sourcebook: The Complete Guide to Renewable Energy Technologies and Sustainable Living,* 12th ed. Gabriola Island, Canada: New Society Publishers, 2005. A reference guide to all the various forms of renewable energy, sustainable home construction, off-the-grid living, and alternatively fueled transportation.

Sweet, William. *Kicking the Carbon Habit: Global Warming and the Case for Renewable and Nuclear Energy.* New York: Columbia University Press, 2006. Citing coal as the primary culprit behind the carbon emissions that are being held responsible for global warming, proposes that wind energy, already a major source of power in European countries, and nuclear energy, of which, he claims, major accidents and safety problems are aberrations, have the capability to revolutionize American power production.

Tickell, Joshua. *From the Fryer to the Fuel Tank: The Complete Guide to Using Vegetable Oil as an Alternative Fuel,* 3d ed. New Orleans: Joshua Tickell Media Productions, 2003. Provides history and technical information on biodiesel and other forms of vegetable-oil–based fuels.

ARTICLES AND PAPERS

Barnham, K. W. J., B. Clive, and M. Mazzer. "Resolving the Energy Crisis: Nuclear or Photovoltaics?" *Nature Materials* 5 (March 2006): 161–164. Using photovoltaics alone as a model, concludes that, despite claims to the contrary, renewable resources are capable of replacing the energy contribution from nuclear energy.

Bates, Todd B. "Nuclear Power Is Regenerating Interest," *Asbury Park Press,* 15 February 2006. Offers a primer on nuclear energy generation and production, nuclear energy policy, and nuclear safety issues, while discussing recent technological developments and a possible revival of nuclear energy in the United States.

Blake, E. Michael, Dick Kovan, and Rick Michal. "Nuclear Technology—a Good Story Untold." *Nuclear News* 49 (January 2006): 50–67. A report on the 2005 winter meeting of the American Nuclear Society, endorsing nuclear energy for a variety of applications, exploring different views on nuclear proliferation, discussing nuclear energy infrastructure needs, and analyzing risk and safety issues from an industry perspective.

Boser, Steve. "Finding Greener Wheels." *Home Power* 112 (April/May 2006): 52–54. Assesses time, cost, availability, build quality, pollution reduction, and dependability features of eight types of alternatively fueled vehicles, from all-electric to gas-electric hybrid to biodiesel. Offers background information on each alternative fuel technology.

de Winter, Francis, and Ronald B. Swenson. "A Wake-Up Call." *Solar Today* 20 (March/April 2006): 15–19. Analyzes the growth of photovoltaic capacity against the backdrop of declining oil supply and uses data to refute various arguments about why solar power will never surpass fossil fuel energy, including objections to cost, efficiency, and inconvenience.

Hammerschlag, Roel. "Ethanol's Return on Investment: A Survey of the Literature 1990–Present." *Environmental Science and Technology, Washington, D.C.* 40 (2006): 1,744–1,750. Considers various conflicting reports on the value of ethanol and whether the level of nonrenewable fuel consumption needed for its manufacture renders it unsuitable as a fossil fuel alternative. Uses mathematical calculations to conclude that ethanol at least produces some renewable energy and consumes less nonrenewable energy in its manufacture than gasoline.

Howland, George, Jr. "Clean-Energy Frenzy," *Seattle Weekly,* 14 December 2005, 18. A report on the state of Washington's successes with initiatives to convert its residents and businesses over to renewable or alternative energies.

Kerr, Andy. "Cost-Effective Hybrids." *Home Power* 113 (June/July 2006): 66–71. Analyzes the purchase prices and expenses associated with driving a hybrid and whether hybrid vehicles make economic sense. Compares hybrid vehicles offered by different manufacturers. Offers background information on hybrid vehicle technology.

Kidd, Stephen W. "Time for a Fresh Look at Nuclear?" *Energy and Environment* 17 (2006): 175–180. Explores a revived interest in nuclear power, as well as a concurrent backlash despite improved operations and safety records. Proposes that nuclear and renewable energies may be seen as complementary rather than competitive energy solutions.

Klara, Scott, John Litynski, Howard McIlvried, and Rameshwar Srivastava. *Climatic Change* 74 (2006): 81–95. Describes U.S. Department of Energy–sponsored research into ways of reducing greenhouse gas emissions, including technologies to capture and sequester carbon dioxide within geologic formations, terrestrial ecosystems through natural processes, or above- and below-ground biomass in forests.

Menicucci, Dave. "Giving Green Credit Where It's Due: Technical Aspects to Consider in Establishing a Renewable Energy Credit Program." *Solar Today* 20 (May/June 2006): 30–33. Defines and gives background on state renewable energy credit programs. Describes technical aspects, such as system power requirements, energy ratings, and monitoring challenges, involved in developing renewable energy credit programs for various renewable technologies.

Ramirez, C. A., and E. Worrell. "Feeding Fossil Fuels to the Soil." *Resources, Conservation and Recycling* 46 (2006): 75–93. Surveys technological improvements in the nitrogen fertilizer industry and the idea that energy is embedded in fertilizer consumption.

Schultz, K. R. "Why Fusion? A Discussion of Energy Alternatives." *IEEE Control Systems Magazine* 26 (April 2006): 32–34. Claims that, if successfully developed, fusion is an attractive, sustainable, environmentally benign, and economic source of energy. Provides a scientific and practical rationale for developing countries to continue to pursue fusion energy despite its safety issues and other challenges.

WEB DOCUMENTS

Aitken, Donald W. *Transitioning to a Renewable Energy Future.* International Solar Energy Society (ISES), 2005. Available online. URL: http://www.whitepaper.ises.org. Accessed on January 29, 2006. An ISES white paper that recommends that governments set, assure, and achieve goals to achieve energy efficiency aggressively, pursuing renewable energy objectives.

American Council on Renewable Energy (ACORE). "Renewable Energy in America: The Policies for Phase II." ACORE Phase II White Paper, October 2005. Available online. URL: http://www.acore.org/download/phaseII_forum_summary.pdf. Accessed on January 29, 2006. Outlines steps for implementing clean energy strategies, policies, and programs. Recommendations are part of "Phase II" of ACORE's 2005 National Policy Forum, a working group whose goal is to increase the utilization of renewable energy in the United States.

Clemmer, Steven, Jeff Deyette, Deborah Donovan, and Alan Nogee. *Clean Energy Blueprint: A Smarter National Energy Policy for Today and the Future.* Union of Concerned Scientists (UCS), October 2001. Available online. URL: http://www.ucsusa.org/assets/documents/jump.jsp?origID=pdf-399. Accessed on January 29, 2006. Examines renewable energy and energy efficiency policies, including renewable portfolio standard, public benefits fund, net metering, production tax credit, increased funding for research and development, and improved efficiency standards. Part of an overall analysis conducted by UCS and the American Council for an Energy-Efficient Economy, with the Tellus Institute, exploring the value of standards and incentives for increasing investment in clean energy and overcoming market barriers to investment.

National Resources Defense Council (NRDC). *A Responsible Energy Plan for America.* National Resources Defense Council, Energy, News & Issue Highlights, April 2005. Available online. URL: http://www.nrdc.org/air/energy/rep/repinx.asp. Accessed on January 29, 2006. Makes recommendations for U.S. energy efficiency measures to help cut energy costs and increase productivity at the same time.

OTHER MEDIA

Pelley, Scott. "Global Warning!" *60 Minutes.* Produced and directed by Bill Owens. CBS Television, February 19, 2006. Television news report. Features an interview with Robert Correll, a senior fellow of the American Meteorological Society and director of the Arctic Climate Impact Assessment, and other scientists regarding melting glaciers, their causes, and their impact on the environment. Explores the possibility that greenhouse gas emissions have resulted in the melting of glacial ice and are disrupting ocean currents and other environmental factors.

U.S. Energy-Related Infrastructure Issues

WEB DOCUMENTS

Connolly, Michael. "Developments in Hydrogen Production, Technology and Use under the Energy Policy Act of 2005." Thelen Reid Publications, *Energy, Utility and Infrastructure,* November 15, 2005. Available online. URL: http://www.thelenreid.com/articles/article/art_264_idx.htm. Accessed on January 20, 2006. Offers an overview of the Energy Policy Act of 2005 and other U.S. legislative initiatives that propose to increase the nation's capacity for producing renewable or alternative energies. Focuses primarily on details of hydrogen-related energy legislation and building of a renewable energy infrastructure.

U.S. Energy Supplies, Crises, and Shortfalls

BOOKS

Alters, Sandra M. *Energy: Supplies, Sustainability, and Costs.* Detroit: Thompson Gale, 2005. Includes source documents, statistical and factual data, and resource information focusing mainly on U.S. energy cost issues and the future availability of sources of energy in the United States.

Annotated Bibliography

Bent, Robert, Randall Baker, and Lloyd Orr, eds. *Energy: Science, Policy, and the Pursuit of Sustainability.* Washington, D.C.: Island Press, 2002. Studies the environmental, societal, political, and economic effects of energy use.

Borowitz, Sidney. *Farewell Fossil Fuels: Renewing America's Energy Policy.* Cambridge, Mass.: Perseus Book Group, 1999. Discusses the limitations and benefits of both conventional and alternative sources of energy.

Boyle, Godfrey, Bob Everett, and Janet Ramage, eds. *Energy Systems and Sustainability: Power for a Sustainable Future.* New York: Oxford University Press, 2003. Studies the history behind the current state of energy production and consumption and explores issues related to energy policy as well as the economic, social, and environmental dimensions of energy. Proposes that technological developments may improve the balance between energy demand and energy supply and lead to greater sustainability.

Darley, Julian. *High Noon for Natural Gas: The New Energy Crisis.* White River Junction, Vt.: Chelsea Green, 2004. Highlights the increasing U.S. dependency on natural gas for electricity generation and other energy activities and argues that it has the potential to cause serious environmental, political, and economic consequences.

Deffeyes, Kenneth S. *Hubbert's Peak: The Impending World Oil Shortage.* Princeton, N.J.: Princeton University Press, 2001. Reviews the findings of M. King Hubbert, who predicted that U.S. oil production would peak in the 1970s and, on the basis of Hubbert's methods and contemporary energy data, theorizes that world oil production will peak in the first decade of the 21st century.

Goodell, Jeff. *Big Coal: The Dirty Secret behind America's Energy Future.* New York: Houghton Mifflin, 2006. Explores the U.S. reliance on coal and its implications for the environment, health, and society. Investigates various dimensions of the coal industry, including environmental effects and mine safety. Presents statistics and facts that, it is claimed, point toward a resurgence in coal energy that is ultimately at the expense of the American people.

Heinberg, Richard. *The Party's Over: Oil, War and the Fate of Industrial Societies.* Gabriola Island, Canada: New Society, 2005. Explores the historical events that shaped modern dependency on oil as a primary source of energy, such as the Industrial Revolution, and calls upon the United States to join other countries to implement a global program of resource conservation and sharing.

Huber, Peter W., and Mark P. Mills. *The Bottomless Well: The Twilight of Fuel, the Virtue of Waste, and Why We Will Never Run Out of Energy.* New York: Basic Books, 2005. An analysis of current U.S. energy consumption and the technologies that hold potential for helping Americans meet their future energy needs.

Kunstler, James Howard. *The Long Emergency: Surviving the Converging Catastrophes of the Twenty-First Century.* New York: Atlantic Monthly Press, 2005. Argues that there are abundant signs that global oil production has peaked and will soon end and that alternative energy sources are incapable of filling oil's role. Predicts that, absent oil, Americans will be forced to adopt a lifestyle reminiscent of that more than 100 years ago, relying on animal and mechanical energy. Also predicts that big cities will collapse and hunger and poverty will result.

Mast, Tom. *Over a Barrel: A Simple Guide to the Oil Shortage.* Austin, Tex.: Hayden, 2005. Concise reference guide covering the basics of energy science and history; oil consumption, reserves, and production; and the facts behind and implication of the prospect of permanent depletion of oil supplies. Includes a special emphasis on the consequences of oil shortfalls in the United States.

Plunkett, Jack. *Plunkett's Energy Industry Almanac 2006.* Houston, Tex.: Plunkett Research, 2005. A complete guide to the energy industry in the United States, with analyses of the latest business trends and statistics, as well as profiles of leading energy companies.

Roberts, Paul. *The End of Oil: On the Edge of a Perilous New World.* Boston: Houghton Mifflin, 2004. Investigates the effects of U.S. dependency on oil as its primary source of energy on the economy, the environment, health, and foreign relations and explores the overall prospects for national energy supply and the future of energy consumption in the United States.

Simon, Julian Lincoln. *The Ultimate Resource 2.* Rev. edition. Princeton, N.J.: Princeton University Press, 1998. Argues against those who propose that the world is running out of resources for its ever-growing population and that it is thus headed for an unparalleled social and economic collapse. Claims that the human imagination is the "ultimate resource" that keeps growing in supply as the world population grows, enabling human civilization to advance and expand beyond any perceived limitations.

Wright, Russell O. *Chronology of Energy in the United States.* Jefferson, N.C.: McFarland, 2003. A chronology tracing the historical development of U.S. energy use and energy technologies since 1752, when Benjamin Franklin discovered that lightning and static electricity represented the same expression of electrical energy.

Yount, Lisa. *Energy Supply.* New York: Facts On File, 2005. Thorough introduction to energy types, sources of energy, as well as U.S. energy history and issues, with additional coverage of global energy issues and the future of energy. Features a special emphasis on energy and the law, with reprints of U.S. legal documents related to energy issues. Provides specific tools for researching energy and energy supply, including biographies, a glossary, a timeline, and a bibliography.

ARTICLES AND PAPERS

Akins, James. "The Oil Crisis: This Time the Wolf Is Here." *Foreign Affairs,* April 1973, 473. Written by a U.S. State Department analyst and chief energy policymaker in response to the 1973–74 Arab oil embargo, recommends increased domestic production and improved relations with Middle Eastern oil-producing countries. Considered by many energy industry analysts to be a watershed article on U.S. oil supply.

Bartlett, Roscoe G. "Transiting to a New Paradigm." *Solar Today* 20 (March/April 2006): 27–29. Written by a U.S. congressional representative, suggests that the United States effect a transition to an economy in which energy needs have been reduced to a level that sustainable energy resources can easily meet and urges that the United States pass an energy bill to form an energy project with other countries to work on solutions to the problem of peak oil.

Campbell, C. J. "The Second Half of the Age of Oil Dawns." *Solar Today* 20 (March/April 2006): 21–23. Discusses the declining supply of oil and gas as cheap, convenient, and abundant sources of energy and proposes five strategies to help the United States plan and prepare for the need to balance its oil demand with reduced oil supply. Contains figures illustrating current oil and gas production and discovery data.

Kerr, Richard A. "USGS Optimistic on World Oil Prospects." *Science* 289 (July 14, 2000): 237. Examines the implications of a U.S. Geological Survey (USGS) 2000 finding that there are more untapped oil reserves in the world than previously thought. Includes data on the amounts of oil already produced and oil reserves yet to be uncovered in various regions of the world and the impact of these data for the United States.

WEB DOCUMENTS

Energy Information Administration. *Annual Energy Review 2004.* Energy Information Administration, August 15, 2005. Available online. URL: http://www.eia.doe.gov/emeu/aer/contents.html. Accessed on November 11, 2006. EIA's primary report on historical annual energy statistics, including data on total energy production, consumption, and trade; overviews of petroleum, natural gas, coal, electricity, nuclear energy, renewable energy, and international energy; and conversion tables.

Energy Information Administration. *Annual Energy Outlook 2006 with Projections to 2030.* Energy Information Administration, August 15, 2005. Available online. URL: http://www.eia.doe.gov/oiaf/aeo/index.html. Accessed on January 29, 2006. EIA's primary forecast and analysis of U.S. energy supply, demand, and prices through 2030.

Energy Information Administration. *International Energy Annual.* Energy Information Administration, July 2005. Available online. URL: http://www.eia.doe.gov/iea. Accessed on January 29, 2006. EIA's primary report on international energy statistics.

Energy Information Administration. *Monthly Energy Review.* Energy Information Administration, May 25, 2006. Available online. URL: http://www.eia.doe.gov/oiaf/aeo/index.html. Accessed on June 2, 2006. EIA's primary forecast and analysis of U.S. energy supply, demand, and prices for each month. Provides an overview for all types of fuels, sources, and sectors.

Helmore, Edward, Oliver Morgan, and Sarah Ryle. "Oil and U.S. Jitters Prompt Fears of Global Slowdown" *The Observer,* August 8, 2004. Available online. URL: http://observer.guardian.co.uk/business/story/0,,1278354,00.html. Accessed on January 29, 2006. Reviews a prediction by leading economists that the global economy is approaching a crisis, caused by reduced consumer confidence and spending in response to soaring oil prices, bleak U.S. unemployment rates, and a potential real estate crash both in the United States and across the Atlantic.

Lynch, Michael C. "Unconventional Oil: Filling in the Gap or Flooding the Market?" Department of Energy National Energy Modeling System, April 12, 2005. Available online. URL: http://www.eia.doe.gov/oiaf/aeo/conf/pdf/lynch.pdf. Accessed on January 29, 2006. Discusses the viability of alternative fuels.

States News Service. "President Bush Signs into Law a National Energy Plan." The White House, News & Policies, August 8, 2005. Available online. URL: http://www.whitehouse.gov/news/releases/2005/08/20050808-4.html. Accessed on January 29, 2006. Outlines the background of and some of the energy policies promoted by the national energy law adopted in the United States in August 2005, the Energy Policy Act of 2005.

U.S. Foreign Oil Dependence/Energy-Related Political Issues

BOOKS

Bloomfield, Lincoln P., Jr., and James A. Kelly, Jr., eds. *Global Markets and National Interests: The New Geopolitics of Energy, Capital, and Information.* Washington, D.C.: Center for Strategic and International Studies, 2002. Discusses the effects of globalization on U.S. foreign policy and energy politics.

Bronson, Rachel. *Thicker than Oil: America's Uneasy Partnership with Saudi Arabia.* New York: Oxford University Press, 2006. Draws on archival material, declassified documents, and interviews with Saudi and American officials to chronicle the relationship between Saudi Arabia and the United States. Concludes that the U.S. part in this relationship was less motivated by a need to secure oil supplies than by a desire to form a strategic political alliance to combat Soviet expansion.

Day, James McDonald. *What Every American Should Know about the Mid East and Oil.* Carson City, Nev.: Bridger House, 2001. Unfavorable commentary on the Middle Eastern policies of the administration of President George W. Bush, suggesting that such policies are unduly influenced by oil.

Everest, Larry. *Oil, Power and Empire: Iraq and the U.S. Global Agenda.* Monroe, Maine: Common Courage Press, 2003. Suggests that there was a relationship between oil interests and the 2003 U.S. invasion of Iraq.

Fitts, Catherine Austin, and Michael C. Ruppert. *Crossing the Rubicon: The Decline of the American Empire at the End of the Age of Oil.* Gabriola Island, Canada: New Society, 2004. Claims that "the war on terror," the U.S. response to the terrorist attacks on America on September 11, 2001, is an orchestrated move through which the United States can sidestep an emerging economic crisis caused by shrinking oil supplies. Suggests that military spending and foreign lending help the United States fund the oil and natural gas industries and continue to sustain economic growth.

Haley, James, ed. *At Issue: Foreign Oil Dependence.* San Diego, Calif.: Greenhaven Press, 2004. Through a compilation of essays and other writings by various authors, presents a debate about the extent of U.S. dependence on oil imports and possible ways of cutting back on these oil imports.

Kalicki, Jan H., and David L. Goldwyn. *Energy and Security: Toward a New Foreign Policy Strategy.* Washington, D.C.: Woodrow Wilson Center Press, 2005. Argues that U.S. foreign policy must incorporate a coherent energy policy.

Klare, Michael T. *Blood and Oil: The Dangers and Consequences of America's Growing Dependency on Imported Petroleum.* New York: Henry Holt, 2004. Suggests that American foreign policy is unduly motivated by dependency on oil, providing a historical overview of the U.S. formation of relationships with oil suppliers in dangerous and unstable parts of the world.

Parra, Francisco. *Oil Politics: A Modern History of Petroleum.* New York: I. B. Tauris, 2004. Commentary on the relationships between Middle Eastern governments, multinational oil companies, and Mideast oil–consuming nations and how these relationships have affected the international petroleum industry.

ARTICLES AND PAPERS

Morton, Peter. "U.S. Energy Industry in For a Jolt: U.S. Drives to Cut Mideast Oil Dependency," *National Post,* 25 October 2005, FP6. Reviews some of the policies that the United States has proposed for reducing its dependency on foreign oil, such as the Alaska National Wildlife Refuge (ANWR) plan.

The Tribune Co. "Foreign Fuel Subsidy Burns Taxpayers," *The Tampa Tribune,* 26 December 2005, 14. Defines biodiesel fuel and describes its benefits but laments the high tax cost associated with biodiesel subsidies.

Webber, Fed. "US Wants to Go It Alone." *Professional Engineering,* 25 May 2005, 39–40. Discusses tax breaks and incentives for automobile manufacturers and drivers related to vehicles that run on alternative fuels or on a combination of electricity and gasoline and the potential impact of these measures on reducing foreign oil dependence.

WEB DOCUMENTS

Shenk, Mark. "Oil May Fall for a 5th Week on U.S. Supply Gains, Survey Shows." Bloomberg News Service, February 17, 2006. Available online. URL: http://www.bloomberg.com/apps/news?pid=10000087&sid=aHvjNtXBNOwE&refer=top_world_news#. Accessed on February 18, 2006. Reports on impact of rising domestic stockpiles of crude oil and oil products. Reviews U.S. and international concern over Iranian nuclear energy program and details effects of the controversy on the U.S. energy market.

WORKS ABOUT ISSUES IN INTERNATIONAL ENERGY SUPPLY AND NATURAL RESOURCES

International Energy Affordability and Economics

BOOKS

Economides, Michael, and Ronald Oligney. *The Color of Oil: The History, the Money and the Politics of the World's Biggest Business.* Katy, Tex.: Round Oak, 2000. Historical overview of the foundations and growth of the oil industry and the role energy has played in the wars of the 20th and 21st centuries. Identifies the oil industry as the biggest business in the world, with energy the most important factor in the histories of countries around the world.

Mitchell, John, Koji Morita, Norman Selley, and Jonathan Stern. *The New Economy of Oil: Impacts on Business, Geopolitics and Society.* London: Earthscan, 2001. Examines the prospective loss of revenue and other consequences that may result in world governments as the global supply of oil, the world's primary energy source, begins to run out.

Tietenberg, Tom. *Environmental Economics and Policy.* 3d ed. Boston: Addison Wesley Longman, 2001. Provides an introduction to environmental economics and policy, with chapters on sustainable development, natural resource economics, and energy.

Yergin, Daniel. *The Prize: The Epic Quest for Oil, Money, and Power.* New York: Free Press, 1993. Winner of the 1992 Pulitzer Prize in nonfiction, chronicles the growth of the oil industry, from the ascent of John D. Rockefeller to the oil crises of the 1970s to the Persian Gulf War.

ARTICLES AND PAPERS

al-Husseini, Sadad. "Why Higher Oil Prices Are Inevitable This Year, Rest of Decade." *Oil & Gas Journal* (August 2, 2004): 14–18. Written by a former executive of Saudi Aramco, Saudi Arabia's national oil company, predicts that oil supply constraints will inevitably lead to higher oil prices through the first decade of the 21st century.

Ghouri, Salman Saif. "Correlation between Energy Usage and the Rate of Economic Development." *OPEC Review* 30 (2006): 41–54. Reviews the correlation between per capita GDP and per capita consumption of different sources of energy in a selection of countries, including OPEC countries. Estimates the ratios for total GDP and total energy consumption of different sources of energy. Concludes that most OPEC countries exhibit negative and weak ratios for all forms of energy, including electricity.

Simms, Andrew. "It's Time to Plug into Renewable Power." *New Scientist* 183 (July 3, 2004): 18. A critique of fossil fuel use that deals with the impact of fossil fuel dependency on economic losses, conflicts, and crises in both industrialized and developing countries. Promotes the use of renewable energies.

WEB DOCUMENTS

AFX News Service. "OPEC Tells European Countries to Cut Oil Taxes." Forbes.com, News & Analysis, November 20, 2005. Available online. URL: http://www.forbes.com/markets/feeds/afx/2005/11/20/afx2347009.html. Accessed on November 20, 2005. Offers information on the gathering of the International Energy Forum (IEF) in November 2005, the first major meeting of oil-producing and -consuming countries after oil prices reached a record high on August 30, 2005. Discusses oil-producing countries' opinion that leading oil consumer countries should cut taxes on oil to alleviate price hikes.

Harrison, Michael. "Shell Pays £83m Fine to Settle Scandal over Reserves." *The Independent,* July 30, 2004. Available online by subscription. URL: http://news.independent.co.uk/business/news/article49633.ece. Accessed on June 2, 2006. Describes an agreement by which the Shell Oil Company would pay a record 83 million British pounds in regulatory fines for misreporting its reserves figures. Explains the details of a continuing UK and U.S. investigation into the actions of the oil company.

Macalister, Terry. "Shell's Shame: FSA Spells Out Abuse." *The Guardian,* August 25, 2004. Available online. URL: http://business.guardian.co.uk/story/0,,1290141,00.html. Accessed on June 2, 2006. Reports that the United Kingdom's Financial Services Authority accused the Shell Oil Company of "unprecedented misconduct" in inflating oil and gas reserves figures in order to conceal unfavorable investment information.

Marquez, Humberto. "Oil: Consumer Worries Rise alongside Soaring Prices." IPS-Inter Press Service/Global Information Network, August 12, 2005. Available online. URL: http://www.redorbit.com/news/business/208907/oil_consumer_worries_rise_alongside_soaring_prices/index.html. Accessed on January 29, 2006. Surveys rising prices in the world oil market and details some of the factors that have contributed to the increase.

Seager, Ashley. "Oil Threat to World Economy." *The Guardian,* August 5, 2004. Available online. URL: http://www.guardian.co.uk/oil/story/0,11319,1276393,00.html. Accessed on January 29, 2006. Discusses sharp rises in the price of oil and petrol.

Seager, Ashley. "Petrol Prices Unlikely to Fall Again as OPEC Warns on Supply." *The Guardian,* August 4, 2004. Available online. URL: http://business.guardian.co.uk/story/0,3604,1275310,00.html. Accessed on January 29, 2006. Reports that U.S. light crude oil reached a record high price and discusses experts' predictions that the current spike may not be followed by reductions later, because of a reduced supply of oil in OPEC member countries.

Smith, David. "High Oil Prices Are Here to Stay." *The Sunday Times,* June 19, 2005. Available online. URL: http://www.redorbit.com/news/business/208907/oil_consumer_worries_rise_alongside_soaring_prices/index.html. Accessed on January 29, 2006. Discusses the outlook for oil prices as demand for oil continues to rise in China and non–Organisation for Economic Cooperation and Development (OECD) countries. Also presents various points of view regarding the theory of peak world oil.

INTERNATIONAL ENERGY-RELATED ENVIRONMENTAL ISSUES/ RENEWABLE AND ALTERNATIVE ENERGIES

BOOKS

Cothran, Helen, ed. *Opposing Viewpoints: Global Resources.* San Diego, Calif.: Greenhaven Press, 2003. A compilation of selections from articles, books, speeches, and other source documents that present opposing viewpoints on a variety of natural resources–related topics, from global oil reserves to genetically engineered foods to sustainable development.

Crane, R. G., L. Kump, and J. F. Kasting. *The Earth System.* Upper Saddle River, N.J.: Prentice Hall, 1999. Contains information on the impact of energy production and consumption on the environment, especially in chapter 3, "Global Energy Balance."

Deutch, John, and Richard K. Lester, eds. *Making Technology Work: Applications in Energy and the Environment.* New York: Cambridge University Press, 2003. Discusses the economic, political, social, and environmental factors that influence whether a new energy technology is practical enough for primary use.

Doig, Alison, Simon Dunnett, and Tim Jackson. *Energy for Rural Livelihoods: A Framework for Sustainable Decision Making.* Rugby, England: ITDG, 2006. Offers ways to evaluate and compare energy technology options and suggestions for ensuring access to sustainable energy in rural communities.

Egendorf, Laura K., ed. *Opposing Viewpoints: Energy Alternatives.* San Diego, Calif.: Greenhaven Press, 2006. A compilation of selections from articles, books, speeches, and other source documents that express opposing opinions on alternative energies—whether they are necessary, whether nuclear power is a viable form of them, which alternative energies should be pursued, among other issues.

Franz, Cecilia M., and Donald G. Kaufman. *Biosphere 2000 . . . Protecting Our Global Environment.* 3d ed. Dubuque, Iowa: Kendall/Hunt, 2000. A scientific textbook covering the fundamentals of biosphere and environmental studies; ecological principles; population, food, and energy concerns; global energy issues; fossil fuel energies; alternative energy sources; management of natural resources; and environmental protection.

Gipe, Paul. *Wind Energy Comes of Age.* Hoboken, N.J.: John Wiley & Sons, 1995. A comprehensive guide to wind power and the development of the wind power industry.

Gore, Al. *An Inconvenient Truth: The Planetary Emergence of Global Warming and What We Can Do About It.* Emmaus, Pa.: Rodale Press, 2006. Data, analysis, graphs, photographs, and figures representing facts and issues related to global warming, with suggestions about ways the world can begin to conserve energy and use fewer fossil fuels. Focuses mainly on the concept of global warming as a result of human energy use, not natural ecological processes.

Hayden, Howard C. *The Solar Fraud: Why Solar Energy Won't Run the World.* 2d ed. Pueblo West, Colo.: Values Lake, 2005. Written by a physics professor, identifies scientific properties of sunlight that greatly limit the ability of solar energy to provide more of the world's energy supply.

Henderson-Sellers, A., and P. J. Robinson. *Contemporary Climatology.* New York: J. Wiley & Sons, 1986. Features fundamental information on climate and the factors that affect it.

Hoffman, Peter, and Tom Harkin. *Tomorrow's Energy: Hydrogen, Fuel Cells, and the Prospects for a Cleaner Planet.* Cambridge, Mass.: The MIT Press, 2001. Primarily concerned with the potential of hydrogen and fuel cells to produce cleaner energy, chronicles efforts that have been made to develop and promote hydrogen and analyzes the successes and failures of these efforts.

Holland, H. D., and U. Peterson. *Living Dangerously: The Earth, Its Resources and the Environment.* Princeton, N.J.: Princeton University Press, 1995. Provides a large-scale overview of the effects of energy use on the environment as well as discussions of natural resources depletion.

Johansson, Thomas B., Henry Kelly, Amulya K. N. Reddy, and Robert H. Williams, eds. *Renewable Energy: Sources for Fuels and Electricity.* Washington, D.C.: Island Press, 1993. Comprehensive reference guide to all the various forms of renewable energy and their applications, benefits, limitations, economic and environmental effects, and emerging technologies.

Kursunogammalu, Behram N., Stephan L. Mintz, and Arnold Perlmutter, eds. *Global Warming and Energy Policy.* Dordrecht, The Netherlands: Springer, 2001. Proceedings of a symposium on global warming and energy policy. Explores the environmental concerns surrounding fossil fuel use, including global warming as a result of carbon dioxide emissions and health hazards caused by pollution caused by combustion products. Offers ideas about what can be done to rectify the problems.

Lavorgna, Gary. *Environment: Issues and Alternatives.* Dubuque, Iowa: Kendall/Hunt, 1998. Explores the history of interactions between human beings and their environment, with particular emphasis on fundamentals of ecology, climate, and natural resources. Focuses mainly on human consumption of natural resources for energy and the effects of this consumption.

Moench, Mel. *Encyclopedia of Renewable Energy for Home and Auto.* Buffalo, Minn.: Osprey Press, 2006. Exhaustive guide to renewable energy sources, including alternatively fueled vehicles and new and emerging technologies. Emphasizes the use of alternative energy in the home.

Pringle, Laurence. *Global Warming.* Des Moines, Iowa: Perfection Learning, 2001. An introduction to the threat of global warming, how it began, the problems it has caused, and theories about how it might be reversed. Features a strong focus on conservationism.

Sørenson, Bent. *Hydrogen and Fuel Cells: Emerging Technologies and Applications.* Burlington, Mass.: Elsevier Academic Press, 2005. Explains the techniques associated with the production, conversion, and setup of utilizing energy from hydrogen fuel cells. Also discusses the economic, social, and environmental implications of the use of hydrogen fuel cells.

Weyant, John. *Energy and Environmental Policy Modeling.* Boston: Kluwer Academic, 1999. Scholarly work containing mathematical models for the environmental aspects of energy development as well as for environmental policies designed to address energy's impact on the environment.

Winteringham, F. Peter W. *Energy Use and the Environment.* Chelsea, Mich.: Lewis, 1992. Offers an overview of the fundamentals of energy production and consumption and their effects on the environment.

ARTICLES AND PAPERS

Bailis, Robert, et al. "Energy Management and Global Health." *Annual Review of Environment and Resources* 29 (2004): 181–204. Reviews current research on the impact of energy use and energy technologies on public health. Emphasizes the risks associated with indoor and outdoor air pollution from energy use. Also explores the link between the local and global environmental health effects of energy use. Suggests that public policy needs to focus more closely on energy-associated health risks.

Christiansen, Atle Christer, and John Birger Skjærseth. "Environmental Policy Instruments and Technological Change in the Energy Sector: Findings from Comparative Empirical Research." *Energy and Environment* 17 (2006): 223–241. Explores the extent to which and the ways in which environmental policy has affected the institution of environmentally oriented technological changes in the energy sector.

Connolly, Sarah, and Dara O'Rourke. "Just Oil? The Distribution of Environmental and Social Impacts of Oil Production and Consumption." *Annual Review of Environment and Resources* 28 (2003): 1–23. Presents data on the impact of oil production and consumption in different parts of the world. Analyzes the environmental, social, and health effects of oil extraction, transport, refining, and consumption. Emphasizes the way these effects are distributed among various socioeconomic and ethnic groups.

Demrba, Ayhan. "Fuel Conversion Aspects of Palm Oil and Sunflower Oil." *Energy Sources* 25 (2003): 457–466. An interesting comparison of the scientific properties of palm oil and palm kernel oil that may render these oils superior to other renewable and clean engine fuel alternatives to fossil fuels.

Ding, J., et al. "Hydrogen Production by Fermentation: Review of a New Approach to Environmentally Safe Energy Production." *Aquatic Ecosystem Health Management* 9 (2006): 39–42. Explores a novel method in hydrogen production technology: production of hydrogen through the fermentation of bacteria in organic wastewater.

Doughton, Sandi. "The Truth about Global Warming," *The Seattle Times,* 9 October 2005, A1. A special report on the scientific evidence of global warming and its human causes.

Glasby, Geoffrey. "Drastic Reductions in Utilizable Fossil Fuel Reserves: An Environmental Imperative." *Environment, Development and Sustainability* 8 (2006): 197–215. Cautions that use of even less than 20 percent of global hydrocarbon reserves will push the concentration of carbon dioxide in the atmosphere to a level that will cause a massive environmental catastrophe. Urges an immediate, large-scale program for carbon sequestration.

Langenkamp, R. Dobie, and Rex J. Zedalis. "International Comparisons of Energy Use and the Environment: Does It Make Sense to Call on Wind Power?" *European Environmental Law Review* 15 (2006): 162–174. Examines the comparative energy use levels, and corresponding contributions to global carbon dioxide emissions, of the world's major energy-consuming regions. Argues that the direct benefit of electrical energy produced by wind power is a reduction in greenhouse gas emissions, but notes that the limitations of wind power must be carefully and candidly addressed by governments before any major wind power program can be deployed.

Mills, David. "Renewable Energy Capability vs. Climate Necessity." *Bulletin of Science, Technology, and Society* 26 (2006): 78–83. Claims that current proposed targets for reducing global carbon dioxide emissions may need to be reduced further and that aggressive energy efficiency and solar electricity are the most practical means for accomplishing this goal.

Nordell, Bo. "Thermal Pollution Causes Global Warming." *Global and Planetary Change* 38 (2003): 305–312. A scientific study concluding that since 1880 the global use of nonrenewable energy sources has resulted in an imbalance between solar heat and geothermal heat (heat reemitted from the Earth back into the atmosphere), causing excess heat in the global environment as a whole.

Scott, Karen N. "Tilting at Offshore Windmills: Regulating Wind Farm Development within the Renewable Energy Zone." *Journal of Environmental Law* 18 (2006): 89–188. Explores the need for a legislative and policy framework to protect and conserve the marine environment where installation of offshore wind farms is concerned.

WEB DOCUMENTS

Diehl, Sarah J., and James Clay Moltz. *Nuclear Weapons and Nonproliferation: A Reference Handbook.* Santa Barbara, Calif.: ABC-CLIO, 2000. E-book. Available online at http://www.abc-clio.com. Offers general information pertaining to nuclear energy and the nuclear energy industry, including an introductory overview of the science, technology, politics, and culture behind nuclear energy production. Accessed January 29, 2006.

International Energy-Related Infrastructure Issues

BOOKS

Barton, Barry, Lila K. Barrera-Hernández, Alastair R. Lucas, and Anita Rønne, eds. *Regulating Energy and Natural Resources.* Oxford: Oxford University Press, 2006. Examines the impact of democratization, globalization, and environmental awareness on governmental regulation of the energy industry and the use of natural resources.

Bradbrook, Adrian, and Rosemary Lyster, eds. *Energy Law and the Environment.* Cambridge: Cambridge University Press, 2005. A compilation of essays from energy industry analysts from across the globe, including the United States and China, about the role of law in sustainable development. Explores energy and environmental laws and legal issues amid current practices in worldwide energy production that have led to environmental problems.

Milani, Brian. *Designing the Green Economy.* Lanham, Md.: Rowman & Littlefield, 2000. Argues that sustainability, social justice, and economic stability cannot be ensured if an economy is not rebuilt on ecological principles. Investigates the manufacturing, energy, and financial sectors of the economy for evidence of the capacity for this restructuring.

Stein, Matthew. *When Technology Fails: A Manual for Self-Reliance and Planetary Survival.* Sante Fe, N.Mex.: Clear Light Books, 2000. Based on the notion that the world is rapidly depleting its supply of fossil fuels and other natural resources while becoming increasingly dependent on a high-technology infrastructure for meeting its needs, offers tips and instructions for sustainable, self-reliant living in the event of a technology breakdown or emergency resource shortage.

ARTICLES AND PAPERS

Agarwal, Sachin G., Douglas F. Barnes, and R. Anil Cabraal. "Productive Uses of Energy for Rural Development." *Annual Review of Environment and Resources* 30 (2005): 39–74. Highlights the impact of energy services on education, health, and gender equality and suggests that a refined understanding of the far-reaching effects of energy use is needed to shape public policy decisions.

Annette, Jenny, José Raúi Díaz López, and Hans Joachim Mosler. "Household Energy Use Patterns and Social Organisation for Optimal Energy Management in a Multiuser Solar Energy System." *Progress in Photovoltaics* 14 (2006): 353–362. Describes the successful implementation of a community-controlled energy distribution system: a central photovoltaic system in a Cuban village managed collaboratively by the residents there.

Farrell, Alexander E., Hadi Dowlatabadi, and Hisham Zerriffi. "Energy Infrastructure and Security." *Annual Review of Environment and Resources* 29 (2004): 181–204. Discusses how, in an age of terrorist-related attacks, energy infrastructure and security have become related, how this relationship differs from energy security issues of the past, and what this may mean for private and public policy decisions.

Tambo, N. "Technology in the High Entropy World." *Water Science and Technology* 53 (2006): 1–8. Describes solar radiation as the basic energy source that drives the global ecosystem. Proposes a high-entropy solar power concept for driving the operations of water and sanitation systems.

Watts, Michael. "Righteous Oil? Human Rights, the Oil Complex, and Corporate Social Responsibility." *Annual Review of Environment and Resources* 30 (2005): 39–74. Reviews oil companies' efforts to develop corporate social responsibility. Also surveys the various regulatory systems put in place by governments and agencies to ensure that the oil industry complies with human, social, political, and environmental rights, especially in less developed countries.

International Energy Supplies, Crises, and Shortfalls

BOOKS

Adelman, M. A. *The Economics of Petroleum Supply: Papers by M. A. Adelman, 1963–1993.* Cambridge, Mass.: MIT Press, 1993. Anthology of works about the depletion of natural resources and the monopoly on world oil written by a prominent authority on energy resource economics. Covers the theory and measurement of the depletion and scarcity of fossil fuel resources, the control of the oil market by the OPEC cartel, and problems and options in public policy.

Brown, Charles E. *World Energy Resources.* New York: Springer Verlag/Telos Press, 2002. Provides an in-depth, country-by-country analysis of global energy resources and international energy markets in the past, present, and future. The book includes discussion of transportation, electricity, and environmental issues.

Campbell, C. J. *Oil Crisis.* Essex, England: Multi-Science, 2005. Written by a former petroleum geologist and the current president of the Association for the Study of

Peak Oil, argues that the world is horribly unprepared for the consequences of a worldwide oil shortage.

Campbell, C. J. *The Coming Oil Crisis.* Essex, England: Multi-Science, 2004. Written by a former petroleum geologist and the current president of the Association for the Study of Peak Oil, concludes that world oil production will reach its peak in the first decade of the 21st century and that grave economic and political consequences will result as nations around the world experience energy scarcity and sustained increases in the price of oil.

Cleveland, Cutler J., ed. *Encyclopedia of Energy.* St. Louis, Mo.: Elsevier Science, 2004. Reference guide about energy supply that covers all the scientific, social, and environmental dimensions of energy. Written by energy experts.

Cordesman, Anthony H., and Khalid R. Al-Rodhan. *The Global Oil Market: Risks and Uncertainties.* Washington, D.C.: Center for Strategic and International Studies, 2006. Studies major oil-producing regions of the world (the Middle East, Africa, Asia and the Pacific, Europe and Eurasia, North America, and South and Central America) and outlines national oil developments in these areas, focusing on risks and uncertainties such as economic instability, geopolitical risks, oil production uncertainties, and the unpredictable nature of resources.

Datamonitor. *Global Energy: Industry Profile.* London: Datamonitor, 2006. Summary of facts and data related to the energy industry, including profiles of leading energy companies, a survey of energy market values, a report on market segmentation, statistics on market shares, in-depth look at the competitive climate, and business forecasts.

Debeir, Jean-Claude, Jean-Paul Deléage, and Daniel Hémery. *In the Servitude of Power: Energy and Civilization through the Ages.* London: Zed Books, 1990. A comprehensive chronology of energy use throughout the world, with descriptions of the sociological, cultural, ecological, and economic repercussions of technological developments in energy, energy crises, and revolutions in the types of resources chosen by various civilizations as primary sources of energy.

Deffeyes, Kenneth S. *Beyond Oil: The View from Hubbert's Peak.* New York: Hill & Wang, 2005. Based on the theories of M. King Hubbert, who predicted that world oil supply would peak in 2000, proposes that oil supply will peak at the end of 2005. Explores the physics of retrieving oil as well as the grim possibilities for the world's economies and societies when oil is absent from the energy picture.

Doyle, Jack. *Riding the Dragon: Royal Dutch/Shell and the Fossil Fire.* Monroe, Maine: Common Courage Press, 2004. Discusses the health and environmental effects of the multinational oil company's projects around the world.

Dupler, Douglas. *Energy: Is There Enough?* Detroit: Information Plus, 2001. Reference book examines range of energy sources and their contribution to the world's energy supply.

The Economist Intelligence Unit. *Energy and Electricity Forecast: World.* London: The Economist Intelligence Unit, 2005. Provides five-year energy forecasts and energy market profiles for each of 60 countries around the world.

Franchi, John R. *Energy in the 21st Century.* Toh Tuck Link, Singapore: World Scientific, 2005. Provides a history of human energy use, a discussion of the energy options available today, and an overview of environmental and economic implications of energy technology developments that have potential in primary energy use of the future.

Goldemberg, Jose, ed. *World Energy Assessment: Energy and the Challenge of Sustainability.* New York: United Nations Press, 2001. Examines ways to encourage energy sustainability by pursuing specific energy policies. Also studies energy-related economic, social, safety, and environmental issues.

Goodstein, David. *Out of Gas: The End of the Age of Oil.* New York: W. W. Norton, 2004. Applies the peak oil theory of the renowned geologist M. King Hubbert to predict that world oil production will reach its peak within the decade and that world oil reserves will be completely exhausted within the century. Explores the possibilities for civilization's future energy choices once oil is depleted.

Heinberg, Richard. *Powerdown: Options and Actions for a Post-Carbon World.* Gabriola Island, Canada: 2004. Puts forward four likely scenarios of the impact of oil and natural gas depletion on industrial societies. Calls for a reduction in resource usage among wealthy countries, the development of alternative energy sources, and more equitable distribution of resources, among other measures, to combat a future energy crisis.

International Energy Agency. *Energy: The Next Fifty Years.* Paris: Organisation for Economic Cooperation and Development (OECD), 2000. From the proceedings of an OECD conference about the outlook of world energy use, forecasts the challenges that the next 50 years is likely to introduce to the technological, economic, environmental, and geopolitical dimensions of world energy consumption and production.

Jaccard, Marc. *Sustainable Fossil Fuels: The Unusual Suspect in the Quest for Clean and Enduring Energy.* Cambridge: Cambridge University Press, 2006. Acknowledges that oil and gas are in limited supply but points to energy efficiency technologies that may possibly extend fossil fuels' availability indefinitely. Argues that modifying the *way* we use fossil fuels is an essential part of any plan to achieve energy sustainability. Especially highlights coal's potential for generating hydrogen, a clean fuel.

Klare, Michael T. *Resource Wars: The New Landscape of Global Conflict.* New York: Henry Holt, 2002. Contends that international power struggles over petroleum, water, timber, and other natural resources will increasingly drive international politics and increase military activities in the coming years.

Leggett, Jeremy. *The Empty Tank: Oil, Gas, Hot Air, and the Coming Global Financial Catastrophe.* New York: Random House, 2005. Studies the world's dependency on oil as a primary energy source and the factors that have caused and continue to sustain this dependency. Predicts that the effects of waning oil supply will have a devastating impact on financial markets and eventually cause an international economic collapse. Also deals with environmental concerns related to fossil fuel consumption.

Leone, Daniel A., ed. *At Issue: Is the World Heading toward an Energy Crisis?* San Diego, Calif.: Greenhaven Press, 2005. Through a compilation of essays and other writings by various authors, presents a debate about how many natural resources are left and the policies that should be pursued to prevent an energy crisis.

Lynch, Michael C. *Oil Supply Security: Does the Song Remain the Same?* Boulder, Colo.: International Research Center for Energy and Economic Development, 2004. Analysis of the world oil and gas market, focusing on the errors inherent in petroleum supply forecasting and refuting the claim that an international oil crisis is on the horizon. Supports the view that oil reserves rise and fall in such a way that sweeping predictions of scarcity cannot be validated.

Marcel, Valerie, and John V. Mitchell. *Oil Titans.* Washington, D.C.: Brookings Institution Press, 2006. Studies five national oil companies of the Middle East: Saudi Aramco, the Kuwait Petroleum Corporation, the National Iranian Oil Company, Sonatrach of Algeria, and the Abu Dhabi National Oil Company. Explains the complex relationship between each state company and its oil company.

Maugeri, Leonardo. *The Age of Oil: The Mythology, History, and Future of the World's Most Controversial Resource.* New York: Praeger, 2006. Written by an executive of an Italian energy company that is the sixth-largest publicly listed oil company in the world, describes controversial events in the history of the oil industry and explains the fundamentals of oil production.

McKillop, Andrew, ed. *The Final Energy Crisis.* London: Pluto Press, 2005. Anthology of internationally authored essays on the consequences of a fossil fuel shortage on aspects of everyday life, from mass-produced food to clothing to international travel to automotive transportation. Offers global perspectives on the prospective dangers of the overconsumption of oil, gas, and coal and weighs the costs of alternatives, such as nuclear power.

McVeigh, J. C., and J. G. Mordue, eds. *Energy Demand and Planning.* New York: Brunner-Routledge 1998. Compilation of essays on issues that world energy policies should address over the next 50 years, such as global warming, population growth, and sustainable development.

Miller, E. Willard, and Ruby M. Miller. *Contemporary World Issues: Energy and American Society.* Santa Barbara, Calif.: ABC-CLIO, 1993. Provides an overview of the state of U.S. energy resource supplies and consumption, with a chronology, biographical sketches, statistical information, and primary-source material.

Odell, Peter R. *Oil and Gas: Crises and Controversies, 1961–2000.* Vol. 1, *Global Issues.* Brentwood, England: Multi-Science, 2001. A series of essays written by Peter Odell from 1961 to 2000 that challenge the widely held notion that there is a scarcity of oil supplies. Cites multiple facts and data regarding technological developments in oil recovery, fluctuations in reserves figures, and steady increases in production figures worldwide.

Ramage, Janet. *Energy: A Guidebook.* London: Oxford University Press, 1983. Offers a general overview of fundamental energy principles and laws and relates them to world energy production and consumption.

Shah, Sonia. *Crude: The Story of Oil.* New York: Seven Stories Press, 2004. Explains the complex geopolitical history behind oil and how this fossil fuel has managed to transform the world.

Simmons, Matthew R. *Twilight in the Desert: The Coming Saudi Oil Shock and the World Economy.* Hoboken, N.J.: John Wiley & Sons, 2005. Provides a historical overview of the U.S.-Saudi Arabia relationship and U.S. reliance on Saudi oil. Highlights discrepancies between Saudi Arabia's actual production capabilities and the reserves claims that it reports. Suggests that the near future hold the prospect of depleted Saudi Arabian oil resources and the increased oil scarcity will have significant consequences around the world, including higher costs of oil.

Smil, Vaclav. *Energy at the Crossroads: Global Perspectives and Uncertainties.* Cambridge, Mass.: MIT Press, 2003. Explores the effect of energy issues on the U.S. economy, quality of life, environment, and political conflict. Discusses the inadequacies of energy forecasting. Examines the pros and cons of fossil fuels, hydropower, biomass energy, wind power, and solar power.

Tertzakian, Peter. *A Thousand Barrels a Second: The Coming Oil Break Point and the Challenges Facing an Energy Dependent World.* New York: McGraw-Hill, 2006. Predicts that the world of the "fossil fuel age" is heading toward an energy "breakpoint," a crisis in which growth in oil demand will outpace supply. Proposes that there is no "magic bullet" solution, only the possibility of taking gradual, incremental steps toward conservation and sustainability.

Vaitheeswaran, Vijay V. *Power to the People: How the Coming Energy Revolution Will Transform an Industry, Change Our Lives, and Maybe Even Save the Planet.* New York: Farrar, Straus, 2003. Discusses today's energy problems, such as reliance on oil, global warming, air pollution, and nuclear energy dangers, and calls for liberalization of energy markets, greater environmental awareness, and innovations in hydrogen fuel-cell technology.

ARTICLES AND PAPERS

Bicker, Alan, Johan Pottier, Paul Sillitoe, and Thomas F. Thornton. "Development and Local Knowledge: New Approaches to Issues in Natural Resources Management, Conservation, and Agriculture." *Journal of Anthropological Research* 62 (2006): 115–166. Identifies a method for managing and conserving resources that utilizes a local community–based approach.

Brennand, Garry, Mohamed Hamel, and Adnan Shihab-Eldin. "Oil Outlook to 2020." *OPEC Review* 27 (2003): 79–128. Applies OPEC's World Energy Model, which assumes a world economic growth averaging 3.3 percent annually, to project oil demand and supply to 2020. Predicts that world oil demand will rise from 76 million barrels a day in 2000 to 89 million barrels a day in 2010 and 107 million barrels a day in 2020, with more than three-quarters of the increase from developing countries. Also suggests that non-OPEC output is expected to grow by 2010 but stabilize thereafter.

Haavind, Robert. "The Future of Energy." *Solid State Technology* 48 (11): 12. Editorial addressing the question of whether the world is running out of and what the world is doing to combat a future shortage of oil. Presents informative international energy data related to the topic.

Laherrère, Jean. "Future Oil Supplies." *Energy Exploration and Exploitation* 21 (2003): 227–267. Reviews various formulae that have been used to determine figures representing global oil reserves. Discusses the lack of consensus on these figures and the faults inherent in each formula. Uses a mathematical model based on past and current trends, human behavior, oil-field declines, and economic criteria to project future global oil reserves.

Rosen, Joe. "Conservation Law." *Encyclopedia of Physics.* New York: Facts On File. Available online at *Science Online.* URL: www.fofweb.com. Accessed January 29, 2006. Provides a scientific definition of *conservation* in terms of laws of energy.

Volti, Rudi. "Energy Efficiency." *The Facts On File Encyclopedia of Science, Technology, and Society.* New York: Facts On File, 1999. Available online at *Science Online.* URL: www.fofweb.com. Accessed January 29, 2006. An overview of the concept of energy efficiency and its relevance in assessing the usefulness of an energy resource.

WEB DOCUMENTS

British Petroleum. *BP Statistical Review of World Energy 2006.* British Petroleum, June 2006. Available online. URL: http://www.bp.com/genericsection.do?categoryId=92&contentId=7005893. Accessed on January 29, 2007. Comprehensive data on the energy industry, with figures on production, consumption, reserves, and projections, all organized by fuel type, from one of the world's largest energy companies.

Energy Information Administration. *Country Analysis Briefs.* Available online. URL: http://www.eia.doe.gov/emeu/cabs/contents.html. Accessed on February 17, 2006. EIA's energy overviews for various countries. In addition to comprehensive energy statistics, provides a broad overview of the country and its socioeconomic and geopolitical climate. Includes briefs on the United States, China, Germany, Iran, Saudi Arabia, Nigeria, and Venezuela.

International Energy Agency. *Key World Energy Statistics,* 2006 edition. International Energy Agency, The Online Bookshop, 2006. Available online. URL: http://www.iea.org/w/bookshop/add.aspx?id=144. Accessed on March 17, 2007. Data and statistics covering the supply, transformation, and consumption of all the major energy sources around the world.

Santa Barbara, Jack. "Peak Oil, Energy Futures and Violent Conflict." *Peace Magazine* 21 (October 21, 2005): 12. Available online. URL: http://www.peacemagazine.org/archive/v21n4p12.htm. Accessed on June 2, 2006. Describes the implications of peak oil theories for the world and the effect of waning global oil reserves on relations between countries.

OTHER MEDIA

Skrebowski, Chris. "Joining the Dots." Speech presented at the 2004 Energy Institute Conference: *Oil Depletion: No Problem, Concern or Crisis?* London: November 10, 2004. Available as audio recording by contacting info@energyinst.org.uk. Outlines challenges that confront the international community in regard to a peak in world oil.

International Energy-Related Political Issues

BOOKS

Claes, Dag Harald. *The Politics of Oil-Producer Cooperation.* Boulder, Colo.: Westview Press, 2001. Study of the international politics behind the activities of the world oil market.

Ebel, Robert, and Rajan Menon, eds. *Energy and Conflict in Central Asia and the Caucasus.* Lanham, Md.: Rowman & Littlefield, 2000. Examines the interplay between security and energy development in the Caspian Sea region. Describes new findings related to competition for energy resources, political and economic development in the region's states, and a propensity for conflict in the area.

Mommer, Bernard, and Ali Rodriguez Araque. *Global Oil and the Nation State.* New York: Oxford University Press, 2002. Studies the relationships among government, politics, and the economy and oil producers, energy investors, and energy consumers.

Noreng, Oystein. *Crude Power: Politics and the Oil Market.* New York: I. B. Tauris, 2002. Explores the world's dependency on oil from the politically volatile region of the Middle East.

ARTICLES AND PAPERS

Glaser, Alexander. "On the Proliferation of Uranium Fuel for Research Reactors at Various Enrichment Levels." *Science and Global Security* 14 (2006): 1–2. Reviews the rationale for limiting the enrichment level of uranium for nuclear research reactor fuel in order to minimize proliferation risks.

Roberts, Paul. "Crude Awakening Indeed to Possible Oil Conflicts," *Times Union,* 2 July 2004, B8. Discusses the international race for energy security as the state of maintaining a level of fuel and electricity production and affordability that will secure a country's economic stability and political power for generations.

Smith, Benjamin. "Oil Wealth and Regime Survival in the Developing World, 1960–1999." *American Journal of Political Science* 48 (April 2004): 232–246. Examines a debate regarding the extent to which a country's oil supply bolsters that country's political power or undermines it. Assesses the effects of oil wealth on regime survival, political unrest, and civil wars. Based on case studies of 105 countries, including China, Iran, Nigeria, Saudi Arabia, and Venezuela.

WORKS ABOUT ISSUES IN ENERGY SUPPLY AND NATURAL RESOURCES IN CHINA

BOOKS

Akiyama, Takamasa, M. Hashem Pesaran, and Ron P. Smith. *Energy Demand in Asian Developing Economies.* Oxford: Oxford University Press, 1998. Developed as part of a World Bank research project on energy demand in Asian developing countries, analyzes the environmental problems, policy needs, and patterns of growth involved in energy demand in developing countries such as China.

Cummings, Sally N., ed. *Oil, Transition and Security in Central Asia.* Richmond, England: Curzon Press, 2003. Analyzes China's energy industry since 1991.

Downs, Erica Strecker. *China's Quest for Energy Security.* Santa Monica, Calif.: Rand Corporation, 2000. Describes steps China is taking to achieve energy security and the motivation behind these actions. Considers China's investment in overseas oil projects and increasing dependency on energy imports.

Kleveman, Lutz. *The New Great Game: Blood and Oil in Central Asia.* New York: Atlantic Monthly Press, 2003. Argues that China is engaged in a power struggle with the United States, Russia, and Iran for control of the Caspian Sea region's vast oil and gas reserves.

Manning, Robert A. *The Asian Energy Factor: Myths and Dilemmas of Energy, Security and the Pacific Future.* New York: Palgrave, 2000. Explores the challenges that the Asia-Pacific region's soaring energy consumption presents to the world, national security, and Asian culture. Includes a case study of China and its search for energy policies that adequately address its energy-related environmental issues and precarious domestic energy supply.

National Research Council, Chinese Academy of Sciences, and Chinese Academy of Engineering. *Cooperation in the Energy Futures of China and the United States.* Washington, D.C.: National Academies Press, 2000. Discusses the connection between the United States and China related to energy consumption and production and evaluates the environmental impact of growing energy demand in both countries.

ARTICLES AND PAPERS

Andrews-Speed, Philip, and Sergei Vinogradov. "China's Involvement in Central Asian Petroleum: Convergent or Divergent Interests?" *Asian Survey* 40 (March–April 2000): 377–397. Assesses China's actions in expanding its source of energy supply. Presents an overview of China's foreign and energy policies and possible conflicts of interest involved in China's extensive international investment in petroleum in Central Asia.

Totten, Michael. "In China, a Climate for Renewables." *Solar Today* 20 (March/April 2006): 33. Reviews the components of China's Renewable Energy Law and analyzes what the United States can learn from them to grow toward more sustainable energy use.

Wallace, William, and Zhongying Wang. "Solar Energy in China: Development Trends for Solar Water Heaters and Photovoltaics in the Urban Environment." *Bulletin of Science, Technology, and Society* 26 (2006): 135–140. Presents trends and requirements for developing solar urban applications in China, currently the world's largest market for solar water heating systems.

WEB DOCUMENTS

The Economist Newspapers Ltd. "No Questions Asked: China and Africa." *The Economist,* January 21, 2006. Available online. URL: http://economist.com/world/asia/

displaystory.cfm?story_id=5425730. Accessed on January 29, 2006. Discusses Western misgivings about China's trade agreements with Africa, by which China has made large-scale investments in African infrastructure projects in exchange for airtight contracts to secure oil from Africa's energy sector—all in spite of the fact of human rights abuses and corruption in Africa.

OTHER MEDIA

Mergent, Inc. *Mergent Industry Reports: Oil and Gas—Asia Pacific.* Charlotte, N.C.: Mergent, 2005. Available as a CD-ROM or PDF file through www.mergent.com. Examines the activities of the Asian-Pacific oil and gas sector and global and regional influences on various industry segments. Features a profile of the energy market in China.

WORKS ABOUT ISSUES IN GERMANY'S ENERGY SUPPLY AND NATURAL RESOURCES

ARTICLES AND PAPERS

Platts. "In Germany, Sweeping Renewables Law Set to Come into Force Next Month." *Renewable Energy Report,* 1 July 2004, 4. Provides details on the incentives offered by Germany's Renewable Energy Sources Act to encourage the development of biomass, geothermal, solar, and offshore wind energy.

Platts. "Renewable Energy's Contribution to Nation's Energy Mix Accelerates." *Renewable Energy Report* 94 (December 12, 2005): 24. Presents a forecast on German renewable energy use that predicts that renewable energy's contribution to Germany's electricity consumption will reach 18.2 percent by 2011.

Platts. "Study of German Exports Finds Obstacles Persist for Renewable Energy Companies." *Renewable Energy Report* 88 (September 19, 2005): 6. Reports difficulties within the renewable energy industry in Germany in selling renewable technologies overseas due to insufficient renewable energy infrastructures in other nations.

Platts. "Wind Power Tops Hydro." *EU Energy* 90 (September 10, 2004): 15. Reports that for the first time 10 percent of Germany's gross power consumption was from renewable energy sources and that more than half of this was supplied by wind power.

WEB DOCUMENTS

AII Data Processing. "Solar Energy Costs to Reach Nonrenewables in 15 Years—ISE." *German News Digest,* May 6, 2004. Available online. URL: http://www.aiidatapro.com. Accessed on January 29, 2006. Discusses Germany's adoption of the Renewable Energy Sources Act and its stimulation of growth in power generation from renewables, particularly in the installation and operation of solar power plants.

WORKS ABOUT ISSUES IN ENERGY SUPPLY AND NATURAL RESOURCES IN IRAN AND SAUDI ARABIA

Iran

BOOKS

Farmanfarmaian, Manucher, and Roxane Farmanfarmaian. *Blood and Oil : A Prince's Memoir of Iran, from the Shah to the Ayatollah.* New York: Random House, 2005. Written by a former Iranian prince and a director of the National Iranian Oil Company, describes the emergence of Middle East oil industry from British domination and the beginnings of the partnership that led to the formation of OPEC.

Peimani, Hooman. *The Caspian Pipeline Dilemma: Political Games and Economic Losses.* New York: Praeger, 2001. Discusses the United States' policy of isolation toward Iran and the impact of this policy on oil supply.

ARTICLES AND PAPERS

Eckel, Mike. "Iran, Georgia Strike Gas Deal amid Georgia Energy Crisis," Associated Press, International News, 28 January 2006. Reports on a deal between Iran and the former Soviet republic Georgia involving Iran's offer of emergency natural gas supplies to Georgia.

Heiskanen, Veijo. "Oil Platforms: Lessons of Dissensus." *Nordic Journal of International Law* 72 (2005): 179–208. Analyzes disagreements among the judges in the International Court of Justice's decision on a case brought against the United States by Iran.

Iran Petroleum Magazine. "Zanganeh stresses increased recovery, collecting associate gas." *Iran Petroleum Magazine* 24, December 2005. Available online. URL: http://www.iranpetroleummag.com. Accessed on January 29, 2006. Describes an announcement by Iran's minister of petroleum about Iranian plans to boost oil and gas production and improve oil and gas recovery concerning the use of oil platforms.

WEB DOCUMENTS

Associated Press. "Oil Prices Jump on Iranian Leader's Threat." *Houston Chronicle,* June 5, 2006. Available online. URL: http://www.chron.com/disp/story.mpl/ap/fn/3932444.html. Accessed on June 5, 2006. Reports that oil prices jumped to $73 a barrel after the announcement by Iran's supreme leader, Ayatollah Ali Khamenei, threatened that the world's oil supply would be jeopardized if Western nations punished Iran for its nuclear program.

Fletcher, Sam. "Hopes for Iranian Settlement Cut Energy Prices." *Oil & Gas Journal,* Market Watch, June 8, 2006. Available online. URL: http://www.ogj.com/articles/article_display.cfm?article_id=257179. Accessed on June 8, 2006. Reports that Iran's announcement of its willingness to negotiate a plan for delaying its nuclear energy program until the United Nations (UN) can determine whether the intent of this program is peaceful had a positive impact on the price of crude oil worldwide.

Murphy, Brian. "Iranians Are Awash in Crude, but Gasoline Is Another Story." *Houston Chronicle,* June 6, 2006, 2. Available online. URL: http://www.chron.com/CDA/archives/archive.mpl?id=2006_4128531. Accessed on June 4, 2006. Discusses Iran's refinery shortage, which has left it heavily dependent on gas imports. Suggests that this shortage is a possible reason that Iran is eager to avoid sanctions due to its nuclear energy program.

Saudi Arabia

BOOKS

Bronson, Rachel. *Thicker than Oil: America's Uneasy Partnership with Saudi Arabia.* New York: Oxford University Press, 2006. Draws on archival material, declassified documents, and interviews with Saudi and American officials to chronicle the relationship between Saudi Arabia and the United States. Concludes that the U.S. part in this relationship was less motivated by a need to secure oil supplies than by a desire to form a strategic political alliance to combat Soviet expansion.

Brown, Anthony Cave. *Oil, God, and Gold: The Story of Aramco and the Saudi Kings.* Boston: Houghton Mifflin, 1999. Provides history on the relationships of the Saudi royal family, multinational oil companies, and the American and British governments since the 1930s. Offers details from documents from the Saudi national oil company Aramco.

Citino, Nathan J. *From Arab Nationalism to OPEC: Eisenhower, King Sa'ud, and the Making of U.S.-Saudi Relations.* Bloomington: Indiana University Press, 2002. Analyzes the ties established between President Dwight Eisenhower and the founder of Saudi Arabia after World War II, the factors that led the United States to replace Britain as leading foreign power in the Middle East, and the foundation of OPEC.

Simmons, Matthew R. *Twilight in the Desert: The Coming Saudi Oil Shock and the World Economy.* Hoboken, N.J.: John Wiley & Sons, 2005. Provides a historical overview of the U.S.-Saudi Arabia relationship and U.S. reliance on Saudi oil. Highlights discrepancies between Saudi Arabia's actual production capabilities and the reserves claims that it reports. Suggests that the near future holds the prospect of depleted Saudi Arabian oil resources and the increased oil scarcity will have significant consequences around the world, including higher costs of oil.

ARTICLES AND PAPERS

al-Husseini, Sadad. "Why Higher Oil Prices Are Inevitable This Year, Rest of Decade." *Oil & Gas Journal,* 2 August 2004, 14–18. Written by a former executive of Saudi Aramco, Saudi Arabia's national oil company, predicts that oil supply constraints will inevitably lead to higher oil prices through the first decade of the 21st century.

Mouawad, Jad. "Saudi News Drives Up Oil Prices." *New York Times* (August 2, 2005): B2. Discusses the impact of the news of the death of King Fahd bin Abdul Aziz al-Saud (1923–2005), king and prime minister of Saudi Arabia. Provides valuable background on Saudi Arabia's plans to bolster its oil supplies from 10.5 million barrels per day to 12.5 million barrels per day by 2009.

Mouawad, Jad. "Saudis Aim for Precision, Not Surplus, in Meeting Global Oil Needs," *New York Times*, 6 December 2005, C1. Describes Saudi Arabia's efforts to monitor daily oil production as well as a visit to the country by the U.S. energy secretary, Samuel Bodman.

WORKS ABOUT ISSUES IN ENERGY SUPPLY AND NATURAL RESOURCES IN NIGERIA

BOOKS

Dube, Ikhupuleng, and Bereket Kebede. *Energy Services for the Urban Poor in Africa*. London: Zed Books, 2004. Studies the limitations experienced in African countries such as Nigeria in relation to energy services and electricity infrastructure. Focuses on three main energy-related problem areas: urban welfare, public finances, and the economy.

Omoweh, Daniel A. *Shell Petroleum Development Company, the State and Underdevelopment of Nigeria's Niger Delta*. Trenton, N.J.: Africa World Press, 2004. Commentary on the activities of Shell Petroleum in Nigeria's Niger Delta region and their economic, environmental, and social effects on the region.

WEB DOCUMENTS

Ahemba, Tume. "Nigeria Can't Stop Flaring Right Now, Industry Says." Reuters News Service, November 11, 2005. Available online. URL: http://www.planetark.com/dailynewsstory.cfm/newsid/33493/story.htm. Accessed on January 29, 2006. Details the Nigerian government's efforts to end gas flaring in Nigeria, the infrastructure problems that have made the government's plans difficult to implement, and the health and environmental risks associated with gas flaring.

Ibiam, Agha. "Man Who Brings Gas to Your Homes." This Day Online, *This Day* (Nigeria), January 12, 2006. Available online by subscription. http://www.thisdayonline.com/nview.php?id=37952. Accessed on January 29, 2006. A commentary on natural gas policy in Nigeria, including efforts to reform natural gas infrastructure and end natural gas flaring.

Parker, Leia. "Nigeria Needs Come First As Oil Majors' Role Shrinks." Dow Jones Newswires, September 30, 2005. Available online by subscription. URL: http://www.djnewswires.com. Accessed on January 29, 2006. Discusses Nigeria's plans to strike deals with multinational oil companies that give a higher priority to Nigeria's needs as a whole, especially its need for infrastructure development.

This Day (Nigeria). "Nigeria, 2005: Forward Strides, Backward Moves." This Day Online, January 10, 2006. Available online by subscription. URL: http://www.thisdayonline.com/nview.php?id=37811. Accessed on January 29, 2006. Describes the inadequacy of public infrastructure in the housing, aviation, tourism, and energy sectors of Nigeria.

WORKS ABOUT ISSUES IN ENERGY SUPPLY AND NATURAL RESOURCES IN VENEZUELA

BOOKS

Tugwell, Franklin. *The Politics of Oil in Venezuela.* Stanford, Calif.: Stanford University Press, 1975. Historical overview of the impact of oil reserves and production on the political climate of Venezuela from 1920 to 1973, focusing on tensions between Venezuela and oil-consuming countries.

ARTICLES

Avery, William P. "Oil, Politics, and Economic Policy Making: Venezuela and the Andean Common Market." *International Organization* 30 (Autumn 1976): 541–571. Offers historical background on Venezuela's oil trade.

WEB DOCUMENTS

Agence France Presse. "Venezuela Set to Take Over Oil Field January 1." TODAYonline, December 22, 2005. Available online. URL: http://www.todayonline.com/articles/91537.asp. Accessed on January 29, 2006. Provides details on a Venezuelan plan to commandeer a U.S.-based oil operation in Venezuela if the U.S. oil firm that controls it refuses to accept the new terms for foreign oil companies under Venezuela's 2001 Hydrocarbons Law, which requires foreign oil investments to be in the form of joint ventures with Venezuela's state-owned oil corporation.

Chestnut, Teddy, and Jessie Gaskell. "U.S. Policy toward Venezuela: Playing with Oil, Playing with Trade, Playing with Fire." Council on Hemispheric Affairs (COHA), News and Opinions on the Situation in Venezuela, August 23, 2005. Available online. URL: http://www.coha.org/NEW_PRESS_RELEASES/New_Press_Releases_2005/05.99_Robertson_Advises_Bush_to_Take_Already_Self-Destructive_Vendetta_Against_Chavez_to_an_Outrageous_Extreme_htm. Accessed on February 18, 2006. Commentary on the Bush administration's discomfort with the administration of the Venezuelan president Hugo Chávez.

Hudson, Saul. "U.S. Says Has 'Inoculation' Strategy to Curb Chávez." Reuters News Service, February 17, 2006. Available online. URL: http://www.reuters.com. Accessed on February 17, 2006. Discusses U.S. strategy to garner the support of European and Latin American countries for the U.S. charge that the Venezuelan president, Hugo Chávez, has abused his power to target other political leaders. Also presents background information on the deteriorating ties between the United States and oil-rich Venezuela.

Weisbrot, Mark. "Is U.S. Policy toward Venezuela Counter-Productive?" Venezuelanalysis.com—Venezuelan Views, News and Analysis, September 21, 2003. Available online. URL: http://www.venezuelanalysis.com/articles.php?artno=1021. Accessed on February 18, 2006. Offers an opinion on the political tensions between Venezuela and the United States, a point of view that is mainly unfavorable toward the United States.

Chronology

440–445 MILLION YEARS AGO

- According to scientists, fossil fuels (coal, oil, and natural gas) are formed during this period.

MORE THAN 15,000 YEARS AGO

- Human beings begin to depend on various forms of energy for their survival and continuation as a species: the energy of their own bodies, the mechanical energy of tools and weapons, the thermal and radiant energy of the Sun and fire, and the energy of burning plant matter, such as wood, for fuel.

CIRCA 4000 B.C.E.

- People begin to settle in the valleys of the Nile, Tigris, and Euphrates Rivers, within the western and southwestern regions of the Asian continent, in regions such as Egypt and Mesopotamia. The waters of the Nile and other rivers provide a natural irrigation system that enables the inhabitants of these regions to grow grains on the land for food. The people also begin to rely on animal energy to supplement their own human energy.

CIRCA 2650 B.C.E.

- Rivers in western and southwestern regions of the Asian continent become major sources of energy to fulfill the need to transport and distribute grains.

770 B.C.E.

- Deforestation becomes a problem in China as a result of a high demand for wood to be used as fuel.

770 TO 453 B.C.E.

- The Chinese use coal as a fuel for cooking and domestic heating, as well as for the thermal energy needed to forge metals to make tools.

ENERGY SUPPLY AND RENEWABLE RESOURCES

CIRCA 600 B.C.E.

- The first oil wells and oil pipelines appear in China.

500–400 B.C.E.

- Greece makes improvements in the structure and shipping capacity of shipping vessels. These improvements lead to significant advancements in maritime transportation, not only for Greece but for other regions of the world.

300–200 B.C.E.

- Rome experiences shortages in its supply of wood, a valuable energy commodity. Human energy from hundreds of thousands of slaves, captured from the many cities Rome has conquered, takes on greater importance as an energy resource.

1–100 C.E.

- North Africa and Egypt ship vast amounts of wheat along the Tiber River. Europe begins using water mills for energy, to complete such tasks as grinding grain and weaving cloth.

400–500

- This period marks the decline of the Roman Empire. Rome makes fewer and fewer conquests and begins to experience a shortage in the human energy it once obtained from slaves procured from pillaging other cities. Rome begins to make greater use of hydropower for energy, relying on water mills to perform the work that slaves once completed.

900–1600

- Water mills are a major energy resource in Europe. They are ubiquitous, cluttering rivers and obstructing boat traffic.

1180

- The first recorded use of windmills for energy dates to northwestern Europe during this year.

1100S-1200S

- ***mid-1000s to late 1200s:*** Europeans clear millions of hectares of forests and swamps in order to obtain wood for fuel.
- ***late 1100s to 1200s:*** Windmills are in widespread use all across Holland.

1346

- England implements a policy to protect forests from further depletion; much of its forestland has been used for wood for thermal energy by this time.

Chronology

1500s

- England uses coal as a fuel for cooking and heating. Many of England's early industries, including glassworks, potteries, and breweries, are also powered by coal.

1700s

- In France unprecedented domestic thermal energy consumption and a growing industrial sector dependent on wood for fuel cause serious deforestation. This trend leads to escalating wood prices and governmental restrictions on woodcutting.

1712

- The first installation of a coal-powered machine for pumping water from coal mines, devised by the British inventor Thomas Newcomen, takes place in England. The machine features an engine that burns coal in a boiler to create high-pressure steam. The steam drives a shaft that can pump water. The machine enables the British to extract higher-quality coal from coal mines, to be used for heating purposes. The invention helps coal to surpass wood as the primary source of thermal energy.

1730s, 1776

- French peasants organize uprisings in response to restrictions on woodcutting and price increases on wood.

1752

- Benjamin Franklin discovers that static electricity and lightning are expressions of one and the same form of energy—electrical energy.

1765

- The Scottish inventor James Watt adds improvements to the coal-powered steam engine invented by Thomas Newcomen. Watt's steam engine could be used for purposes other than pumping water, and it soon becomes the energy powerhouse behind the operation of silk mills and other textile factories, first in England in the 18th century and then in the United States in the 19th century. The textile industry, which previously had depended on energy from water mills and windmills, is transformed. Watt's coal-powered steam engine is credited with setting in motion the Industrial Revolution.

1781

- The horse-drawn carriage is the standard method of passenger travel worldwide.

ENERGY SUPPLY AND RENEWABLE RESOURCES

EARLY 1800s

- Gas, a product derived from the distillation of oil, is used as a fuel for lighting. Gas lighting is the main source of light in workshops and factories in London and other industrial cities until the late 1800s.

1812

- The first gas company is founded.

1830s

- The British scientist Michael Faraday discovers that an electric current can flow in a wire within a magnetic field. His work demonstrates basic principles of electricity production: electromagnetism, induction, generation, and transmission. It also leads to new technologies using electrical energy, such as the telegraph and electroplating.

1850–1980

- The average amount of energy used by each person increases worldwide.

1857

- Kerosene is widely used for lamp lighting, replacing whale oil, a more expensive and scarce oil.

1859

- The U.S. entrepreneur "Colonel" Edwin Drake strikes a large oil deposit 69 feet beneath the earth in Titusville, Pennsylvania. It is the first U.S. oil well, and its discovery soon leads to the development of oil fields, places where oil is extracted and produced for energy. Within six years the first U.S. oil pipeline, a line of pipe with pumps and valves for conveying oil, is constructed. The oil industry is born.

1860s

- James Clark Maxwell, a Scottish mathematical physicist, demonstrates that electric and magnetic fields travel in waves through space. Maxwell develops a mathematical theory that expresses four basic laws of electricity and magnetism. His equations become the basis for many uses of electrical power in the late 1800s and later for the development of radio and television.

LATE 1800s

- A burgeoning Chinese rice-farming industry cuts into pastures and forests, depleting valuable forestland. By the mid-1900s deforestation in China leads to soil erosion, a lack of cultivable land, and finally a food shortage and famine.

- The Industrial Revolution also begins in England during this time, driven by the introduction of the coal-powered steam engine. The coal-powered steam engine train becomes the standard method of passenger travel worldwide.

1863

- U.S. petroleum producers form their first professional association, the Association of Petroleum Producers.

1870

- Oil is the United States' second-largest export.
- John D. Rockefeller forms the Standard Oil Company of Ohio, an oil production company that begins to dominate the oil industry. Rockefeller's company seizes control of several oil refining, transportation, and distribution businesses and soon develops a monopoly over oil production and distribution in the United States.

1878

- The first electric railway is introduced. Electricity is also found to have applications in the heavy chemicals and metallurgical industries. The American inventor Thomas Edison founds the Edison Electric Light Company.

1879

- Thomas Edison creates the first practical incandescent electric light. The lightbulb, in which a filament gives off light when heated to incandescence by an electric current, is modeled on an 1878 prototype created by Edison. With this introduction of artificial lighting, one of the remaining limitations on industrial productivity—the onset of evening—diminishes. Lightbulbs begin to replace the oil and gas lamps used in industrialized nations such as the United States and England.

1880–1914

- There is an annual growth of more than 10 percent in electrical consumption in the world's major industrialized countries.

1881

- Thomas Edison enters into a partnership with Werner and William Siemens to design Europe's first public lighting network in London.

1882

- Thomas Edison establishes the first electricity generating plants in London and New York. They are powered by coal.

ENERGY SUPPLY AND RENEWABLE RESOURCES

1885–1950

- The most widely used fuel in the United States is coal.

1889

- Thomas Edison consolidates his various electric businesses into the Edison General Electric Company.

1890

- The first automobiles utilizing an internal combustion engine are produced in large quantities in the United States, creating a greater demand for the gasoline that is needed to fuel them.
- Oil is struck in southeast Asia, and a petroleum company from Britain called British Shell sets up operations there, along with another company, the Royal Dutch Company for the Exploitation of Oil Wells of the Dutch Indies.

1890s

- George Westinghouse, a U.S. inventor and manufacturer, promotes alternating current (AC), an electric current that, unlike direct current (DC), flows back and forth instead of in one direction. It can also be transmitted over long distances with minimal energy loss. It becomes the main electric current used in electricity generation plants, replacing direct current, which cannot be transmitted over long distances.

1892

- Edison General Electric merges with another U.S. electrical power distribution company, forming General Electric.
- The U.S. government dissolves John D. Rockefeller's Standard Oil Trust.

1896

- Antoine Henri Becquerel, a French professor and scientist, makes a discovery that is foundational to the development of nuclear energy: the radioactivity of uranium. Radioactivity is a property possessed by some elements, by which energy can be spontaneously emitted during disintegration of these elements' atomic nuclei.

1899

- A machinist and engineer employed by the Edison Company, Henry Ford, successfully designs his first car and launches the Detroit Automobile Company (which later becomes the Ford Motor Company).

Chronology

1900

- John D. Rockefeller's Standard Oil Company controls greater than half of the world's sales of oil.
- Oil companies during this time routinely rely on surface geology, or a basic evaluation of land formations, to find likely spots for oil.

1900–1930s

- Monopolies form within the electric utility industry, with companies such as General Electric controlling the related functions of generating electricity, transmitting it over distances, and distributing it to customers.
- The Middle East becomes the locus of various competing oil companies, as well as governmental interests, with Germans, Russians, Britons, and Americans all seeking control of the Persian oil fields. Oil-related contracts between countries and oil companies multiply, setting the stage for an interdependence among governments, the economy, and the energy industry that still shapes world politics today.
- Production and consumption of oil skyrocket. Great Britain has control of most world oil production during this time.

1901

- In the United States oil is struck in Spindletop, the first large oil field in Texas, near the town of Beaumont. The Spindletop oil well opens with a "gusher," or a large burst of oil. The oil shoots more than 150 feet into the air. The Spindletop field begins producing almost 100,000 barrels of oil a day, more than all the other oil fields in the United States combined. The Spindletop event sets off a veritable race for oil, with oil exploration and drilling taking place all across Texas and other parts of the United States.

1903

- Antoine Henri Becquerel shares the Nobel Prize in physics with colleagues Marie Curie and Pierre Curie for the discovery and further study of the radioactivity of uranium.

1904

- The Standard Oil Company processes more than 84 percent of all U.S. crude oil in its refineries.

1908

- Henry Ford produces his first Model T automobiles. Their popularity results in soaring sales of gasoline, and the number of motor vehicles in the United States increases from 8,000 in 1900 to about 1.3 million in 1913.

1910

- A former assistant to the U.S. inventor Thomas A. Edison and president of the Commonwealth Edison Company, Samuel Insull, develops what is the largest electricity distribution network at the time. It is capable of transmitting electricity over long distances. In his network, electricity is generated as DC at a few central power stations, changed to the more efficient high-voltage AC, and transmitted throughout a city via a network of wires. In substations in different parts of the city, transformers and converters change the electricity back to low-voltage DC for nearby homes and businesses. Eventually, under this system, electricity is sent through entire states and then multistate regions. It becomes the standard way to distribute electricity.

1911

- A U.S. Supreme Court ruling breaks up the holding company to which John D. Rockefeller's Standard Oil Company transferred its assets. Yet Rockefeller retains his enormous fortune, later making donations to many philanthropic causes.

1913

- The German mechanical engineer Rudolf Diesel dies. After his death an engine he invented in 1895 is modified to run on petroleum-based fuel, even though this "diesel engine" had originally been developed to run on a variety of fuels, including vegetable oil. Oil companies begin labeling one of the by-products of gasoline distillation diesel fuel.

1920

- The Ford Motor Company begins to produce Model Ts by the millions, manufacturing nearly 17 million cars before discontinuing the model in 1928.

1920–1929

- Electricity output increases 106 percent in the United States and 116 percent in Europe.

1921

- Oil production begins in Venezuela.

1925

- Petroleum geologists are routinely using seismic exploration to find oil. This method involves the sending of high-intensity sound waves into earth or water (at first, mainly through the use of explosives) and then interpreting the resulting echoes for evidence of the presence of oil reservoirs.

Chronology

1928

- Major oil corporations of various countries sign an agreement known as the Achnacarry Accords. This agreement sets standards for pricing and competition among oil companies from Britain and from Western European nations, including Dutch, German, French, Belgian, and Russian corporations.

1930

- On October 5 of this year, Columbus Marion ("Dad") Joiner strikes possibly the largest pool of oil ever found in America, near Tyler, Texas.

1930s

- The administration of the U.S. president Franklin D. Roosevelt (1932–45) helps to reform the electric sector. His New Deal administration takes a direct role in developing U.S. natural resources by establishing the Tennessee Valley Authority in 1933 and the Rural Electrification Administration in 1935. Throughout the 1930s the New Deal administration enforces laws to address problems caused by monopolies and to dissolve public utility holding companies.
- During this decade Roosevelt also establishes ties with the founder of Saudi Arabia, securing U.S. access to Saudi oil. Major U.S. oil companies, including Chevron, Texaco, Exxon, and Mobil, buy the rights to extensive oil fields in Saudi Arabia. Saudi Arabia is one of the top three suppliers of crude oil to the United States today.

1939

- Scientists discover that it is possible to maintain a fission reaction capable of releasing enormous amounts of energy. Fission is the splitting apart of the atomic nucleus of a radioactive element, such as uranium, by the impact of a subatomic particle, or neutron, which results in the release of large amounts of energy.
- Also at the onset of World War II during this year, production in Venezuela reaches almost 30 million tons of oil annually. U.S. companies begin to step up their investment in Venezuela's oil, setting the stage for an eventual U.S. dependency on oil imports from Venezuela.

1939–1945

- During World War II supplies of U.S. oil are sufficient for the country's operations both home and abroad. Around this time, however, American habits, especially in relation to transportation, begin to require large amounts of oil, and consumption eventually begins to outpace production.

1942

- U.S. scientists produce nuclear energy in a sustained nuclear reaction. This new technology is used in the creation of atomic bombs in the 1940s.

1945

- In August of this year, the final year of World War II, U.S. forces drop atomic bombs on the cities of Hiroshima and Nagasaki in Japan. This event marks the first use of nuclear energy in warfare.
- By the end of World War II in September of this year, seven large multinational oil companies, based in the United States or in Europe, have control of 92 percent of world oil reserves and 88 percent of world oil production, which they retain until the 1950s. Known as the Seven Sisters, these mostly privately owned companies become partial owners or controllers of the assets of oil-producing countries. They control three-fourths of the world's oil refinery capacity and distribution, nearly one-third of the oil tanker fleets, and a good share of maritime traffic, and they are able to increase prices on all their oil-related activities without much competition.

1946

- Oil replaces coal as the world's primary energy source.

1950s

- To meet increased domestic oil demand, the United States begins to import more of its oil.

1951

- Electricity generated by a nuclear reactor is produced for the first time—in the United States at an experimental station in Idaho.

1956

- A renowned U.S. geophysicist named M. King Hubbert projects that U.S. oil production will reach its peak in the 1970s and decline thereafter, basing this projection on his conclusion that oil production rises and falls according to a bell curve. This projection turns out to be accurate.

1957

- The first commercial nuclear power plant begins operating in England. In nuclear power plants electrical energy is produced by controlled fission reactions that result in the production of steam that turns turbines.

Chronology

1960

- At a meeting in Baghdad, Iraq, on September 14, Iran, Iraq, Kuwait, Saudi Arabia, and Venezuela form the Organization of Oil Exporting Countries (OPEC), a multinational organization aimed at coordinating oil production policies and pricing among oil-producing nations. At the time of its founding OPEC's highest priority is to establish arrangements between major oil companies and oil-producing countries that are more favorable to the latter. OPEC standardizes the amount of oil exported by its members, thus regulating the price of oil. Today OPEC member nations, which produce 40 percent of the world's oil, are Algeria, Indonesia, Iran, Iraq, Kuwait, Libya, Nigeria, Qatar, Saudi Arabia, the United Arab Emirates, and Venezuela.

1970

- The Middle East owns most of the world's oil reserves, with the Seven Sisters controlling 70 percent of the Western world's oil production. The Seven Sisters reduce their investments in Middle Eastern oil production to keep prices down and deflect competition. Sudden oil shortages in the United States in the winter of 1970–71 results.
- Also during this year U.S. production of petroleum (including crude oil and natural gas) reaches its highest level, at 9.4 million barrels per day, as reported by the Energy Information Administration (EIA). Since that time oil production in the lower 48 states has been declining overall.
- The United States also passes a major law designed to improve air quality in the nation during this year. The Clean Air Act, whose predecessor was the Air Pollution Control Act of 1955, establishes regulations whose purpose is to minimize the impact of harmful pollutants in the United States, including those emitted by electric power plants and automobiles. It is later revised (in 1977, 1990, and 1997). Under the Clean Air Act the U.S. Environmental Protection Agency (EPA) sets limits on the amount of emissions allowable anywhere in the United States from pollutants considered to be primary causes of ozone depletion and harm to the environment.
- In 1970 Iran ratifies the Nuclear Non-Proliferation Treaty (NPT), which binds member countries to an agreement to develop, research, produce, and use nuclear energy for peaceful purposes only.

1973

- Oil commands nearly half of the world's total primary energy supply, double its share in 1949.

- In October of this year a coalition of Middle Eastern Arab nations led by Egypt and Syria battle Israel in a war known as the Yom Kippur War.

October 1973–March 1974

- Arab member nations of the Organization of Oil Exporting Countries (OPEC) stop shipping oil to the United States and other Western nations because of their support of Israel during the Arab-Israeli Yom Kippur War. OPEC decides to carry out a coup on oil prices, increasing them from $3.0111 a barrel in October 1973, at the start of the Arab oil embargo, to $11.651 in January 1974. More than a third of America's oil is from the Arab countries of the Middle East around this time, so the Arab oil embargo, combined with the OPEC price increases, cause a severe energy crisis in the United States.

1975

- The United States passes the Energy Policy and Conservation Act in response to the 1973–74 oil crisis. Its purpose is to reduce U.S. dependency on high-priced oil imported from politically unstable countries, to prepare the United States for energy shortage conditions, and to improve energy efficiency and conservation. It establishes the Strategic Petroleum Reserve (SPR), a stockpile of up to 1 billion barrels of petroleum to tap into during energy emergencies. It also creates the Corporate Average Fuel Economy (CAFE) program, which sets minimal requirements for the number of miles per gallon of gas that cars and light trucks should be able to achieve. Under the CAFE program automobile manufacturers are required to average at least a minimal specified miles per gallon (mpg) for each category of cars they make.

1977

- The U.S. passes the Federal Surface Mining Control and Reclamation Act, which contains regulations whose purpose is to restore topsoil and vegetation to areas damaged by coal surface mining.

1978

- The United States passes the Public Utility Regulatory Policies Act (PURPA) of 1978 as part of a larger law known as the National Energy Act of 1978 (NEA). NEA attempts to address the instability in U.S. energy supplies and prices that came to light in the 1970s after the Arab oil embargo. PURPA establishes the Federal Energy Regulatory Commission (FERC) and gives it the authority to enforce many measures aimed at promoting conservation of electrical energy and diversifying U.S. sources of this energy. One of these measures is a requirement that utilities buy power from nonutility, small-power producers that use renewable sources such as wind and solar power to generate electricity, or from "qualifying facilities," known as QFs, which produce power through cogeneration, a method of electricity production that uses waste heat from factory processes. PURPA is credited with helping to supplement electric utility generation with more efficiently produced energy, as well as with increasing competition in the energy market.

1979

- ***early to mid-1979:*** The United States again experiences an oil crisis, in the wake of political instability in the Arab world. Islamic fundamentalists who resent the pro-Western policies of the shah of Iran, Mohammad Reza Pahlavi, drive Pahlavi out during the Iranian Revolution. A new Islamic republic emerges in Iran, based on fundamentalist Islamic principles and led by the national spiritual leader, Ayatollah Khomeini. This new regime manages oil exports inconsistently—at a lower volume and at higher prices than during Pahlavi's regime. The United States experiences widespread panic-driven oil demand, which drives the price of oil up even higher and causes long lines at the gas pumps.
- ***March 28:*** A U.S. nuclear power plant, in Three Mile Island, Pennsylvania, leaks a small amount of radioactive material into the air and water as a result of equipment failure and human error.

1979–1982

- Overall energy consumption in the United States declines 10 percent, and consumption in the industrial sector decreases by 20 percent. Some of the decline is attributed to U.S. policies implemented to increase energy efficiency and conservation.

1980

- Under the leadership of Saddam Hussein, Iraq's president from 1979 to 2003, Iraq invades Iran. The conflict nearly drives Iranian oil production to a halt and greatly slashes oil production in Iraq. The situation affects oil imports to the United States.

1986

- A reactor in a nuclear power plant, in Chernobyl, Ukraine (then part of the Soviet Union), overheats. The reactor's lid blows off, spewing a cloud of radioactive waste into the atmosphere that affects all of Europe. The accident causes 30 fatalities.

EARLY 1990S

- The United States for the first time imports more oil and refined oil products than it produces. Since this time greater quantities of imports have been added each year as U.S. demand for petroleum grows and domestic production declines.

1991

- The U.S. president George H. W. Bush unveils an energy policy that promises to scale down foreign oil dependence by increasing domestic oil production, producing oil from environmentally sensitive areas in the United States, and encouraging domestic pipeline construction. His proposals meet with opposition from environmentalists and renewable energy advocates.
- ***January 16:*** A coalition of U.S. and international military forces launches Operation Desert Storm, or the Persian Gulf War, to force Iraq out of Kuwait. Iraq had invaded Kuwait in 1990, possibly driven by a desire to control Kuwaiti oil fields. The Persian Gulf War threatens the supply of oil imports to the United States. During the war Iraqi soldiers also set oil fields ablaze in a gesture of defiance of Western countries.

  Also on this date, the first emergency drawdown of the SPR is authorized. In the middle of the effort to counter Iraq's invasion of Kuwait, the U.S. Department of Energy (DOE) implements a plan to draw down and sell 33.75 million barrels of crude oil from the SPR because of the Persian Gulf conflict's threat to oil supplies. The drawdown proceeds on schedule, but world oil supplies and prices stabilize before the final sale, so the U.S. government reduces the drawdown to 17.3 million barrels and sells them to 13 oil companies.

1992

- The United States passes the Energy Policy Act (EPAct) of 1992 to increase the diversification and competition in the electricity sector that began as a result of the Public Utilities Regulatory Policies Act (PURPA) of 1978. EPAct creates a category of power producers called exempt wholesale generators (EWGs), thereby enabling nonutility generators that are exclusively in the business of wholesale electricity to produce and sell their wholesale power without being regulated as utilities. EPAct also authorizes the FERC to require that utilities owning transmission lines carry power from wholesale electricity producers. EPAct also allows EWGs to build generating plants. In these ways EPAct helps to open up the national electricity transmission system to nonutility businesses.

LATE 1999

- The United States connects Canadian natural gas supplies to U.S. consumers through the construction of a pipeline from Canada to Chicago through the upper Midwest.

2000

- According to the U.S. Census Bureau, the United States has a trade deficit of $449 billion. According to the EIA, 20 percent of this deficit is the value of imported oil.

- ***January:*** Natural gas pipelines, running from Canada's Sable Island to New England, become operational.
- ***August 19:*** A corroding natural gas pipeline near the Pecos River in southeastern New Mexico ruptures, causing an explosion so large that 11 nearby campers are killed by the blast and a fireball can be seen as far as 20 miles to the north in Carlsbad, New Mexico.

2001

- ***early 2001:*** The administration of President George W. Bush refuses to ratify the Kyoto Protocol, an international treaty that calls for the world's 38 most industrialized countries to reduce fossil fuel emissions by an average of 5 percent below 1990 levels by 2012. The treaty developed out of the work of the United Nations Framework Convention on Climate Change (UNFCCC), which aimed to find ways to minimize greenhouse gas emissions in order to reduce global climate change. The United States had helped to draft the protocol in 1997 and even signed it in 1998, agreeing, at least initially, to a 7 percent cut in emissions, but refused to ratify it on the grounds that it gave certain developing nations an unfair economic advantage by excluding them from its emissions reduction requirements.

  Also during early 2006 California experiences rolling electrical blackouts. Gray Davis, governor of California from 1999 to 2003, declares a state of emergency, and the California state government puts the brakes on its plans to deregulate power utilities. In an attempt to guarantee steady and reliable electric power in the future, Davis appoints the California Department of Water Resources to replace the state's utility companies as state buyer of wholesale electricity and signs long-term, expensive contracts with other energy suppliers.
- ***May:*** The National Energy Policy Development (NEPD) Group, spearheaded by the U.S. vice president, Dick Cheney, a former oil industry executive, releases a report making recommendations to the president regarding a new national energy policy. The group's overall findings are controversial because environmentalists and watchdog agencies, pointing to Cheney's oil industry background, suspect that the interests of the fossil fuel industry had too much influence on the NEPD's recommendations.
- ***September 11:*** Members of a group of Islamic fundamentalist extremists execute a terrorist attack on the United States, hijacking U.S. domestic flights and using the airplanes to destroy the World Trade Center in New York City and strike the Pentagon in Washington, D.C. More than 3,000 Americans are killed. The event has a profound effect on U.S. oil supply and affordability, with the United States experiencing record-high oil prices and a general insecurity over oil imported from the Middle Eastern region.

There are also mounting fears of possible terrorist strikes on U.S. nuclear power facilities.

- ***November 2:*** Hugo Chávez, president of the Bolivarian Republic of Venezuela, signs Venezuela's Hydrocarbons Law. This law increases the royalties that private companies must pay to the government for producing oil in Venezuela and requires foreign oil investments to be in the form of joint ventures with Venezuela's state-owned oil company, Petroleos de Venezuela, SA (PdVSA). The law is controversial because it guarantees PdVSA a majority share of any new projects, while there are many doubts about PdVSA's ability to fund investment in expanding crude oil production sufficiently.
- ***November 13:*** The Bush administration orders the SPR filled to capacity in response to the threat to oil security posed by the September 11 terrorist attacks that year. After gas prices soar, however, 50 members of the U.S. House of Representatives request that the president stop the filling of the SPR so that the amount of oil available for current use will increase and gas prices will stabilize. The president refuses their request.

2002

- ***July 26:*** Iran and Russia indicate that they are working together on a plan to construct nuclear power plants in Iran, raising global concerns that Iran is pursuing nuclear energy for the purpose of weapon development. Yet Iran claims its pursuit of nuclear power is a part of an attempt to diversify its energy assets to bolster its largely oil-dependent economy while making good use of its significant reserves of uranium ore.
- ***December 2:*** Venezuelan citizens opposed to the policies of their president, Hugo Chávez, lead a nationwide revolt against him. The country's oil workers go on strike, and all oil operations cease, hurting Venezuela's economy as well as leading to a spike in U.S. oil prices in early 2003. The Venezuelan economy enters a recession at the end of 2002 because of the strike. Later Chávez dismisses nearly half of the country's oil workforce.
- ***December 17:*** The Pipeline Safety Improvement Act (PSIA) of 2002 is signed into law in the United States. It establishes new requirements for the way the natural gas industry ensures the safety and integrity of its pipelines.

2003

- President George W. Bush waives the new source review requirement of the Clean Air Act, a requirement that had called for new or remodeled coal-powered electricity plants to use the cleanest-burning technologies possible. President Bush's action draws criticism from environmentalists.

- ***February:*** Crude oil prices in the United States reach a 29-month high due to a cold winter and the strike by Venezuelan oil workers at the end of 2002. The price of oil reaches $49 a barrel, almost double the previous February's price.
- ***March 20:*** After Iraq's president, Saddam Hussein, does not cooperate with weapons inspections supervised by the UN, the United States leads an invasion of Iraq, contending that Iraq possesses weapons of mass destruction. The U.S. invasion of Iraq puts Iraqi oil production on hold. Saudi Arabia and other oil-producing countries, however, step up production to circumvent the shortfall. Still, the remainder of 2003 is marked by reduced economic activity in the United States, with a stagnant stock market, especially during the year's early months.
- ***August 14:*** Defects in the North American electricity transmission system become apparent when New York, much of the Midwest, and Ontario, Canada, experience the worst blackout in U.S. history. The blackout is attributed to the failure of an electrical utility supplier, FirstEnergy Corporation, to trim trees in part of its Ohio service area. It is claimed that when the trees had contact with overloaded transmission lines, a cascading series of power outages was triggered.

2004

- The U.S. Department of Transportation (DOT) reports that 429 hazardous natural gas pipeline incidents occurred during the year.
- The 21.4 billion barrels of proven oil reserves in the United States during this year represent a 46 percent decline in reserves since 1970. According to EIA data, 29 percent of all energy consumed in the United States is imported from other countries.
- ***March 1:*** Nigeria passes the Electric Power Sector Reform (EPSR) Act. The purpose of this law is to reform Nigeria's unreliable and disorganized electricity generation system. The law enables private power companies to participate in electricity generation, transmission, and distribution and separates Nigeria's National Electric Power Authority (NEPA) into 11 electricity distribution firms, six electricity-generating companies, and an electricity transmission company. The law calls for all of these operations to be privatized. Nigeria's passage of the EPSR Act sets in motion the privatization of NEPA and a long-awaited reform of the nation's electricity generation system.
- ***July 21:*** Germany signs the Renewable Energy Sources Act (Act Revising the Legislation on Renewable Energy Sources in the Electricity Sector). The law's purpose is to increase the percentage of Germany's electricity that is produced through renewable energy sources—from 6.7 percent in 2000 to 12.5 percent by 2010, 20 percent by 2020, and 50 percent by 2050. The law calls for improved financial incentives for the continuing development of not only wind

power for electricity, but also hydropower, now the second-largest renewable source of electricity in Germany after wind. It also calls for more electricity derived from solar, biomass, and geothermal power.

- ***November:*** The Arctic Climate Impact Assessment, a study conducted by the Arctic Council and the International Arctic Science Committee (IASC), is released at the Arctic Climate Impact Assessment (ACIA) International Scientific Symposium in Reykjavik, Iceland. It reports that rapidly melting glacial ice in the arctic region, believed to have been caused by global warming, has harmed the food chain and indigenous life in the region and is disrupting ocean currents and storm systems.

  Also during this month Venezuela's president, Hugo Chávez, announces a new royalty rate for oil companies that is the highest rate allowable under all of Venezuela's prior laws regulating the hydrocarbons sector. The new rate renders Venezuela's oil business particularly unfriendly to foreign investment.

2005

The U.S. DOT reports that 396 hazardous natural gas pipeline incidents occurred during the year.

- ***June:*** Mahmoud Ahmadinejad, a leader known for his militancy and radicalism, becomes president of Iran. Five days after his election Iran resumes a discontinued project to enrich uranium at a facility in Isfahan, Iran. Ahmadinejad maintains that the research is for the purpose of generating nuclear power to help his country, which, he claims, is running short on energy.
- ***August 8:*** President George W. Bush signs into law the Energy Policy Act of 2005. Among its provisions are an authorization to fill the SPR to capacity, new reliability standards for electricity utilities for the purpose of modernizing the electrical grid, incentives for expanding the use of nuclear energy for electricity, and a commitment to invest \$2 billion over 10 years to support research into environmentally "cleaner" ways of using coal to generate electricity. It also includes measures for promoting production and use of ethanol, a renewable fuel; institutes a tax credit for owners of hybrid vehicles; and extends daylight savings time in the United States by four weeks on the presumption that Americans will use less electricity if daylight is available for a longer period. The act also authorizes subsidies for producers of wind energy and other alternative energies, \$50 million in an annual biomass grant, and tax breaks for energy conservation improvements made to homes.

  The original bill for this law includes a controversial plan to drill for oil within a portion of the Arctic National Wildlife Refuge (ANWR), an environmentally sensitive, 19-million-acre tract of land near an oil-rich site in Alaska. As passed, however, the Energy Policy Act of 2005 omits this plan, mainly because of lack of support. The plan becomes a separate bill in the Senate, which is later rejected in early 2006, but analysts expect that U.S. lawmakers will continue to consider the ANWR plan.

After the Energy Policy Act of 2005 is passed there remains a great deal of uncertainty about the extent of the impact of its provisions on U.S. energy supply.

- ***August 29:*** The eye of Hurricane Katrina makes landfall in the U.S. Gulf Coast region, where a quarter of all domestic oil production is situated. Oil production facilities, terminals, pipelines, and refineries off the coasts of Texas and Louisiana are damaged. Disruptions in the U.S. supply of gasoline and other refined oil products cause oil prices to spike nationwide.
- ***August 30:*** The price of oil reaches a record high of $70.82 per barrel in the United States after Hurricane Katrina's destruction to U.S. oil production facilities along the Gulf of Mexico.
- ***September:*** Iranian leaders threaten to make the investment climate of Iran unfavorable to countries that attempt to hinder its access to nuclear energy technology.
- ***September 2:*** The U.S. president, George W. Bush, acting in conjunction with the International Energy Agency (IEA), authorizes a drawdown of crude oil from the SPR in response to the oil production disruption caused by Hurricane Katrina. Secretary of Energy Samuel W. Bodman immediately authorizes the sale of 30 million barrels of crude oil to U.S. markets. The DOE accepts the bids of five companies for a total sale of 11 million barrels.
- ***November:*** Iran rejects a plan whereby it would abstain from uranium enrichment and accept enriched uranium from Russia instead.

2006

- ***early 2006:*** Quantities of oil produced in the United States are nearly 60 percent below 1970 levels.
- ***January 1:*** A U.S. federal tax credit is made available to individuals or businesses who have solar and other energy-saving systems installed and operating on a home or a commercial building between January 1, 2006, and December 31, 2007.

  Also on this date, China's Renewable Energy Law, signed into law in 2005, becomes effective. The law earns praise from some U.S. environmentalists, who say it is a good model for the United States to follow in formulating future U.S. energy policies. The law is China's attempt to establish measures to protect the environment, prevent energy shortages, and reduce dependence on energy imports. It stipulates that electricity power grid operators purchase resources from approved renewable energy producers. The law includes in its definition of renewable energy hydroelectricity, wind power, solar energy, geothermal energy, and marine energy. It also requires that national financial incentives be available to foster state and local development of renewable energy resources, including solar electricity, solar water heating, and renewable energy fuels; that loan and tax discounts be given for renewable energy projects, such as the construction of commercial renewable energy facilities; and that specific penalties be imposed for noncompliance with the law. China projects that the law will boost its capacity to use renewable energy to 10 percent by the year 2020.

- ***January 12:*** The Charter of the Asia-Pacific Partnership on Clean Development and Climate (AP6) is ratified at the inaugural ministerial meeting of the Asia-Pacific Partnership on Clean Development and Climate in Sydney, Australia. The AP6 is a nontreaty agreement of the United States, Australia, India, Japan, China, and South Korea to cooperate on developing and implementing technology to reduce greenhouse gas emissions. The AP6 agreement allows countries to set their own goals for reducing emissions, and there is no mechanism for mandating compliance. Because AP6, unlike the Kyoto Protocol, imposes no mandatory limits on greenhouse gas emissions, environmentalists and nations that have ratified the Kyoto Protocol have called AP6 meaningless.
- ***January 31:*** During his State of the Union address the U.S. president, George W. Bush, announces the Advanced Energy Initiative, calling for a 22 percent increase in clean energy research at the U.S. DOE, with investment both in coal-fired power plants with reduced polluting capacity and in advanced nuclear energy designs focused on "clean and safe" technologies. The plan includes other proposals that, in the words of the president, could help to end America's "addiction" to oil "imported from unstable parts of the world."
- ***February:*** Iran is reported to the UN Security Council over charges that its nuclear program violates the Nuclear Non-Proliferation Treaty. The Iranian president, Mahmoud Ahmadinejad, threatens that Iran will "revise its policies" if the rights of the Iranian people are violated.
- ***February 16:*** The Kyoto Protocol, calling industrialized countries to achieve, by 2012, at least a 5 percent reduction in greenhouse gas emissions below 1990 levels, goes into effect. The agreement is ratified by more than 165 countries as of this writing, but not by the United States or Australia.
- ***April:*** The world's five largest oil companies, including Exxon Mobil, and BP Plc, report record profits for the first quarter of 2006—about $29 billion, or $4.46 for every person on Earth. At the same time by mid-2006 gasoline prices reach about three dollars per gallon and oil costs nearly $65 per barrel.

  Also in April the Iranian president, Mahmoud Ahmadinejad, announces that Iran has successfully enriched uranium. The uranium is enriched by using more than 100 centrifuges, devices that separate substances of different densities. This quantity of centrifuges would make the uranium capable of being used in a nuclear reactor, but several thousands of centrifuges would be required for a nuclear bomb. However, that same month the Institute for

Science and International Security (ISIS) publishes satellite images that allegedly show new nuclear sites under construction in Iran. In the meantime the United States, Great Britain, France, and other nations press for UN sanctions against Iran if it continues to pursue uranium enrichment. These countries attempt to institute a series of negotiations to obtain Iran's cooperation in nuclear nonproliferation.

- ***May 12:*** The House of Representatives Energy and Commerce Committee gives the Bush administration the authority to set per-gallon mileage targets for passenger cars in a new system that is based on vehicles' size and weight. Some lawmakers call for this authorization to include a stipulation that higher CAFE standards be established by 2015 and a fleetwide average of 33 mpg for light trucks and passenger cars be set by 2015, but the Energy and Commerce Committee rejects that proposal.
- ***May 19:*** The U.S. House of Representatives approves a Democratic plan to renegotiate contracts granted to oil companies in 1998 and 1999 for leasing land in the Gulf of Mexico for oil production. The original leases excused several oil companies, such as Exxon Mobil Corp. and ConocoPhillips, from any additional royalty fees they would have had to pay to account for increases in oil prices. (The agency responsible for the contracts claimed that the price thresholds for royalty relief were accidentally omitted.) The Democratic plan also proposes to take back $10 billion in tax incentives granted to oil companies. The approval of this plan marks the first time that either house of Congress passed a measure designed to address tax incentives and subsidies granted to the oil industry. However, another proposal from Congress, aimed at taxing oil companies for their record profits, meets with opposition from the president around the same time that this particular plan is approved.
- ***June:*** Iranian officials indicate that they will consider proposals to delay their nuclear power program until UN officials determine that it will be used for peaceful purposes only.
- ***July:*** China receives its first shipments of crude oil from a new pipeline connecting it with Kazakhstan.
- ***August 18:*** The president of Nigeria, Olusegun Obasanjo, orders police and security forces to crack down on militants suspected of orchestrating kidnappings and attacks on employees of foreign energy companies. Hundreds of arrests ensue, and several foreign oil companies must temporarily halt operations.
- ***August 25:*** China's government releases data showing that the country's demand for oil increased 12.2 percent in July alone.
- ***August 31:*** Iran fails to meet a UN Security Council deadline to either suspend its nuclear program or receive sanctions.

- ***September 11:*** At its 142nd meeting OPEC decides to uphold its daily oil production ceiling of 28 million barrels, despite a sustained decrease in the price of crude oil following a peak in mid-July.
- ***November:*** Venezuela's president Hugo Chávez more than doubles the number of U.S. households and states included in a program by which Venezuela provides cheaper heating oil to low-income families in the United States.
- ***December:*** Chávez wins reelection in his country.

  Also during this month, the UN Security Council votes to impose sanctions on Iran's nuclear materials and technology trade. In response Iran vows to further increase its uranium enrichment activities.

2007

- ***January:*** Imports supply more than 65 percent of U.S. daily oil demand. During his State of the Union address on January 23, U.S. President George W. Bush asks Congress to double the capacity of the SPR. He also calls for a mandatory fuels standard requiring 35 billion gallons of renewable and alternative fuels by 2017. Venezuela's president, Hugo Chávez, begins his third term in office and takes initial steps toward nationalizing key sectors of the economy, including the electric power sector.
- ***February:*** The U.S. DOE announces SPR expansion plans that include the development of a new SPR storage location.
- ***March:*** Iran fails to meet another UN Security Council deadline for halting uranium enrichment. Germany and five permanent council nations, the United States, Russia, China, Britain, and France, draft a resolution to embargo arms exports and impose financial sanctions on Iran's Revolutionary Guards and one of its banks, in an effort to apply pressure to Iran to cause it to stop enriching uranium.

Glossary

alternating current *(AC)* an electric current that, unlike direct current (DC), flows back and forth. It can also be transmitted over long distances with minimal energy loss.

anticlines porous rock formations in the ground where oil is commonly found.

atmosphere the thick layer of colorless, odorless gases, or air, that surrounds the Earth. The atmosphere consists of carbon dioxide, plús nitrogen, oxygen, argon, and other gases in trace amounts.

biomass organic matter that is burned, including wood and other plant matter, as well as animal waste. Biomass is considered a renewable resource since plants can always be regrown and since what living things take in for nourishment they eliminate as waste.

British thermal unit (Btu) a measurement of energy that is gauged on the amount of heat required to raise the temperature of one pound of water by one degree Fahrenheit (1°F).

carbon dioxide (CO_2) a colorless, odorless gas that is emitted whenever humans and animals breathe, forests are burned, land is cleared, vegetation is left to decompose, or fossil fuels are burned. Plants absorb carbon dioxide in a process called photosynthesis, which helps to keep them alive. Of all the gases emitted when fossil fuels are burned, carbon dioxide is generally considered to be one of the most harmful to the environment. This is because carbon dioxide collects high up in the atmosphere and reduces the amount of heat from the Sun that the Earth can reradiate back into space. Retaining this additional fraction of the Sun's heat can cause the Earth's temperature to rise gradually. See *global warming.*

coal a fossil fuel formed when increased pressure, temperature, and moisture act upon masses of plant matter that have become compacted beneath marshes and lakes. It exists as a layer of brownish black sedimentary rock in the earth and must be extracted by mining, which traditionally consists of driving shafts into the ground to dig it up. Coal was the first of all fossil fuels to be used extensively. Coal is a solid, combustible substance consisting primarily of carbon, with sulfur and other impurities mixed in.

coal-powered steam engine an engine that burned coal in a boiler to create high-pressure steam, which drove a shaft that could do work, such as cloth weaving or spinning. This machine was one of the driving forces behind the Industrial Revolution.

combustion a rapid chemical process that generates heat (which is energy).

conversion the process whereby the energy of a given energy source is changed, or converted, into *actual* work. Conversion generally depends on an the interaction between the source of energy and a specific process or device, usually known as a converter. The ease with which the potential energy of oil can be converted into the actual work of, say, powering a vehicle helps oil to maintain its position as a major source of energy, whereas alternative types of fuels, which might involve a more complex conversion process, are slow to enter widespread use, even though they may present fewer safety and environmental hazards.

deforestation the cutting down or clearing away of forests. Soil erosion and the displacement of forest species from their natural habitat are two of the negative effects of deforestation.

diesel engine an engine created by Rudolf Diesel in 1895 and originally designed to run on a variety of fuels, including vegetable oil. Aside from being able to run on cheap fuels, his engine had 30 percent efficiency—a 15 percent greater efficiency than the standard combustible engine. After Diesel died in 1913, his engine was modified to run on petroleum—the cheapest fuel available—and oil companies labeled one of the by-products of gasoline distillation *diesel fuel.*

direct current *(DC)* an electric current that flows in only one direction and cannot easily be transmitted over long distances.

ecosystems communities of organisms within a geographical area that depend for their survival and proper functioning on an intricate network of interactions with the environment.

efficiency the degree to which the energy produced by a particular energy source, such as a windmill, surpasses the energy and cost that have been invested in order to use it for energy.

electrical energy energy released from the movement and interaction of negatively charged elemental particles, called electrons, and positively charged elemental particles, called protons. It can occur naturally, such as in lightning or in static electricity, or may be deliberately produced, as through a generator.

Glossary

embargo a stoppage or a ceasing of a specific activity; generally refers to a stoppage of the import or export of a particular commodity, such as oil.

energy crisis an immediate shortfall in energy supply and/or a sudden, sharp increase in the cost of energy resources.

energy the ability to do work, the capacity to cause matter to move or change. Energy takes several forms: mechanical energy, light (or radiant) energy, heat (or thermal) energy, electrical energy, chemical energy, and nuclear energy.

ethanol an alcohol that can be used as a fuel for internal combustion engines that have been modified to run on it. It is usually made from biomass.

fission the splitting apart of the atomic nucleus of a radioactive element, such as uranium, by the impact of a subatomic particle, or neutron, resulting in the release of large amounts of energy.

fossil fuels substances consisting of compounds of carbon or hydrocarbon that are called fossil fuels because scientists believe that they were formed in the earth by the decomposition of plants and animals that died millions of years ago. This decomposing organic material in the Earth's crust resulted in naturally occurring compounds of carbon or hydrocarbon that include coal, petroleum, natural gas, oil shale, and tar sands. When fossil fuels are burned for energy, the gas carbon dioxide is released, along with other gases.

gasoline a liquid fuel composed of a mixture of small, light hydrocarbons. Gasoline is one of the products of a process that crude oil undergoes in refineries, where it is heated and broken up into smaller molecules to remove its impurities. Gasoline is the world's primary fuel for powering vehicles and other forms of transportation.

global warming (or global climate change) a process whereby the release of gases such as carbon dioxide, sulfur dioxide, nitrogen oxides, and ash particles is thought by scientists to result in an increase in the average overall temperature of the Earth. Some scientists believe that even a slight change in average global temperature can radically alter weather and climate patterns, as the warmer air melts the ice of the arctic regions, heats up the oceans, and causes more frequent and violent storms. The problem of global warming has often been blamed on the use of fossil fuels for energy, since burning these fuels releases carbon dioxide and other gases, which collect high up in the atmosphere and reduce the amount of heat from the Sun that the Earth can reradiate back into space.

hybrid vehicles cars that are powered by both an electric motor and a gasoline engine.

hydrocarbon a combination of the gases hydrogen and carbon.

hydropower the energy that moving water provides. It is so called because water is two atoms of hydrogen plus one atom of oxygen. Energy from moving water is considered renewable energy.

Industrial Revolution a time of major socioeconomic change that began in England at the end of the 18th century.

infrastructure the foundation or framework required to allow something to operate properly.

internal combustion engine an engine powered by combustion, a rapid chemical process that generates heat. Unlike in the steam engine, where combustion takes place inside a separate furnace chamber, combustion in the internal combustion occurs inside the engine itself. Because of this feature, the internal combustion engine is lighter than the steam engine; thus it is more portable and suitable for a greater variety of energy needs. Vehicles and other forms of transportation utilize an internal combustion engine that is fueled by gasoline.

kerosene a fuel derived from the distillation of petroleum. It was used in oil lamps from the medieval period onward.

kilowatt hour (kWh) a measure of a unit of electricity expended per hour.

kinetic energy energy being released to do work.

lightbulb a device for lighting, in which a filament gives off light when heated to incandescence by an electric current.

liquefied natural gas (LNG) a form of natural gas that has been chilled until it liquefies so that it can be transported by shipping vessels overseas.

mechanical energy the energy that an object possesses because of its motion or position. Mechanical energy may be either kinetic energy or potential energy, and it is usually associated with a physical object, as opposed to a chemical or biological change or process.

natural gas a fossil fuel that is usually found above oil-bearing rock and is usually extracted along with oil so that it too can be used as a fuel. Natural gas is a simple molecule made up of four hydrogen atoms around one atom of carbon; it consists mainly of a gas called methane.

natural resources matter that is supplied by the environment, such as trees or fossil fuels, which may act as a source of energy or have the potential to provide for human existence in some other way.

New Deal administration the administration of the U.S. president Franklin D. Roosevelt (1932–45), which took a direct role in developing U.S. natural resources, addressing problems caused by monopolies in the energy industry, and dissolving public utility holding companies.

nonrenewable resources natural resources that are limited in supply and/or cannot be replenished more quickly than they are consumed. Fossil fuels are one type of nonrenewable resource.

nuclear energy energy derived from fission. See *fission.*

nuclear power plants facilities that produce electrical energy by controlled fission reactions that result in the production of steam that turns turbines.

oil the fossil fuel petroleum, or the other fuels that petroleum yields when it is purified and distilled. Petroleum is an oily, flammable, liquid mixture of hydrocarbons and traces of other substances. It may range in color from almost clear to black. It is also called crude oil. It lies under pressure within rock formations known as anticlines.

Organization of Petroleum Exporting Countries (OPEC) a multinational organization aimed at coordinating oil production policies and pricing among oil-producing nations. Its member nations, which produce 40 percent of the world's oil, are Algeria, Indonesia, Iran, Iraq, Kuwait, Libya, Nigeria, Qatar, Saudi Arabia, the United Arab Emirates, and Venezuela.

petroleum See *oil.*

pipeline a line of pipe with pumps and valves for conveying oil or natural gas. Pipelines facilitate the process of distributing oil or transporting it across various (nonoverseas) regions.

potential energy the energy that all living things (and many nonliving things) have stored within them; when used, it becomes actual, or kinetic, energy.

proven oil reserves quantities of oil that can actually be recovered under existing economic conditions.

radiant energy energy that travels as electromagnetic waves and that may produce light as it does so.

radioactivity a property possessed by some elements, by which energy can be spontaneously emitted during disintegration of these elements' atomic nuclei.

refinery a facility wherein crude oil is heated and broken up into smaller molecules to remove its impurities.

renewable resources natural resources that are not limited in supply but are essentially inexhaustible, or at least constantly "renewed," or replaced, by the environment. As long as renewable resources are not used up more quickly than they can be restored, their supply is plentiful. Aside from the Sun, wind and water are renewable resources.

rotating coupler a device that can group various electricity systems.

seismic exploration a method for finding likely spots for oil, whereby high-intensity sound waves are sent into earth or water and the resulting echoes interpreted for evidence of the presence of oil reservoirs.

Seven Sisters seven large multinational oil companies, based in the United States or Europe, that held a near monopoly over deliveries of oil in the industrialized world until the 1950s. The Seven Sisters were Royal Dutch/Shell, Exxon, Gulf, Texaco, BP, Mobil, and Standard Oil of California (Chevron).

solar energy energy from the Sun.

Strategic Petroleum Reserve (SPR) an emergency stockpile of oil in the United States that was established by the Energy Policy and Conservation Act of 1975. It has a 727-million-barrel capacity; the Energy Policy Act of 2005 called for an expansion to a 1-billion-barrel capacity.

subsidies government funds allocated to lower the price of a particular good or service, usually with public interests in mind. Many people criticize U.S. oil subsidies for their potential to mask the true cost of oil and artificially keep it just low enough so that demand for oil will continue at the present level or even increase.

surface geology a method for evaluation of landmasses to find likely reservoirs of oil, whereby the Earth's surface is studied for visible features such as seeping oil or natural gas, craters caused by escaping gas, or the inverted-cup–shaped hills that are characteristic of anticlines—all common features of land that hold reservoirs of oil.

thermal energy energy that produces heat.

water mill a wooden wheel to which are attached horizontal spoons; flowing water collects in the spoons and causes the wheel to spin. The spinning motion powers a drive shaft that can rotate a stone upon another stone to complete work such as grinding grain. Water mills are thought to date back to antiquity.

wind power energy harnessed from the movement of winds. Wind is itself directly derived from another source of energy, the Sun, whose unequal warming of the Earth's surface and atmosphere causes shifts in air pressure that result in the formation of wind.

windmill (or wind turbine) a device that operates via the action of wind blowing upon vanes or sails attached to a long vertical structure. The spinning motion of the vanes or sails then turns a shaft that can grind grain, pump water, or, as in the wind turbines that are in use today, generate electricity. The first recorded use of windmills for energy dates back to 1180 in northwestern Europe.

Index

Page numbers in **boldface** indicate major treatment of a subject. Page numbers followed by *f* indicate figures. Page numbers followed by *b* indicate biographical entries. Page numbers followed by *c* indicate chronology entries. Page numbers followed by *g* indicate glossary entries.

Index

Index

Index